Springer Geography

The Springer Geography series seeks to publish a broad portfolio of scientific books, aiming at researchers, students, and everyone interested in geographical research. The series includes peer-reviewed monographs, edited volumes, textbooks, and conference proceedings. It covers the entire research area of geography including, but not limited to, Economic Geography, Physical Geography, Quantitative Geography, and Regional/Urban Planning.

More information about this series at http://www.springer.com/series/10180

Dénes Lóczy
Editor

The Drava River

Environmental Problems and Solutions

Springer

Editor
Dénes Lóczy
Institute of Geography and Earth Sciences
University of Pécs
Pécs, Hungary

ISSN 2194-315X ISSN 2194-3168 (electronic)
Springer Geography
ISBN 978-3-030-06523-2 ISBN 978-3-319-92816-6 (eBook)
https://doi.org/10.1007/978-3-319-92816-6

Printed on acid-free paper

This Springer imprint is published by the registered company Springer Nature Switzerland AG
The registered company address is: Gewerbestrasse 11, 6330 Cham, Switzerland

This book is dedicated to the memory of Prof. György Lovász (1931–2016), a committed researcher of the Drava and Mura Rivers.

Foreword

As an Italian scientist, it was my pleasure to accept the invitation to introduce the international readers to this valuable book on *The Drava River* and its environmental aspects, since this water course is not only one of the most interesting in Europe, from both the environmental and historical viewpoints, but also because its source is located in Italy, which is generally unknown. Actually, the Drava River crosses the Italian region of South Tyrol just for 10 km, before reaching Austria and flowing then in Slovenia, Croatia and Hungary with a total length of more than 700 km.

It is easy to understand how the Drava River, as a remarkable physical element, played a major role in the history and geopolitics of Central and Eastern Europe through time. As a matter of fact, it still marks the border between Croatia and Hungary for a long distance. On the other hand, the river's location in politically and militarily sensitive areas has prevented detailed scientific research to take place there until the 1990s.

The book shows the results of a 4-year research project on the environmental rehabilitation of the Hungarian Drava floodplain, providing, however, a wider overview of the environmental aspects of the watercourse and its surrounding areas. The Drava river is one of the most exploited in the World for hydroelectric purposes and, therefore, the effects of human pressure are tangible, but it still holds elements of remarkable environmental interest that deserve rehabilitation, preservation and enhancement.

As President of the International Association of Geomorphologists (IAG), I would like to underline the scientific value and originality of the book and how its contents match with the interest of presently active IAG Working Groups dealing with 'Geomorphological hazard', 'Geomorphosites' and 'Landform Assessment for Geodiversity'. The topics of the book were taken into account by a previous IAG Working Group on 'Human Impact on the Landscape' which was led between 2001 and 2009 by the editor of this book, Prof. Dénes Lóczy, who has long-standing experience in the field of human-induced transformation of the fluvial environment.

Finally, I would like to congratulate the Editor and all the Authors on this valuable publication since it provides the international community with a precious wide-spectrum contribution on an outstanding European fluvial environment.

Modena, Italy

Mauro Soldati
Full Professor of Geomorphology
President of the International Association
of Geomorphologists (IAG)
University of Modena and Reggio Emilia

Contents

Contributors

Gábor Andrási Department of Physical Geography and Geoinformatics, University of Szeged, Szeged, Hungary

Ognjen Bonacci Faculty of Civil Engineering, Architecture and Geodesy, University of Split, Split, Croatia

Tamara Brleković Faculty of Civil Engineering, University of Osijek, Osijek, Croatia

Alajos Burián South Transdanubian Water Management Directorate, Pécs, Hungary

Rok Ciglič Department of Geographic Information System, Anton Melik Geographical Institute, Research Centre of the Slovenian Academy of Sciences and Arts, Ljubljana, Slovenia

Zoltán Csabai Department of Hydrobiology, Institute of Biology, Faculty of Sciences, University of Pécs, Pécs, Hungary

Szabolcs Czigány Department of Physical and Environmental Geography, Institute of Geography and Earth Sciences, Faculty of Sciences, University of Pécs, Pécs, Hungary

József Dezső Department of Physical and Environmental Geography, Institute of Geography and Earth Sciences, Faculty of Sciences, University of Pécs, Pécs, Hungary

Anita Dolgosné Kovács Department of Environmental Engineering, Faculty of Engineering and Informatics, University of Pécs, Pécs, Hungary

Péter Gyenizse Department of Cartography and Geoinformatics, Institute of Geography and Earth Sciences, Faculty of Sciences, University of Pécs, Pécs, Hungary

Gábor Horváth South Transdanubian Water Management Directorate, Pécs, Hungary

Balázs Kevey Department of Ecology, Institute of Biology, Faculty of Sciences, University of Pécs, Pécs, Hungary

Tímea Kiss Department of Physical Geography and Geoinformatics, University of Szeged, Szeged, Hungary

Gerhard Karl Lieb Institute of Geography and Regional Sciences, University of Graz, Graz, Austria

Dénes Lóczy Department of Physical and Environmental Geography, Institute of Geography and Earth Sciences, Faculty of Sciences, University of Pécs, Pécs, Hungary

László Márk South Transdanubian Water Management Directorate, Pécs, Hungary

Arnold Móra Department of Hydrobiology, Institute of Biology, Faculty of Sciences, University of Pécs, Pécs, Hungary

Gábor Nagy Earth Sciences Doctoral School, University of Pécs, Pécs, Hungary

Adrienne Ortmann-Ajkai Department of Hydrobiology, Institute of Biology, Faculty of Sciences, University of Pécs, Pécs, Hungary

Dijana Oskoruš Hydrometeorological Service of Republic Croatia, Zagreb, Croatia

Tibor Parrag Danube-Drava National Park Directorate, Pécs, Hungary

Hrvoje Petrić History Department, Faculty of Humanities and Social Sciences, University of Zagreb, Zagreb, Croatia

Ali Mohamed Salem Earth Sciences Doctoral School, University of Pécs, Pécs, Hungary

Ulrich Schwarz FLUVIUS, Floodplain Ecology and River Basin Management, Vienna, Austria

Wolfgang Sulzer Institute of Geography and Regional Sciences, University of Graz, Graz, Austria

Péter Sály Department of Hydrobiology, Institute of Biology, Faculty of Sciences, University of Pécs, Pécs, Hungary

Lidija Tadić Faculty of Civil Engineering, University of Osijek, Osijek, Croatia

Enikő Anna Tamás Institute for Hydraulic Engineering and Water Management, Faculty of Water Sciences, National University of Public Service, Baja, Hungary

Gabriella Tóth Department of Physical and Environmental Geography, Institute of Geography and Earth Sciences, Faculty of Sciences, University of Pécs, Pécs, Hungary

Chapter 1
Introduction

Dénes Lóczy

Abstract The first-ever comprehensive physical geographical monograph on the transboundary Drava River tackles diverse environmental problems from the viewpoints of hydraulic engineers, hydrobiologists, and ecologists active in the countries on the drainage basin. The impacts of some rehabilitation actions are also assessed. The book is dedicated to the memory of the late Professor György Lovász, a dedicated researcher of the Drava and Mura Rivers.

Keywords Hydrogeomorphology · Water and sediment regime
River and floodplain ecology · Nature conservation

Issuing at 1,228-m elevation in the Dolomites, the Drava (Drau, Dráva) is an important right-bank tributary of the Danube in southern Central Europe. The 725-km-long river connects countries and cultures from the Italian Alps in South Tyrol, the rivers and large lakes in Carinthia, and the Slovenian Alps all the way to the Carpathian Basin. Near Legrad, Croatia, it is joined by the Mura (Mur) River and forms part of the Croatian–Hungarian border in the Carpathian (Pannonian) Basin. This peripheral and militarily sensitive location explains why this river section and its floodplain had been so neglected in scientific research until the opening of the border zones after the political transformations in 1989–1990 and the end of the Serbo-Croatian war in 1995. The geographical diversity of the Drava catchment is well demonstrated by the fact that it forms a corridor from the Alpine area to the Pannonian biogeographical region.

The rapid-flowing Drava River has been harnessed by 22 hydroelectric power plants in Austria, Slovenia, and Croatia. In Slovenia there are two artificially constructed side channels. In spite being considerably regulated, the Drava River has preserved natural aquatic and wetland habitats along the middle and lower segments and hosts unique assemblages of flora and fauna, including several

D. Lóczy (✉)
Department of Physical and Environmental Geography, Institute of Geography and Earth Sciences, Faculty of Sciences, University of Pécs, Ifjúság útja 6, Pécs 7624, Hungary
e-mail: loczyd@gamma.ttk.pte.hu

D. Lóczy (ed.), *The Drava River*, Springer Geography,
https://doi.org/10.1007/978-3-319-92816-6_1

endemic species. Because of the former political divisions of Europe, however, no comprehensive analysis of all components of the river environment has ever been published.

The present volume summarizes the environmental issues concerning the river channel and the floodplain. It describes the diverse forms of human pressure (river regulation, damming and reservoirs, dredging, intensive agricultural use of the floodplain, etc.) and the environmental changes of hydrological regime, sediment transport, bank stability, biodiversity, etc. involved. Based on the findings of biomonitoring (Ábrahám 2005; Purger 2008) and the SEE River European pilot project (Bizjak et al. 2014), the transboundary nature of the river is emphasized and the resulting problems outlined. Special attention is devoted to the evolution and present conditions of the floodplain (drainage pattern, connectivity, oxbow lakes, and their vegetation). The significance of the main channel and cut-off meanders (oxbow lakes) of the Drava River for nature conservation (UNESCO Transboundary Biosphere Reserve and Ramsar and Natura 2000 sites) and ecotourism are described (expanding on research reports such as UNESCO 2012).

A review of ongoing and planned rehabilitation measures illustrates the difficulties in finding solutions to environmental problems. Relying on environmental monitoring, the difficulties of ensuring proper water availability for the floodplain are studied in detail. The benefits and deficiencies of ongoing rehabilitation projects are outlined. Since a collection of papers have been published recently in three languages (Hungarian, Croatian, and English) on the achievements of the revitalization of side-arms along the lower Drava River (Purger 2013), the biological aspects of rehabilitation are not treated here in detail. The reader is kindly asked to refer to that book for further information.

The international team of authors (representing the countries crossed by the Drava River: Austria, Slovenia, Croatia, and Hungary) approach the topics of the individual chapters from different aspects (including regional physical geography, fluvial geomorphology, water management, hydraulic engineering, forest ecology, hydrobiology, nature conservation). Each chapter summarizes the findings of national research on the Drava—often only available in the native language of the particular author and made public in English here for the first time. The need for international cooperation in solving the environmental problems is often emphasized. The contributors hope that from the chapters of the book supply the reader with a many-sided picture of this European river, which has been neglected in scientific research for a long time.

In the structure of the book, there is some bias in favor of the Drava floodplain in Hungary, the problems of which are treated in seven chapters. The reason is that this publication is mainly the outcome of four years of research within the framework of the project "Rehabilitation potential of the Hungarian Drava floodplain", which was supported by the Hungarian Scientific Research Fund (OTKA, contract no K 104552), now managed by the National Office for Research, Development and Innovation (NKFIH). The project leader was the editor of the book. He is grateful for the funding between 2012 and 2017.

The volume is dedicated to the memory of Prof. György Lovász (1931–2016), who was among the first researchers to study the water regime of the Drava and Mura Rivers from a hydrogeographical aspect as early as the 1960s (Lovász 1972).

References

Ábrahám L (ed) (2005) Biomonitoring along the River Drava in Hungary, 2000–2004. Natura Somogyiensis 7:1–207 http://www.smmi.hu/termtud/ns/ns7.htm

Bizjak A, Cunder M, Marovt L, Skrt P, Softič M, Zinke A (2014) The SEE River Toolkit for Facilitating Cross-Sectoral Management of River Corridors. Booklet 2: Practical Applications on the Drava River. Institute for Water of the Republic of Slovenia, Ljubljana. 62 pp

Lovász Gy (1972) A Dráva-Mura vízrendszer vízjárási és lefolyási viszonyai (River regime and runoff in the Drava-Mura water system). Akadémiai Kiadó, Budapest. 158 pp (in Hungarian)

Purger JJ (ed) (2008) Biodiversity studies along the Drava River. University of Pécs, Pécs, p 328

Purger JJ (ed) (2013) Biota and rehabilitation of four Drava River side-branches in Hungary. BioRes Bt, Pécs, p 164

UNESCO (2012) Mura-Drava-Danube. UNESCO, Paris. http://www.unesco.org/new/en/natural-sciences/environment/ecological-sciences/biosphere-reserves/europe-north-america/croatiahungary/mura-drava-danube/

Chapter 2
The Drava Basin: Geological and Geomorphological Setting

Dénes Lóczy

Abstract The Drava catchment comprises three major geological units of the Eastern Alps and their southeastern foreland:

1. the Austroalpine Nappe System (with a small portion of the High Tauern Window),
2. the Southalpine Nappe System (with the eastern Dolomites) and
3. the southwestern margin of the Pannonian (Carpathian) Basin.

Mountain building in the Alps started in the Cretaceous driven by lithospheric plate movements: the northward drift of the African plate and its microplates. The plate tectonic evolution was also controlled by the opening and closure of two oceans: from the Triassic to the Middle Jurassic the Neotethys extended from the east to the west; the Piemont-Penninic Ocean existed from the Middle Jurassic to the Late Cretaceous and evolved parallel with the opening up of the Atlantic Ocean. The principal tectonic units of the upper Drava catchment are divided by the marked Peri-Adriatic Lineament system. Although with lower intensity than in the Western and Central Alps, the orogeny still goes on today on the eastern end of the mountain arc. Huge horizontal displacements of blocks are observed in the form of strike-slip faults, which control the overall drainage pattern (e.g., in the Gail Valley). According to age, the rocks of the Eastern Alps range from metamorphosed Paleozoics (e.g. in the Tauern Window, Austria) to the late Holocene alluvia of the Lower Drava Valley (e.g. at the Kopački rit in Croatia). Rock variability is also great with regard to resistance to erosion. The most spectacular landforms are the high-mountain glacial and karst assemblages, the dolomite cliffs and pinnacles, earth pyramids, deep ravines, marked landslide features, narrow gorges with waterfalls, and caves in limestone. There are overdeveloped cutoff meanders and undercut bluffs in the lowland section.

Keywords Geological evolution · Tectonics · Nappe systems · Lithology
Geomorphic processes · Landforms · Eastern Alps · Pannonian basin

D. Lóczy (✉)
Department of Physical and Environmental Geography, Institute of Geography and Earth Sciences, Faculty of Sciences, University of Pécs, Ifjúság útja 6, Pécs 7624, Hungary
e-mail: loczyd@gamma.ttk.pte.hu

D. Lóczy (ed.), *The Drava River*, Springer Geography,
https://doi.org/10.1007/978-3-319-92816-6_2

2.1 Introduction

Ever since the Austrian mountaineer and geologist Otto Ampferer (1875–1947) proposed the so-called "undercurrent theory" (Unterströmungstheorie) in 1906 (Ampferer 1906), intensive studies on the mobility of the Earth's crust in the Alpine region have been conducted and led to the identification of nappe systems. The hypothesis, at first neglected by the academic world, became fully accepted in the 20th century, when the tectonical, petrographical, and mineralogical properties of Alpine nappe systems were disclosed. In Austrian geology, the nappes of the Eastern Alps were subdivided into the lowest Penninic Nappe Complex (exposed in the High Tauern window) and overlying Austroalpine Nappe Complex (Bögel and Schmidt 1976; Scarascia and Cassinis 1997; Stüwe and Schuster 2010). In the 21st century the radiometric ages for subduction (e.g. Thöni 2006) and nappe metamorphism (Froitzheim and Schmid 2008) began to be determined as well as the pre-Triassic plate tectonic configurations were reconstructed (Pfiffner 2014). Recently, information on the deep structure of the Eastern Alps is available from the findings of the German-Austrian-Italian TRANSALP seismic reflection profiling project completed in 2003 (Lüschen et al. 2004). Among other results, the project revealed tectonic shear zones in the lower crust at the contact of the European and the Adriatic plate as well as the steeply dipping structures of the Tauern Window (Fig. 2.1).

The mountainous upper Drava basin mostly occupies the southern zone of the Austroalpine Nappe System, which overlies the Penninicum in the Eastern Alps. The main units are the crystalline Central Eastern Alps with the tectonic windows, the Gailtal Alps, the Karawanks and the eastern margin of the Dolomites with the Lienz Dolomites (East Tyrol) in the source area of the river. The 700-km-long Peri-Adriatic Lineament with its branches is the most significant lineament system of the whole Alps (Bartel et al. 2014). Its straight alignment controls the longitudinal valleys typical of the Austrian Alps (the Puster and Gail valleys). To the south, the Dinaric Nappes with southern vergence (opposed to the northern vergence of other nappes) show moderate displacements.

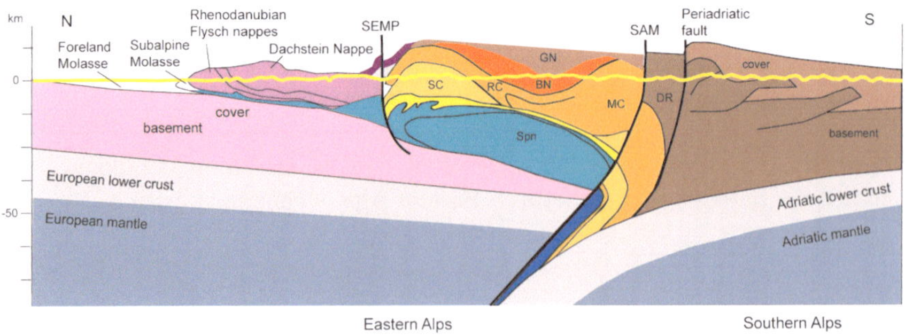

Fig. 2.1 Profile across the Eastern Alps based on findings of the TRANSALP Project (after Schmid et al. 2004, 2013)

The Drava crosses the Klagenfurt Basin and the Mura the Styrian Hills (or Graz Basin) (Ebner and Sachsenhofer 1991) and together enter the extensional Pannonian (Carpathian) Basin (Fodor et al. 1999; Haas 2012).

2.2 Geological Evolution of the Eastern Alps

The Alps have several 'forerunners'. Traces of mountain building, Carboniferous flysch-like clastics, sandy, and pelitic turbidites have been identified in the Carnic Alps at 300–350 m depth (Schönlaub and Heinisch 1993), and Variscan orogeny is assumed to have affected the crystalline basement ('Altkristallin') of the Saualpe-Koralpe and in the Gurktal block. An early orogenic phase is also evidenced in Late Permian conglomerates and breccias (Krainer 1993).

The early history of the present-day Alps begins with the break-up of the megacontinent Pangaea in the Permian (Oberhauser 1980). The rock masses which build up the mountain arc mostly derived from sediments deposited in the Neotethys (Meliata) Ocean (Frisch et al. 2011), a major Mesozoic sedimentation basin. For instance, the dolostone series of the Middle to Late Triassic carbonate platform of the Dolomites reaches more than 1,000 m thickness (Bosellini et al. 2003). Also more than 3,000 m of Permo-Mesozoic sediments were deposited on top of the thermally subsiding Adriatic microcontinent, a broad carbonate shelf (Schuster et al. 2013).

During the Mesozoic, the present Alpine region showed variable and rather complicated patterns of oceanic basins and land masses. A variety of sedimentary environments have been reconstructed: rift zones, deep-sea basins, shelf seas, and submarine rises (Pfiffner 2014). The resulting rock assemblages are called mélange in the Alpine region. Rifting was accompanied with collisions (Frisch 1979). The Adriatic (or Apulian) microplate was detached from Africa on the Jurassic/Cretaceous boundary (Schuster and Stüwe 2010). Ophiolite zones (sutures) were created on its margins when in the Middle Cretaceous continental collision started and the Austroalpine sediments were overthrust over the Penninic unit (Siegesmund et al. 2008). The compression had lasted to the early Paleogene, when the Penninic Ocean closed and was consumed by subduction underneath the Adriatic plate (Schuster and Stüwe 2010). Between 135 and 15 Ma, the nappe systems of the Alps formed in connection with south- to southeast-dipping subduction (Froitzheim et al. 2008). Through ^{40}Ar dating high-pressure metamorphism the time of subduction reaching to over 20 km depths can be estimated at 90 Ma BP (Berger and Bousquet 2008). Intraoceanic subduction and the emplacement of ophiolite nappes onto the Adriatic margin are documented since the Middle Jurassic (Schuster et al. 2013) and continued even after the closure of the last ocean remnants (in the Eocene, ca. 40 Ma ago). Exposed in the Tauern Window, the youngest eclogites, attesting to these processes, are only about 32 Ma old (Froitzheim et al. 2008).

As a result of the TRANSALP project (Lüschen et al. 2004), two different concepts of continental collision in the area of the present Eastern Alps have emerged (Fig. 2.2):

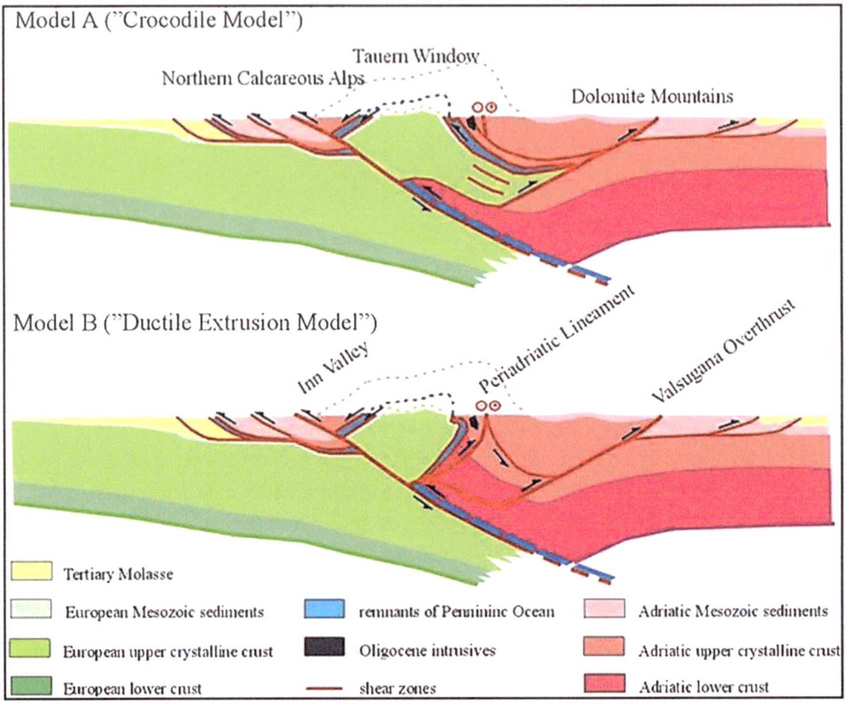

Fig. 2.2 The Ductile Extrusion and the Crocodile Models for continental collision (after Lüschen et al. 2004)

1. the Ductile Extrusion Model (Castelarin et al. 2001) is founded on the decisive role of the Peri-Adriatic Fault in the collision process: the fault of rigidly straight alignment controls the shape of the collision front and the exhumation of the Tauern Window, while
2. the Crocodile Model (Lammerer et al. 2001) does not attach much significance to the Peri-Adriatic Fault and assumes a z-shaped collision front ("double crocodile" structure).

The grade of Cretaceous metamorphism for the Upper Central Austroalpine Nappe (UCAN, comprising the Northern Graywacke Zone, the Graz Paleozoic, the Gurktal Nappe and the small Steinach Nappe) was significantly lower than in the Lower Central Austroalpine units (Schuster et al. 2013). The UCAN incorporates some slightly metamorphosed Paleozoic rocks in the Northern Karawanks, with pillow basalts and deep sea sediments. Cretaceous metamorphic rocks in the Eastern Alps include a wide range of schists with mica, garnet, staurolite, disthene, and felspar minerals (Schuster et al. 2004).

The beginning of the uplift of the southern mountain ranges (Ötztal, Gurktal Alps, and the 'Drauzug' range) along thrust faults, is dated to ca 92 Ma ago

(Schuster and Stüwe 2010). Accompanied by major lateral displacements, uplift can be reconstructed from the regional distribution of metamorphic rocks (Oberhänsli and Goffé 2004). The resulting late Cretaceous landscape resembled the present Adriatic coastal archipelago of Croatia.

After the closure of the Penninic Ocean, huge shearing forces, related to the final stage of collision, resulted in the slab break-off of the subducting oceanic lithosphere from the continental one and 43–24 Ma ago tonalitic-granodioritic plutons (Rieserferner, Ibisten-Unterplanken-Kandellen/Tesido-Planca-Gandella, Lesachtal, Hollbruck) formed at 10–15 km depth along the Peri-Adriatic Lineament and its branching faults (Nemčok et al. 1998; Schuster et al. 2013), in the zone of the lowest-grade metamorphism. The Karawanks tonalitic pluton (32–28 Ma ago) represents the easternmost outcrop of the Oligocene intrusions along the Peri-Adriatic Lineament (Fodor et al. 2008).

The Paleogene saw the beginning of a rapid west to east extension of the lithosphere in the Eastern Alpine area (Schuster and Stüwe 2010). Parallel to the subduction of the Carpathians, this led to marked lateral extrusion and gradual thinning of the Alps between 20 and 10 Ma ago (Froitzheim et al. 1994). At that time, the eastern foreland of the Eastern Alps began to take shape. Recently, field evidence (scratched flowstone in a cave disrupted by a fault) was found for the Holocene activity of the Salzachtal-Ennstal-Mariazell-Puchberg (SEMP) fault, a major strike-slip system responsible for this extrusion (Plan et al. 2010).

In the Late Oligocene, a prominent basement uplift of crystalline rocks emerged in the Tauern Window (Lammerer and Weger 1998). On its southern side, the steep Pustertal Fault, a section of the Peri-Adriatic Lineament, separates the Eastern Alps from the Dolomites. In the Tauern Massif, crustal thickening generated massive uplift between normal faults, the Brenner Fault in the west and the Katchberg Fault in the east (Pfiffner 2014). During the Miocene and Pliocene, the Penninic and even the Helvetic nappes were exhumed over a 20 × 150 km area (Bousquet et al. 2008). The rate of uplift here is still among the highest in the whole Alps, manifested in the rapid incision of the Isel and Möll river systems.

The tectonic evolution of the Dinarides (which started in the Eocene) influenced the Miocene deformations of the Dolomites and the Karawanks (Frisch et al. 2000). Further, plutons formed (e.g. the Pohorje Pluton, Fodor et al. 2008) near to the Peri-Adriatic Lineament. Between 10 and 7 Ma ago, intense basalt volcanic activity took place in Styria. Oligocene to Miocene flysch sedimentation (clay, marl, conglomerate, cross-bedded sandstones) occurred in the intermontane (Klagenfurt and Styrian/Graz Basins) and foreland basins. In the Oligocene and Miocene, deposits of similar character, molasse (e.g. 'Tonmergel' clay marl, 'Nagelfluh' conglomerate, marine turbidites), accumulated above the flysch, but this time with less tectonic deformation.

The next step in the evolution of the Alps includes the development of drainage patterns in the Miocene and Pliocene. The uplift of the Alps, manifested in large-scale thrusts and faults, generated weathering and intense (fluvial) erosion.

The ancestral rivers kept their courses on the substratum, preserved their gradient and stream power and incised deeply into the surface (Pfiffner 2014). The relative relief between mountain peaks and valley floors 5–15 Ma ago was probably similar to that of today.

In the Pleistocene, however, at least 15 cold periods and thousand-meter thick ice sheets brought about major transformation of the topography (van Husen 2000). Overdeepening may have affected the upper Drava Valley, too. Glacial and melt-water processes built a series of landforms of subglacial and glaciofluvial (melt-water) origin. Today such a landform assemblage, which prevailed in the Little Ice Age for the last time, is replaced by a variety of periglacial processes and landforms (Seppi et al. 2015). Mass movements are also very active geomorphic agents in the Alps (Székely et al. 2002; Proske and Bauer 2016).

From the analysis of Alpine river systems it was pointed out that the ice-free Mura and Drava catchments are characterized by channels in morphological equilibrium and only few evidence of strong tectonic activity is shown (Robl et al. 2008). Over the past 4 Ma uplift along the eastern edge of the Alps was estimated to less than 1 mm y^{-1} (Brückl et al. 2010) or 0.1–0.15 mm y^{-1} (Wagner et al. 2011)—one order of magnitude lower than geodetically observed uplift rates. Recently, it was proven that about 90% of the geodetically measured ground uplift in the Alps can be explained by post-glacial rebound in response to deglaciation after the Last Glacial Maximum. An interesting finding of recent research (Tesauro et al. 2005) is that horizontal movements in the Southern Alps also continue to our days: there is an average displacement in north-northwestern direction at a rate of 1.2 mm y^{-1}.

2.3 Geological Evolution of the Lowland Catchment

The Lower Drava Basin has a basement of medium-grade metamorphic rocks and carbonates which is overlain by a more than 6000-m thick sedimentary succession: Lower Miocene terrestrial conglomerates, sandstones, and marls; Middle Miocene deep marine marls and clays and shallow marine limestones. A major regional unconformity is observed between Middle Miocene synrift and Upper Miocene (Pannonian) postrift sediments (Kókai and Pogácsás 1991). The sediments derived from lacustrine deposition in Lake Pannon until 6.8 Ma ago; from this time a south-eastward prograding delta system controlled sedimentation (Haas 2012). In the Pleistocene, the Drava already followed the west-northwest to east-southeast directed axis of the depression, which was gradually shifting in southwestern direction.

2.4 Geological Units and Remarkable Landforms

To understand the problems concerning the Drava River, the physical environment of its drainage area has to be presented to the reader. Therefore, brief summaries of the geographical location, geological buildup, topographic character, and typical landforms of the landscape units the river crosses are provided in this section.

2.4.1 *Dolomites*

The Drava rises on the northern slopes of the Dolomites above the Puster Valley, near Dobbiaco/Toblach (Italy), at 1,450 m elevation (Sailer and Scaglione 2011). The major magmatic and tectonic events which influenced the evolution of the Dolomites include Permian and Triassic volcanism and rifting; rifting associated with the opening of the western Neotethys and Neogene compression and thrusting, resulting in uplift over the past 10 Ma (Bosellini 1998; Bosellini et al. 2003). Permian volcanics, evaporates, and carbonates are overlain by Lower to Middle Triassic shallow and deep marine carbonates and volcanics (Schlager and Thalmann 1962; Hauser 1995). The Middle and Upper Triassic dolostones (*Dolomia Principale*) are the most characteristic formations.

Typical landforms of this high-mountain relief are the dolomite towers (Neukirchen 2011) with subvertical slopes, of which the best known is the Tre Cime di Lavaredo/Drei Zinnen (2,999 m). Rapid mechanical weathering produces extensive debris slopes at mountain feet. Only a small eastern section of the Dolomites belongs to the Drava catchment.

2.4.2 *Carnic Alps*

The geology of the main ridge of the Carnic Alps (highest peak: Hohe Warte, 2,780 m), which runs from west to east along the Lesachtal and Gail Valleys, is very complex (Schönlaub 2012). The valleys (in fact, a single valley with two names) are part of the Peri-Adriatic Lineament system, the geological boundary between the South Alpine and the Austroalpine nappe systems (Krainer 1995). With the fault running along the axis of the valley in both Pustertal, the valley of the Drava, and the Gail Valley, rocks on both sides are quite different. While to the north the Zillertal Alps and Hohe Tauern are composed of crystalline rocks (granite, gneiss), to the south carbonates (limestones, dolostones) prevail. In the Kellerwand and Hohe Warte region 1,300-m thick Paleozoic (Silurian to Carboniferous) carbonates rise to 2,000 m elevation (Buchenauer 1986).

In addition to the richness in rock types, the mountains are made even more interesting by geomorphic features: the 1,500-m-high cliffs (Kellerwand), narrow gorges (Garnitzenklamm), an ice cave (at Obstans, with entrance at 2,300 m elevation), finger lakes (Weiβensee), landslide-dammed lakes (Bodensee), and the richness in Paleozoic and Mesozoic fossils (graptolites, ammonites, sea urchins, sea lilies, bivalves, gastropodes).

2.4.3 High Tauern

The High Tauern Mountains are the highest not only in the Drava catchment, but also in the entire Eastern Alps (Grossglockner, 3,798 m). Only the southern (Carinthian) slopes belong to the catchment. The foremost geological curiosity is the exhumation of the Penninic and Helvetic Nappe Systems from below the Austroalpine Nappes (the Tauern Window) (Krainer 2005). Its severe Miocene overprint by doming and lateral extrusion was most probably triggered by high pressure related to the indentation of a microplate from the south (Schmid et al. 2013).

The Mura, the most important tributary to the Drava, rises in the High Tauern, near Mur, at 1,898 m elevation. Sights of geomorphological interest in the southern foreland of the High Tauern include the earth pyramids at Stronach, carved out by erosion from morainic deposits and the gorges of Daberklamm (incised in calcareous mica schicht) and the Iselschlucht (formed along a north-to-south fault).

2.4.4 Gailtal Alps

The Gailtal Alps is a narrow mountain range wedged between the valleys of the Drava and Gail Rivers. From the west the range starts with the anticline of the Lienz Dolomites (Fig. 2.1), which rise at the confluence of the Drava with the Isel. The Isel is a glacier outlet with average discharge (39 $m^3\ s^{-1}$) almost three times larger than the Drava (14 $m^3\ s^{-1}$). The highest peak, Große Sandspitze (2,770 m) rises more than 2,000 m above the Drava Valley. The mountains are mainly built of Upper Triassic Hauptdolomit. A spectacular gorge is the about 2-km-long Galitzenklamm, where, in addition to main dolomite, sandstone and reddish limestone strata are also exposed.

The Gailtal Alps in a narrow sense is a 65-km-long and 15-km-wide mountain range (Reiβkofel, 2,371 m), part of the so-called 'Drauzug' ranges of west to east strike along the Peri-Adriatic Lineament. Their mostly carbonate rocks range in age from Permian to Mesozoic (high-mountain topography) in the north with some Paleozoic magmatics and quartz phyllites (middle-mountain topography) in the south (Figs. 2.3 and 2.4).

Fig. 2.3 View of the Lienz Dolomites from Ainet, near Lienz, East Tyrol (photo by D. Lóczy)

Fig. 2.4 View of the crest of the Southern Krawanks from Pyramidenkogel. *Source* summitpost.org

2.4.5 Gurktal Alps

The Gurktal Alps are mountain ridges rising between the Mura and Gurk Rivers above 2,000 m (Eisenhut, 2,441 m) (Schuster and Stüwe 2010). In the east, the Metnitz Valley splits the range into two ridges, mostly built of numerous varieties of Paleozoic metavolcanic rocks, phyllites, mica schists, siliceous schists (constituting the Alpine 'Altkristallin') as well as Triassic marbles and dolomites. The peaks are typically rounded knocks, locally called Nockberge. At higher elevations glacial landforms dominate: tarn lakes occupy glacial cirques.

2.4.6 Klagenfurt Basin

Surrounded by the phyllites and greenschists of the Gurktal Nappe, the Klagenfurt basin has a basement of slightly metamorphosed Paleozoic sediments and volcanics. With 1,750 km^2 area, it is the largest intramontane basin of the Alps. The west-to-east stretching Sattnitz Range in the south is composed of Tertiary flysch conglomerates and coal-bearing fine sediments. The basin fill is thick morainic and glaciofluvial gravelly-sandy deposits, resulting from the 700-m-thick Pleistocene ice sheet and its melting.

2.4.7 Karawanks

The Karawanks are a double range of 120 km length and 20–40 km width. The Southern Karawanks of Triassic limestones (highest point: Hochstuhl 2,237 m) rise in the eastern continuation of the Carnic Alps, while the more dissected Northern Karawanks (Hochobir, 2,139 m) belong to the 'Drauzug' ranges (granite and tonalite overlain by Mesozoic carbonates) (Poltnig and Herlec 2012). Orogeny is also manifested in a lateral shift over a distance of 250 km to the east (dextral displacement) along the Peri-Adriatic Lineament, which separates both ranges. Tectonics also brought about several 'flower structures' as well as Paleogene gabbro and syenite and Oligocene tonalite intrusions and volcanic pyroclastics in the south. Various grades of metamorphism can be observed in the suture zone of continental plates. The sediments of the Palaeo- and Neotethys Oceans (Ordovician to Lower Cretaceous platform carbonates) are predominant. Ore formation produced lead molybdate (wulfenite) and zinc deposits at Topla. Topla is the type locality of the rare tourmaline variety, dravite, named after the Drava River.

The high-mountain topography is shaped by glacial, glaciofluvial, and karst processes. The best known river gorge is the Tscheppaschlucht, a collapsed cave with a natural arch. Potholes and plunge pools below waterfalls are both carved by intense evorsion.

2.4.8 Seetal Alps–Saualpe

Together with its northern continuation, the Seetal Alps (highest point: Zirbitzkogel, 2,396 m) and the Saualpe (highest point: Ladinger Spitz, 2,079 m) are part of the crystalline Eastern Alps stretching between the valleys of the Mura and the Drava (Schönenberg and Weissenbach 1975). The most common rock types, according to the grade of metamorphism, are mica schist, gneiss, phyllite, quartzite, and different volcanics (Fritsch 1964). The Permo-Mesozoic carbonate cover has been stripped from the basement during the Cretaceous subduction, but some detached limestone or dolomite klippen (like Ulrichsberg and the 175-m-high hill with the castle of Hochosterwitz) rise above the middle-mountain ridges with subdued relief.

2.4.9 Koralpe

The Koralpe rise between the Drava and Sulm Valleys. The highest peak is Großer Speikkogel (2,140 m), which is built of magmatic rocks (gabbro—Schuster and Stüwe 2010). The same pluton is also exhumed in the Slovenian Pohorje. The Permian to Mesozoic rocks were completely stripped off during an early Alpine orogenic event in the Lower Cretaceous, by the Pleistocene glaciations and post-glacial erosion. Thus, the metamorphic (gneiss) ridges of the Koralpe-Wölz Nappe were heavily denuded into flat surfaces (Rantitsch et al. 2009). The occurrence of eclogite and pegmatite (with spodumen, a lithium aluminium inosilicate) makes the Koralpe a mountain range of geological and economic interest.

2.4.10 Pohorje

South of the Drava, the Pohorje is an Alpine mountain block of ca 50 km east-to-west and 30 km north-to-south extension (840 km^2). The highest peak is Črni Vrh, 1,943 m (in German: Schwarzkogel). Its core is a late Early Miocene tonalitic pluton (with granodiorite and dacite) surrounded with Paleozoic metamorphic rock (Fodor et al. 2008). (The only known occurrence of cizlakite, quartz monzogabbro, a green plutonic rock, in the world is located near the village Cezlak of the Pohorje.) In the southern Pohorje white marble was already quarried in Roman times. “The highest pressures of the Cretaceous metamorphism of the entire Alps are reached in the eclogites of the Pohorje massif (3.0–3.1 GPa, 760–825 °C)” (Fodor et al. 2008).

2.4.11 Low Tauern

The Low Tauern (Niedere Tauern) is the northern watershed of the Mura basin. It embraces three mountain blocks

- the Schladming Tauern (highest point: Hochgolling, 2,863 m) in the west;
- the Wölz Tauern in the centre (Greim, 2,474 m) and
- the Rottenmann Tauern (Großer Bösenstein, 2,449 m) in the east.

The rocks are mostly metamorphic (gneiss, mica schist, phyllite, marble), greywacke, and quartzite. The topography is typically glacial with troughs, cirques, tarns, and morainic landforms.

All sections of the mountains abound in picturesque tarns. In the foreland of the Wölz Tauern the Puxerloch is an opened-up cave with castle ruins.

2.4.12 Eisenerz Alps

Between the Low Tauern and the Hochschwab, on the northernmost boundary of the greywacke zone (Schönlaub 1982), the Eisenerz mountains rise to 2,165 m (Eisenerzer Reichenstein). The constituent rocks are mostly Paleozoic greywacke, Devonian metamorphosed calcareous rocks and Triassic schists. The Mesozoic of the carbonate nappe (Kalkalpendecke) only occur on the northern slopes, outside the Mura catchment.

2.4.13 Hochschwab

To the east the greywacke zone continues in the Hochschwab massif (elevation: 2,277 m). This mountain block of 590 km^2 area (mostly outside the Mura catchment), however, is built, in addition to Paleozoic rocks in the southern part, of Triassic shales and carbonates (limestones and dolomites) in the north. Hochschwab is a good example of crested glaciokarst. The massif is infamous of major landslides and famous for its caves (including the 20.2-km-long Frauenmauerhöhle in Middle Triassic limestone).

2.4.14 Schneealpe

The Schneealpe (1,903 m) is another heavily karstified massif of Triassic limestones with hundreds of caves and dolines. It is among the areas from where the First Vienna Water Main gains drinking water for the Austrian capital.

2.4.15 Fischbach Alps

The Fischbach Alps (highest summit: Stuhleck, 1,782 m) stretch from Bruck an der Mur to the Semmering Pass in the east. Mostly built of schists, Mesozoic carbonates and quartzites occur in the east. The 1.3-km-long Bärenschützklamm gorge of the Mixnitz stream in Almenland is a popular touristic destination.

2.4.16 Styrian (Graz) Basin

Subsidence in this major intramontane basin of the Eastern Alps is due to a single phase of synrift extension in the early Middle Miocene (Ebner and Sachsenhofer 1991). The 4,000-m-thick sequence of marine, lacustrine, and fluvial sediments is interrupted by volcanics of 17–13 Ma age. Late Miocene basin fill mostly consists of limnic-fluviatile deposits, which were uplifted during the Pliocene (Pfiffner 2014). The present hilly topography is formed by erosional processes on these sediments and a second suite of basaltic volcanics. A good example of the 40 basalt volcanoes with tuff rings is Riegersburg.

2.4.17 Pannonian (Carpathian) Basin

Leaving the Styrian Basin, the Drava and the Mura enter the Pannonian Basin. It is a classical back-arc basin, developed as a result of the extension and related thinning of continental lithosphere formed in the Miocene in response to the rapid rollback of a continental slab (Balla 1986; Horváth et al. 2006; Matenco and Radivojević 2012). The extension probably began ca 20 Ma ago, followed by a peak tectonic activity along normal faults during Middle Miocene times. From the Miocene to the Quaternary boundary, the counterclockwise rotation of the indenting Adriatic subcontinent determined tectonic evolution and the inversion of the whole basin (Fodor et al. 1999).

The Drava crosses the southwestern portion of the basin, the Drava Graben, in physical geography: the Drava Plain, with a long subsidence history and a thick sedimentary fill accumulated in Lake Pannon and later by river deltas.

2.4.18 Mura Hills (Goričko and Prekmurje)

The Mura Hills (404 m) are carved by water erosion from marine and fluvial sediments deposited in the Mura Graben. After the subsidence stopped, parallel river channels were oriented to flow more and more towards the south. In the

Pleistocene, the deposits of Alpine glaciers (pebbles and sands) were transported here and accumulated in terraces covered by finer materials.

2.4.19 Zala Hills

Another typical Pannonian hilly landscape is found in the Zala Hills (highest point: Kandikó, 302 m). The Paleozoic-Mesozoic metamorphic basement is overlain by Mesozoic and Paleogene marls and limestones. In the Miocene terrestrial (fluvial gravels) and marine siltstone, clay) also accumulated in the sedimentary basin (Hámor 1998). Uplift began after Lake Pannon filled up. The present-day hilly landscape was formed by Pleistocene fluvial erosion. Mineral oil reserves gave economic importance to the hills.

2.4.20 Ivanščica

Part of the Croatian Zagorje region, the Ivanščica Mountains rise to 1,061 m above the neighboring plains. Prevalently composed of Triassic carbonates with some clastic sediments, remnants of obducted ophiolite derived from the Neotethys Ocean are also exposed here. The landforms are of karstic origin. Volcanic eruptions 22 Ma ago produced the Lepoglava agate, a semiprecious stone.

2.4.21 Kalnik

In Mt. Kalnik (highest point: 642 m) large Triassic basalt blocks incorporated into the Jurassic ophiolite mélange are exposed. Middle Triassic to Lower Cretaceous interlayered chert, siliceous shale, mudstone, and limestone strata also occur.

2.4.22 Bilo

The Bilo Hills (Rajčevica, 309 m) of northwest to southeast strike stretch parallel with the lower Drava along 80 km length in Croatia. The low and wide ridges are built up of Quaternary clastic sediments.

2.4.23 Southern Zselic

This hilly region is a system of alluvial fans of Miocene-Pliocene sands and clays, covered with Pleistocene loess and heavily dissected by fluvial erosion. The highest point is 265 m.

2.4.24 Inner Somogy Hills

Inner Somogy is built up of 100–400 m of Neogene sandy-gravelly sediments and 5–8 m of Pleistocene blown sand. In the centre the north-to-south Marcali loess ridge rising to 192 m. The Somogy Hills are actively affected by compression and strike-slip type movements along lineaments related to the Peri-Adriatic fault system. Where the Drava undercut the hilly ridge reaching the channel at right angles, the high bluff of Heresznye was created (Fig. 2.5).

Fig. 2.5 The high bluff of Heresznye above the Drava, Hungary (photo by D. Lóczy)

2.4.25 Mecsek

The southern slopes of the Mecsek Mountains (highest point: Zengő, 682 m), is among the most varied in rock composition in Hungary (Budai and Gyalog 2010), also belong to the Drava catchment. The oldest formations are Paleozoic metamorphics and Carboniferous granite, found in the east. The western part is a huge anticline of Permian–Lower Triassic terrestrial siliciclastics and Middle Triassic marine carbonates. The eastern Mecsek is syncline dominated by Jurassic shallow to deep marine marls, siltstones, sandstones, and limestones overlain by a Lower Cretaceous basalt volcanic complex. In the middle section of the mountains, Miocene andesite occurs and Miocene sediments are also exposed.

Sites of geomorphological heritage are the Abaliget cave in Middle Triassic limestone and the 'Puppet' stones of Permian–Lower Triassic sandstones and conglomerates.

2.4.26 Villány Hills and Baranya Hills

South of the Mecsek Range, the Villány Hills (highest point: Szársomlyó, 442 m) show an imbricate structure consisting mainly of Mesozoic carbonates; Triassic shallow marine dolomites and limestones, a condensed and discontinuous marine Jurassic succession and a thick Lower Cretaceous shallow-marine limestone sequence. The most important sites for fossils are Csarnóta for Pliocene small mammals and the Templom Hill of Villány for Jurassic ammonites and belemnites.

The area between the Mecsek and the Villány Hills is a dissected, loess-mantled hilly region of Miocene-Pliocene marine sediments, the Baranya Hills (highest point of the western part: 260 m).

2.4.27 Papuk

In the Papuk (highest point: 953 m), the rocks represent various geological environments, spanning 600 Ma of earth history. Some of the oldest rocks in Croatia are found here. The core is built of Lower Paleozoic igneous rocks (granite) and their metamorphic zone (phyllite, chlorite schists, gneiss, migmatites, amphibolite), overlain on the margins by Permo-Triassic carbonates. There are numerous karst features: swallow holes, sinkholes, caves, pits. Sediments abound in fossils (e.g. Carboniferous ferns, Miocene bivalves and gastropodes). Miocene volcanics are also present. Large amounts of debris removed from the Papuk by erosion over millions of years filled up the Sava and Drava valleys.

2.4.28 Drava Plain (Podravina) and Kopački Rit

In the lower section of its valley, the Drava built a broad floodplain (Podravlje in Slovenia, Podravina in Croatia). Meanders at various stages of their development and heavily affected by river regulation works are typical, particularly along the Croatian-Hungarian course.

Finally, the lowermost area of the Drava catchment has to be mentioned. The confluence with the Danube is at 80 m elevation. Kopački Rit (Kopačevo/Kopács swamp) is one of the largest alluvial wetlands in Europe. The dynamics and intensity of flooding controls the environment: the mosaical marshes, bogs and open water surfaces, ponds, and canals (Fig. 2.6).

2.5 Conclusions

The geology of the Drava drainage basin is rather complicated. The complex structure of Alpine nappes, the tectonic processes, the diverse lithology and the resulting variety in relief are impossible to describe and explain within a short chapter. The above presented glaciated high mountains, the karst landforms of the carbonate massifs, the middle-mountain topography on metamorphic rocks and exhumed intrusions locally with isolated klippen, the volcanic cones, the hilly regions in Tertiary sedimentary basins and mountain forelands and the alluvial and terrace surfaces make the landscape of the Drava catchment extremely varied and picturesque.

On the Drava-Mura catchment numerous examples are found for the adjustment of the drainage network to the geological buildup. Major rivers follow zones of weakness (lineaments or rock stripes less resistant to erosion), best attested by the pattern of parallel ranges and river valleys of the ‘Drauzug’. The intricate tectonic pattern of uplifting and subsiding surfaces also exerts a major control not only on the alignment of drainage lines, but also on the longitudinal profile of the Drava and its tributaries.

The weathering and erosion of the highly variable rocks of the mountains contributed to thick accumulations of sediment in the valleys and mountain forelands, the youngest geological formations, also exerting major control on the hydromorphology of rivers in the Drava catchment. Uplift and climatic changes resulted in the incision of channels of the Drava and tributaries, creating a sequence of fluvial terraces.

Fig. 2.6 The wetland of Kopački rit, Croatia (photo by D. Lóczy)

References

Ampferer O (1906) Über das Bewegungsbild von Faltengebirgen. Jahrbuch der Geologischen Reichsanstalt, Wien 56:539–622

Balla Z (1986) Paleotectonic reconstruction of the central Alpine-Mediterranean belt for the Neogene. Tectonophysics 127:213–243

Bartel EM, Neubauer F, Genser J, Heberer B (2014) States of paleostress north and south of the Periadriatic fault: comparison of the Drau Range and the Friuli Southalpine wedge. Tectonophysics 637:305–327. https://doi.org/10.1016/j.tecto.2014.10.019

Berger A, Bousquet R (2008) Subduction-related metamorphism in the Alps: a review of isotopic ages based on petrology and their geodynamic consequences. In: Siegesmund S, Fügenschuh B, Froitzheim N (eds) Tectonic aspects of the Alpine-Dinaride-Carpathian system, vol 298. Geological Society London, Special Publications, pp 117–144

Bosellini A (1998) Geologie der Dolomiten. Athesia, Bozen/Bolzano, 92 pp

Bosellini A, Gianolla P, Stefani M (2003) Geology of the Dolomites. Episodes 26:181–185

Bögel H, Schmidt K (1976) Kleine Geologie der Ostalpen. Ott Verlag, Thun. 231 pp

Bousquet R, Oberhänsli R, Goffé B, Wiederkehr M, Koller F, Schmid SM, Schuster R, Engi M, Berger A, Martinotti G (2008) Metamorphism of metasediments in the scale of an orogen: a key to the tertiary geodynamic evolution of the Alps. In: Siegesmund S, Fügenschuh B, Froitzheim N (eds) Tectonic aspects of the Alpine-Dinaride-Carpathian system, vol 298. Geological Society, London, Special Publications, pp 393–412

Brückl E, Behm M, Decker K, Grad M, Guterch A, Keller GR, Thybo H (2010) Crustal structure and active tectonics in the Eastern Alps. Tectonics 29:TC2011. https://doi.org/10.1029/2009tc002491

Buchenauer L (1986) Karnische Alpen. Leykam, Graz – Wien

Budai T, Gyalog L (eds) (2010) Geological map of Hungary for tourists, 2nd, revised and enlarged edition. Geological Institute of Hungary, Budapest, 276 pp

Castelarin A, Dal Piaz GV, Battista Vai G, TRANSALP Working Group (2001) The TRANSALP seismic profile: remarks and interpretations from the Italian side. Geologische und Paläontologische Mitteilungen, Innsbruck 25:62–64

Ebner F, Sachsenhofer RF (1991) Die Entwicklungsgeschichte der Steirischen Tertiärbeckens. Miteilungen der Abteilung der Geologie und Paläontologie, Landesmuseum Joanneum 49:167–262

Fodor L, Csontos L, Bada G, Győrfi I, Benkovics L (1999) Tertiary tectonic evolution of the Pannonian Basin system and neighbouring orogens: a new synthesis of palaeostress data. Geological Society, London, Special Publications 156:295–334. https://doi.org/10.1144/GSL.SP.1999.156.01.15

Fodor LI, Gerdes A, Dunkl I, Koroknay B, Pécskay Z, Trajanova M, Horváth P, Vrabec M, Jelen B, Balogh K, Frisch W (2008) Miocene emplacement and rapid cooling of the Pohorje pluton at the Alpine-Pannonian-Dinaridic junction, Slovenia. Swiss J Geosci 101(Supplement 1):S255–S271

Frisch W (1979) Tectonic progradation and plate tectonic evolution of the Alps. Tectonics 60:11–15

Frisch W, Székely B, Kuhlemann J, Dunkl I (2000) Geomorphological evolution of the Eastern Alps in response to Miocene tectonics. Zeitschrift für Geomorphologie 44(1):103–138

Frisch W, Meschede M, Blakey RC (2011) Plate tectonics: continental rift and mountain building. Springer, Berlin–Heidelberg, pp 176–179

Fritsch W (1964) Mittelkärnten. Das Kristallin der Saualpe und die Oberkreide (Eozän) des Krappfeldes. Mitteilungen der Geologischen Gesellschaft in Wien 57(1):331–351

Froitzheim N, Schmid SM, Conti P (1994) Repeated change from crustal shortening to orogen-parallel extension in the Austroalpine units of Graubünden. Eclogae Geol Helv 87 (2):559–612

Froitzheim N, Plasienka D, Schuster R (2008) Alpine tectonics of the Alps and Western Carpathians. In: McCann T (ed) The geology of Central Europe. Geological Society, London, pp 1141–1232

Froitzheim N, Schmid SM (eds) (2008) Orogenic processes in the Alpine collision zone. Swiss J Geosci 101(Supplement 1):S1–S310

Haas J (2012) Geology of Hungary. Springer Science & Business Media, Dordrecht, 246 pp

Hámor G (1998) Magyarország miocén rétegtana (Miocene stratigraphy of Hungary). In: Magyarország geológiai képződményeinek rétegtana (Stratigraphy of the geologicalformations in Hungary). MOL Rt – MÁFI, Budapest, pp 437–453 (in Hungarian)

Hauser C (Hrsg) (1995) Geologie von Osttirol. Arbeitstagung 2.-6. Oktober 1995, Lienz, Osttirol. Geologische Bundesanstalt, Wien. 182 pp

Horváth F, Bada G, Szafian P, Tari G, Adam A, Cloetingh S (2006) Formation and deformation of the Pannonian Basin: constraints from observational data. In: Gee DG, Stephenson RA (eds) European lithosphere dynamics. Geological Society Memoirs 32(1):191–206. https://doi.org/10.1144/gsl.mem.2006.032.01.11

Kókai J, Pogácsás Gy (1991) Tectono-stratigraphical evolution and hydrocarbon habitat of the Pannonian basin. In: Spencer AM (ed) Generation, accumulation and production of Europe's hydrocarbons. Oxford University Press, Oxford, pp 307–317 (Special Publication of the European Association of Petroleum Geoscientists 1)

Krainer K (1993) Late and post-Variscan sediments of the Eastern and Southern Alps. In: von Raumer J, Neubauer F (eds) Pre-Mesozoic geology of the Alps. Springer Verlag, Heidelberg, pp 537–564

Krainer K (1995) Der Geologische Aufbau Kärntens. In: Vom Reichtum der Erde. Landesausstellung "Grubenhunt & Ofensau" Hüttenberg, Kärnten, 29. April–29. Oktober 1995. Bd. 2. Beiträge. Kärntner Landesausstellungsbüro, Klagenfurt, pp 25–39

Krainer K (2005) Nationalpark Hohe Tauern – Geologie. Universitätsverlag, Klagenfurt

Lammerer B, Weger M (1998) Footwall uplift in an orogenic wedge: the Tauern Window in the Eastern Alps of Europe. Tectonophysics 285(3–4):213–230
Lammerer B, Lüschen E, Schönrock J, TRANSALP Working Group (2001) Interpretation of the TRANSALP section according to the "Crocodile Model". Geologische und Paläontologische Mitteilungen, Innsbruck 25 pp
Lüschen E, Lammerer B, Gebrande H, Millahn K, Nicolich R, TRANSALP Working Group (2004) Orogenic structure of the Eastern Alps, Europe, from TRANSALP deep seismic reflection profiling. Tectonophysics 249:85–102
Matenco L, Radivojević D (2012) On the formation and evolution of the Pannonian Basin: constraints derived from the structure of the junction area between the Carpathians and Dinarides. Tectonics 31:TC6007. https://doi.org/10.1029/2012tc003206
Nemčok M, Pospisil L, Lexa J, Donelick RA (1998) Tertiary subduction and slab break-off model of the Carpathian-Pannonian region. Tectonophysics 295(3–4):307–340
Neukirchen F (2011) Bewegte Bergwelt. Gebirge und wie sie entstehen. Spektrum Akademischer Verlag, 232 S
Oberhänsli R, Goffé B (2004) Explanatory notes to the map: metamorphic structure of the Alps: introduction. Mitteilungen der Österreichischen Mineralogischen Gesellschaft 149:175–199
Oberhauser R (ed) (1980) Der geologische Aufbau Österreichs. Springer Verlag, Wien, 703 pp
Pfiffner OA (2014) Geology of the Alps. Revised and updated translation of Geologie der Alpen, vol 2. Wiley Blackwell, Chichester, 376 pp
Plan L, Grasemann B, Spötl C, Decker K, Boch R, Kramers J (2010) Neotectonic extrusion of the Eastern Alps: constraints from U/Th dating of tectonically damaged speleothems. Geology 38 (6):483–486. https://doi.org/10.1130/G30854.1
Poltnig W, Herlec U (2012) Geologisch-naturschutzfachliche Grundlagen des Geoparks Karawanken. Geopark Karawanken, Bad Eisenkappel, Carinthia, Austria. http://www.geopark-karawanken.at/files/2012_geopark_karawanken_naturschutzfachliche_grundlagen.pdf
Proske H, Bauer Ch (2016) Indicative hazard maps for landslides in Styria, Austria. Acta Geobalcanica 2(2):93–101. https://doi.org/10.18509/AGB.2016.10
Rantitsch G, Pischinger G, Kurz W (2009) Stream profile analysis of the Koralm range (Eastern Alps). Swiss J Geosci 102(1):31–41. https://doi.org/10.1007/s00015-009-1305-5
Robl J, Hergarten S, Stüwe K (2008) Morphological analysis of the drainage system in the Eastern Alps. Tectonophysics 460(1–4):263–277. https://doi.org/10.1016/j.tecto.2008.08.024
Sailer G, Scaglione, P (2011) Dolomiti Geoscape: Geography + Geology = Landscape
Scarascia S, Cassinis R (1997) Crustal structures in the central-eastern Alpine sector: a revision of the available DSS data. Tectonophysics 271:157–188
Schlager W, Thalmann F (1962) Zur Geologie Der Östlichen Lienzer Dolomiten. Geologische Neuaufnahme Des Kammzuges Zwischen Mur-und Zederhaustal, Gesellschaft Der Geologie-und Bergbaustudenten in Wien, Wien
Schmid SM, Fugenschuh B, Kissling E, Schuster R (2004) Tectonic map and overall architecture of the Alpine orogen. Eclogae Geol Helv 97(1):93–117
Schmid SM, Scharf A, Handy MR, Rosenberg CL (2013) The Tauern Window (Eastern Alps, Austria): a new tectonic map, with cross-sections and a tectonometamorphic synthesis. Swiss J Geosci 106:1–32. https://doi.org/10.1007/s00015-013-0123-y
Schönenberg R, Weissenbach N (1975) Geologie der Saualpe. Sonderband, Clausthaler geologische Abhandlungen, 232 pp
Schönlaub HP (1982) Die Grauwackenzone in den Eisenerzer Alpen (Österreich). Jahrbuch der Geologischen Bundesanstalt 124:361–423
Schönlaub HP (2012) Linking geology between the Geoparks Carnic and Karavanke Alps across the Periadriatic Line. Institute of the Republic of Slovenia for Nature Conservation, Ljubljana. http://www.zrsvn.si/dokumenti/64/2/2012/Geology_Geoparks_Carnic_and_Karavanke_Alps_across_Periadriatic_Line_TEXT_3094.pdf
Schönlaub HP, Heinisch H (1993) The classic fossiliferous Paleozoic units of the Eastern and Southern Alps. In: von Raumer J, Neubauer F (eds) Pre-Mesozoic geology of the Alps. Springer Verlag, Heidelberg, pp 395–422

Schuster R, Koller F, Hoeck V, Hoinkes G, Bousquet R (2004) Explanatory notes to the map: metamorphic structure of the Alps—metamorphic evolution of the Eastern Alps. Mitteilungen der Österreichischen Mineralogischen Gesellschaft 149:175–199

Schuster R, Kurz W, Krenn K, Fritz H (2013) Introduction to the Geology of the Eastern Alps. Berichte der Geologischen Bundesanstalt 99:121–133 (11th Workshop on Alpine Geological Studies and 7th IFAA)

Schuster R, Stüwe K (2010) Die Geologie der Alpen im Zeitraffer. Mitteilungen des naturwissenschaftlichen Vereines für Steiermark 140:5–21

Seppi R, Zanoner T, Carton A, Bondesan A, Francese R, Carturan L, Zumiani M, Giorgi M, Ninfo A (2015) Current transition from glacial to periglacial processes in the Dolomites (South-Eastern Alps). Geomorphology 228:71–86. https://doi.org/10.1016/j.geomorph.2014.08.025

Siegesmund S, Fügenschuh B, Froitzheim N (eds) (2008) Tectonic Aspects of the Alpine-Dinaride-Carpathian system. Geological Society, London. 448 pp (Special Publications 298)

Stüwe K, Schuster R (2010) Initiation of subduction in the Alps: continent or ocean? Geology 38(2):175–178

Székely B, Reinecker J, Dunkl I, Frisch W, Kuhlemann J (2002) Neotectonic movements and their geomorphic response as reflected in surface parameters and stress patterns in the Eastern Alps. EGU Stephan Mueller Special Publication Series 3:149–166

Tesauro M, Hollenstein C, Egli R, Geiger A, Kahle HG (2005) Continuous GDP and broad-scale deformation across the Rhine Graben and the Alps. Int J Earth Sci 94:525–537

Thöni M (2006) Dating eclogite-facies metamorphism in the Eastern Alps - approaches, results, interpretations: a review. Miner Petrol 88:123–148

van Husen D (2000) Geological processes during the Quaternary. In: Neubauer F, Höck V (eds) Aspects of geology in Austria. Mitteilungen der Österreichische Geologische Gesellschaft 92:135–256

Wagner T, Fritz H, Stüwe K, Nestroy O, Rodnight H, Hellstrom J, Benischke R (2011) Correlations of cave levels, stream terraces and planation surfaces along the River Mur: timing of landscape evolution along the eastern margin of the Alps. Geomorphology 134:62–78. https://doi.org/10.1016/j.geomorph.2011.04.024

Chapter 3
Land Use in the Drava Basin: Past and Present

Gerhard Karl Lieb and Wolfgang Sulzer

Abstract Current land use in the Drava River Basin (DRB) and its recent changes are detected from the data of the CORINE Land Cover (CLC) inventory, which allows a multitemporal analysis for the time period 1990–2012. Further data, like a DEM, were taken from the European Environmental Agency database. In order to provide a good overview, 11 land use classes were defined. Their distribution as well as their changes are shown in maps and analyzed statistically. The most striking results are (i) the clear dominance of the land use class forests (46% of the DRB), (ii) a pronounced differentiation of land use between the eastern and the western part of the catchment, both for natural and cultural reasons, and (iii) a relatively satisfactory environmental status, reflected among others by a high proportion of protected areas.

Keywords Land use change · Drava river basin · Protected areas CORINE landcover

3.1 Introduction

The Drava River Basin (DRB) is a second-order sub-catchment of the Danube River Basin (Somogyi et al. 1983) and comprises more than 40.000 km^2, i.e. 5.0% of the Danube River Basin (Sommerwerk et al. 2009). Comparing the different second-order tributaries of the Danube with one another, the DRB is of special interest according, among others, to the following aspects (data taken from Sommerwerk et al. 2009).

G. K. Lieb · W. Sulzer (✉)
Institute of Geography and Regional Sciences, University of Graz, Heinrichstrasse 36, Graz 8010, Austria
e-mail: wolfgang.sulzer@uni-graz.at

G. K. Lieb
e-mail: gerhard.lieb@uni-graz.at

D. Lóczy (ed.), *The Drava River*, Springer Geography,
https://doi.org/10.1007/978-3-319-92816-6_3

- The mean elevation of 760 m a.s.l. is exceeded only by another second-order catchment, that of the Inn River. This indicates the predominance of mountainous relief.
- Also with its mean annual precipitation of 1,121 mm, the DRB ranks second following the Inn catchment. This is due to the high precipitation in the alpine part of the catchment.
- The annual gross domestic product of 15,832 $ per inhabitant is the second highest, too, once again following the Inn catchment. However, the DRB's value is only the half of the Inn's value but twice the value of the entire Danube River Basin.

Naturally, these data hide the internal differentiation of the DRB which will be discussed in Sect. 3.3. However, economic welfare and especially mountainous relief show strong interdependencies with land use even on the entire catchment scale. Taking again the dataset of Sommerwerk et al. (2009) into account, the DRB shows outstanding data compared with other Danube tributaries in the following land use classes (which are named differently from those in the following text):

- Agricultural land covers only 28.7% of the basin, only the value of the Inn catchment is lower.
- Forests cover 45.8% of the DRB, which is the highest rate of all catchments (followed by the Sava catchment).
- Also the high values of natural grassland (9.0%; third highest) and sparse vegetation (3.9%, second highest) indicate once again mountainous relief.

In summary, the hydrological and land use characteristics seem to be largely influenced by the Alps. This is why this study first focuses on the topographical background and then on the historical developments which have strongly influenced land use patterns over time. Because land use and its change are considered important factors of river management, a closer look is taken into the most recent land use changes (Sect. 3.4), using the CORINE land cover dataset. Finally, land use is discussed with regard of nature protection and future perspectives.

3.2 Methods

As part of the CORINE program (Coordination of Information on the Environment), the European Union established a Europe-wide mapping of land cover and/or land use within the framework of the CORINE Land Cover project. A coordinated collection of information on the environment guarantees both a comparability of the data between the individual member states, as well as the possibility of addressing environmentally relevant questions and statements in the all-European context. Therefore, the result of this ongoing initiative provides comparable land use data from each of the DRB countries.

The main European Environmental Agency (EEA) data source is the Copernicus Land Monitoring Service which includes the CORINE Land Cover data set

(European Environmental Agency 2017a). The CORINE Land Cover (CLC) inventory was initiated in 1985; the data sets themselves were first elaborated in 1990. Updates have been produced in 2000, 2006, and 2012 and are based on the cooperation between EEA members, collaborating countries, and the Copernicus Programme (2017a). In 2018, a new inventory is planned. The concept and nomenclature of CLC is used as the quasi-standard for land cover and land use mapping in Europe. The data consists of an inventory of 44 classes. CLC uses a Minimum Mapping Unit (MMU) of 25 hectares (ha) for areal phenomena and a minimum width of 100 m for linear phenomena. The time series are complemented by change layers, which highlight changes in land cover with an MMU of 5 ha. The CORINE data sets of the DRB were provided by the Copernicus Land Monitoring Service (Copernicus Programme 2017b).

As already mentioned, the CLC nomenclature includes 44 land cover classes in a three-level hierarchy (European Environmental Agency 1999). The five "level 1" classes are "*artificial surfaces*", "*agricultural areas*", "*forests and semi-natural areas*", "*wetlands*" and "*water bodies*". In this study, specific CLC classes were merged to new land use classes in order to provide a special focus on catchment properties. This guarantees a better cartographic representation and clear (simplified) tables of specific CLC classes in the DRB. Finally 11 classes (Fig. 3.1) were used in the present CLC analysis:

1. *Built-up areas*: The original term is "artificial surfaces" which includes "urban pattern", "industrial, commercial and transport units", "mine dump and construction sites" and "artificial non-agricultural vegetated areas" of level 2.
2. *Arable land*: Instead of the level 1 class "Agricultural areas" the level 2 differentiation was used. All arable land in the DRB is "non-irrigated arable land".
3. *Permanent crops*: This class includes "vineyards", "fruit trees and berry plantations".
4. *Pastures*.
5. *Heterogeneous agricultural areas*: This class includes areas with "complex cultivation patterns" and "land principally occupied by agriculture, with significant areas of natural vegetation".
6. *Forests*: This class comprises "broad-leaved forest", "coniferous forest," and "mixed forest".
7. *Scrub and/or herbaceous vegetation associations*: Within this class "natural grassland", "moors and heathland," and "transitional woodland shrub" occur in the DRB.
8. *Open spaces with little or no vegetation*: This class consists of small parts of "beaches, dunes, sands" and largely of "bare rocks" and "sparsely vegetated areas".
9. *Glaciers and perpetual snow*: Widespread in the high mountain areas.
10. *Wetlands*: The level 1 CLC term was used and includes "inland marshes" and "peat bogs".
11. *Water bodies*: The class consists of "water courses" and "water bodies".

Fig. 3.1 Visual impressions of land use classes in the DRB (photos by Gerhad Karl Lieb)

The CLC data were analyzed together with a DEM (25 m) which was downloaded from the European Environmental Agency (2017b).

As far as the terms "land cover" and "land use" are concerned, the authors will use the latter one, being aware of the fact that land *cover* is the observed (bio)-physical substrate on the earth's surface. In contrast, land *use* is characterized by the arrangements, activities, and inputs people undertake in a certain land cover type to produce, change or maintain it. Definition of land use in this way establishes a direct link between land cover and the actions of people in their environment (FAO 2017).

3.3 Current Status and Background of Land Use

3.3.1 *Physical-Geographical Background of Land Use*

Land use patterns largely depend on natural conditions. Using the macro-scale European biogeographical division provided by the European Environmental Agency (2016) the DRB belongs to the "Alpine", the "Continental," and the "Pannonian Region". A similar subdivision was also used by Sommerwerk et al. (2009). Another approach, which allows a more detailed insight, was used by Mazúr et al. (1985) providing natural landscape types. According to their map, the entire DRB belongs to the temperate climate zone and has a share of two relief classes, lowlands and mountains. The latter class is further divided into the sub-classes "high mountains" (Alps) and "isolated mountain groups and mountains". The lowlands comprise a broad variety of different types of hilly regions (most widespread type: "accumulational-erosional hilly regions") and "floodplains" along the main rivers. Also within the mountainous region, there are a lot of subtypes ranging from "landscapes of the glacier and névé region" to "basins and wide intra-montane valleys" (which for tectonical reasons are of special significance for the southeastern Alps drained by the Drava River).

Going into more detail is, naturally, far beyond the scope of this chapter. Hence, further considerations are based on the rough distinction between the "alpine" and "pre-alpine region". To divide these two macroregions from each other, we used the Alpine perimeter information which is provided by the Alpine Convention (2017). This is reasonable because this political delineation is very close to the natural one (proved by Figs. 3.2 and 3.3) and is used most frequently in research on the Alps. Accordingly, the alpine area covers 55.3% of the DRB. On this basis, Table 3.1 shows the five land use classes which cover the largest portions of the two macroregions. The table clearly reflects the influence of physical conditions, especially by the limited occurrence of agricultural classes in the Alps. This is due to climatic and geomorphological restrictions (Fig. 3.2) which are also indicated by the large amount of open spaces with little or no vegetation. The latter are very widespread in the western part of the DRB with pronounced high-mountain character—glaciation is still widespread in the Hohe Tauern Range (glaciers and perpetual snow in 2012: 78 km^2) with Großglockner (3,798 m) as the highest summit in the catchment. It may be surprising that in both macro-regions forests are the dominant land use type. This cannot be explained by natural conditions, but mainly by the land ownership situation (Sect. 3.2).

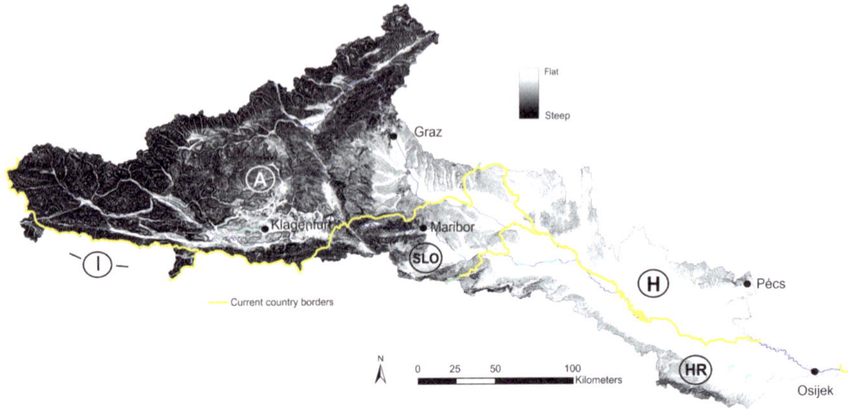

Fig. 3.2 Slope map of the DRB. *Data source* European Environmental Agency (2017b)

Table 3.1 The five most widespread land use classes in the alpine and pre-alpine regions of the DRB in 2012 (explanations in the text)

Alpine region			Pre-alpine region		
Land use class	km^2	%	Land use class	km^2	%
Forests	12,672.74	57.6	Forests	5,435.83	30.6
Scrub[a]	2,826.82	12.9	Arable land	4,871.89	27.4
Pastures	2,010.95	9.1	Heterog. agr.[a]	4,304.29	24.2
Open spaces[a]	1,536.26	7.0	Built-up area	916.23	5.2
Heterog. agr.[a]	1,314.37	6.0	Pastures	904.68	5.1
Others	1,629.10	7.4	Others	1,333.22	7.5
Total	21,989.24	100.0	Total	17,766.14	100.0

[a]*Scrub* scrub and/or herbaceous vegetation associations; *heterog. agr.* heterogeneous agricultural areas; *open spaces* open spaces with little or no vegetation

3.3.2 Historical Background of Land Use

As elsewhere in Central Europe, human presence dates back to prehistoric times in almost the entire DRB and the Romans were the first to establish infrastructure like roads and settlements to a larger extent. From the viewpoint of the present-day ethnicity, the immigration of Slavic tribes (earliest Middle Ages) from the East, the Hungarians (10th century) also from the East and the Bavarians (from the 8th century onwards) from the West was of great importance. Beginning with the turn of the first millennium, human settlement in the western and eastern parts of the DRB was different. Whereas the western part saw a quite continuous establishment of the Habsburg Empire with only a few interruptions of the development of cultural landscapes by wars, the lower eastern part can be labeled as a "region of borders and wars" (Schneider-Jacoby 1996). This is primarily due to the Ottoman expansion, which reached the DRB around 1500 and made its lowest section (approximately up to the mouth of the Mura River) part of the Ottoman Empire. At least until the end of the 17th century, vast areas on both sides of the Drava River remained largely uninhabited and even deserted (Schneider-Jacoby 1996). The re-conquest of the area driven by the Habsburgs brought back agricultural activities which, however, were regionally subordinated to the military tasks the farmers had to fulfill. As a consequence, large parts of the woodland encroached on former arable land during the Ottoman rule has remained until today which is one reason for the large amount of forested areas in the lower DRB and the importance of the region for nature conservation (Sect. 3.5). It was not before the 18th century that large clearings of the forests started again under the rule of mostly Hungarian nobility.

Industrialization in the 19th century was more intensive in the western part of the DRB where first industrial activities date back even to the 17th century based on (iron) ore mining. In the Austrian part of the Austro-Hungarian Empire, economic policy was much more in favor of industry than in Hungary, where agriculture played a major role. This is the reason why the border between Hungary and Austria, which was established as early as the 11th century, further remained an important boundary in terms of land use.

Finally, in the second half of the 20th century the eastern part of the DRB belonged to states which were governed by communist regimes: Yugoslavia (areas of present-day Croatia and Slovenia) created a moderate type of communism. Hungary was situated beyond the Iron Curtain with a political and economic system orientated closely to that of the Soviet Union. From a present day's view, the retarded economic development during the Ottoman period, the focus on agriculture in the Hungarian Kingdom and finally the communist past explain the fact that the eastern part of the DRB is still less developed than the western, although since 2013 the entire DRB belongs to the EU.

In summary, the historical retrospect makes it clear that over time several political borders (Fig. 3.3) influenced the long-term development of land use according to aspects such as:

- The large amount of forests in the eastern DRB—also in areas where physical conditions would allow other types of agriculture—are the consequence (i) of the occupation and subsequent expulsion of the Ottomans and (ii) of the establishment of large estates owned by nobility in the Hungarian Kingdom.
- The transformation of the traditionally widespread heterogeneous agricultural areas to large-scale arable land with huge plots predominantly took place during the communist period, when agriculture was organized by state-owned companies. Although the ownership structure has shifted back to private enterprises, land use itself has remained similar.
- In the former Yugoslav part of the DRB collectivization of arable land took place in limited areas (especially floodplains and fluvial terraces) only. This is why the land use pattern in Slovenia and Croatia is more similar to that of Austria than of Hungary.

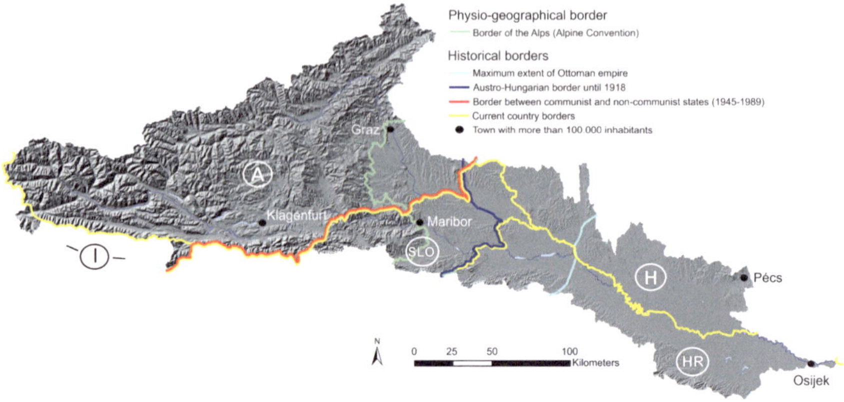

Fig. 3.3 Physical-geographical and historical boundaries (digitized manually from the maps in Bruckmüller and Hartmann 2011) influencing land use (for explanations see text)

- The difference in the economic development of the eastern and the western part of the DRB—mentioned in Sect. 3.1 and emphasized also by Sommerwerk et al. (2009)—is a consequence of the above historical processes. However, it is not directly visible in land use patterns but in environmental aspects which depend on land use (Sect. 3.5).

Summing up, the historical facts again roughly divide the DRB into a western and an eastern part. However, the dividing line is situated more to the east of the alpine boundary (Fig. 3.3).

3.3.3 Regionalization of Land Use

As already indicated, land use shows a quite clear pattern in the DRB. On the one hand, in the alpine region intensive land use is restricted due to high elevation, steep slopes, and gravitational natural hazards. The result of these restrictions is the widespread occurrence of semi-natural land use classes such as scrub and/or herbaceous vegetation associations and open spaces with little or no vegetation (Fig. 3.4), which show a pronounced concentration in the northern and western parts of the alpine region with large areas above the timberline. However, according to Table 3.1, the predominant land use type in the alpine region is forests, which also represent the most striking visual element in Fig. 3.4. Furthermore, built-up areas and agricultural land use classes are concentrated in the valley bottoms and inner-alpine basins, both limited in extension.

On the other hand, the pre-alpine part of the DRB offers favorable conditions for nearly all agricultural activities (Sect. 3.3.1)—although the relief is dominated by hills and broad valley bottoms, but not huge plains. Thus, classes indicating

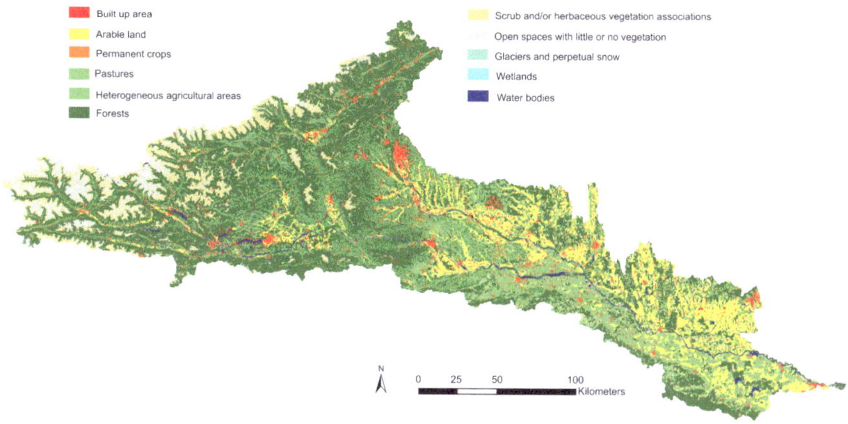

Fig. 3.4 Land use of the DRB in 2012 according to CLC. *Data source* European Environmental Agency (2017b)

agricultural activities (especially arable land and heterogeneous agricultural areas) dominate the visual appearance of Fig. 3.4. However, is has to be pointed out that in the lower eastern part of the DRB forests are also widespread (for reasons see Sect. 3.3.2). As an example of a single land use class, a closer look into forests is possible in Fig. 3.5, which shows a subdivision of forests into three subclasses. In the alpine region, coniferous forests prevail, whereas at the margin of the Alps and in their southern part mixed forests can be found. In contrast, forests in the pre-alpine region are mainly composed of deciduous trees because of the warmer

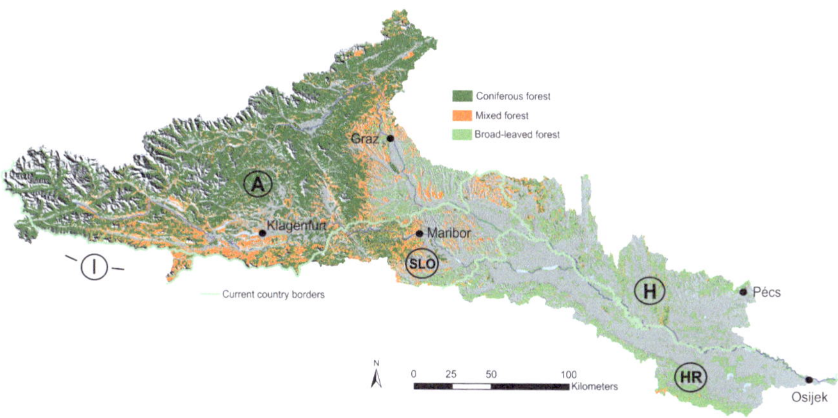

Fig. 3.5 Differentiation of forests in the DRB. *Data source* European Environmental Agency (2017b)

climatic conditions—except in some areas where mixed forests occur due to edaphic reasons.

Summing up, the land use pattern of the DRB can shortly be characterized by a western part with predominance of forests and an eastern section with predominance of agricultural areas. This distribution is the result of both physical conditions (Sect. 3.3.1) and historical processes with the ownerships of real estate's linked to them (Sect. 3.3.2).

3.4 Land Use Changes Since the Last Decade of the 20th Century

Land cover changes reflect the consumption of land of a given type and the formation of another type resulting from the use of land, from natural drivers and in combination with human drivers (Weber 2009). Changes in land use and land cover are key factors for global environmental change (Bürgi 1999). In addition, changes in technology (e.g. construction of traffic networks), culture, power, and political/economic institutions can influence land use/land cover change (Reid et al. 2000). Some major processes influencing land use (change) in the DRB have already been discussed in Sect. 3.3. Despite these changes in the natural, cultural and political framework, variations in land cover reflect a limited number of basic processes such as (Weber 2009):

- Dense and diffuse urban extension (sprawl) over agriculture and natural land;
- Urban land restructuring;
- Extension of agriculture over natural land (deforestation, drainage of wetlands, cultivation of marginal land);
- Intensification of agriculture resulting in internal conversion from pasture and mosaics to arable land;
- Crop rotations;
- Withdrawal of farming;
- Deforestation (if forest is replaced by a land use type other than agriculture);
- Forest rotations with felling and replantation;
- Extension of water bodies;
- Changes in natural land cover due to natural or multiple causes.

Land cover change identification for the DRB as presented in Table 3.2 and Fig. 3.6 is based on the CLC datasets mentioned in Sect. 3.2. Because of methodological restrictions changes in land use are only mapped if they affect an

Table 3.2 Absolute and relative areas of land use classes in the DRB in 1990, 2000, 2006 and 2012

	Land use class	1990		2000		2006		2012	
		km^2	%	km^2	%	km^2	%	km^2	%
1	Built-up area	1,387.89	3.49	1,490.42	3.75	1,603.18	4.03	1,658.27	4.17
2	Arable land	5,378.95	13.53	5,484.59	13.80	5,604.96	14.10	5,520.69	13.89
3	Permanent crops	215.24	0.54	218.06	0.55	181.63	0.46	178.35	0.45
4	Pastures	3,168.32	7.97	3,028.86	7.62	2,925.13	7.36	2,915.63	7.33
5	Heterogeneous agricultural areas	5,762.18	14.49	5,745.57	14.45	5,654.66	14.22	5,618.66	14.13
6	Forests	18,256.34	45.92	18,332.15	46.11	18,219.61	45.83	18,107.57	45.55
7	Scrub and/or herbaceous vegetation associations	3,578.23	9.00	3,423.16	8.61	3,531.57	8.88	3,719.09	9.35
8	Open spaces with little or no vegetation	1,442.48	3.63	1,516.06	3.81	1,531.44	3.85	1,536.98	3.87
9	Glaciers and perpetual snow	134.82	0.34	98.35	0.25	83.02	0.21	77.56	0.20
10	Wetlands	94.65	0.24	79.15	0.20	76.38	0.19	76.77	0.19
11	Water bodies	336.28	0.85	339.01	0.85	343.80	0.86	345.81	0.87
	Total	39,755.38	100.00	39,755.38	100.00	39,755.38	100.00	39,755.38	100.00

According to CLC, *data source* European Environmental Agency (2017b) area larger than 5 ha and with a width of more than 100 m. Figure 3.6 shows the areas in which changes of the specific land use classes occurred.

An area of 1,400 km^2 of land cover change was identified for the period 1990–2012 in the DRB. According to the defined classes, the following statements can be made:

- Built-up areas show an increase from 1990 to 2012 by 20%—with a striking difference between the alpine (+30%) and the pre-alpine region (+12%).
- Arable land (slight increase) and heterogeneous agricultural areas (slight decrease) in total are quite stable land use classes in the entire DRB catchment—though an increase of arable land (24%) in the alpine region can be recognized.
- Fruit tree/berry plantation and vineyard areas are decreasing in each period from 1990 to 2012 (in total −37%), whereas in the alpine area this land use class has remained stable (due to the predominance of fruit trees). In general it is remarkable, that large areas with fruit trees (e.g. in Eastern Styria) are not recorded in the CORINE data.
- In the alpine region the areal extent of pastures, which are located in the alpine foot zone and on clearings at moderate elevations, is stable whereas the pre-alpine pastures show fluctuations with an overall decrease of 20%.
- Forests (46% of DRB) show no or only small changes in their spatial extent from 1990 to 2012. Deciduous and coniferous forests slightly increased whereas mixed forests have decreased.
- Open spaces with little or no vegetation (about 4% of DRB) are dominant in the alpine region (slight increase of 7%); this land use class mainly comprises bare rocks and sparsely vegetated areas above the timberline in the high mountains.
- The summit zones of high mountains show glaciers and perpetual snow with a remarkable decrease of 42% (see below).
- From 1990 to 2012 wetlands (0.2% of the DRB) decreased by 20%, whereas water bodies (water courses: 0.6% and water bodies: 4.5%) slightly increased.

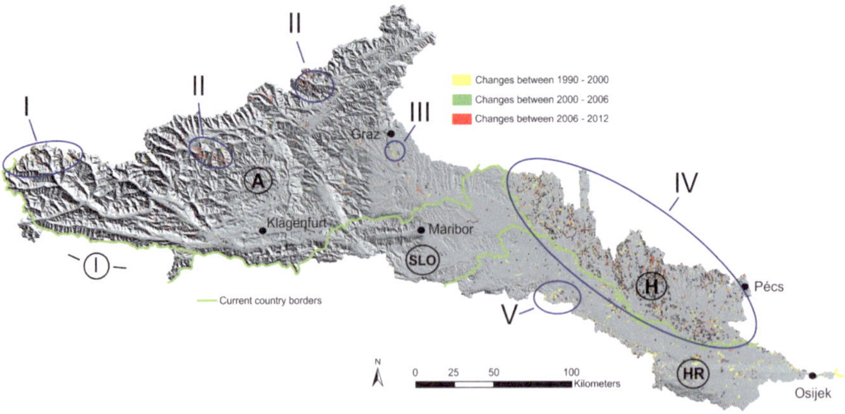

Fig. 3.6 Land use change in the DRB 1990–2012. *Data source* European Environmental Agency (2017b)

In Fig. 3.6, some exemplary areas with striking changes of land use are marked (I–V). They have been chosen because they are conspicuous in Fig. 3.6 and different processes underlie the changes:

- Area I (High Tauern Range) is primarily affected by changes in the land use class glaciers and perpetual snow. This is a consequence of the ongoing accelerated glacier retreat caused by global warming.
- Area II (Low Tauern Range) consists of two sub-areas in which the land use class forests significantly changed. As in area I, there was no deliberate change in land use, but some heavy winter storm events (e.g. "Paula" in January 2008) cleared large tracts of forest. Because all the areas affected are still characterized by intensive forestry, this is, strictly speaking, an example of land *cover* but not land *use* change.
- Area III documents changes in the vicinity south of Graz, the largest of the few large cities in the DRB. Detected changes here mainly refer to the land use class built-up areas and were caused by the expansion of gravel pits and settlements. Hence, this is an example of changes connected with urban sprawl.
- Area IV represents intensive changes in the land use classes linked to agriculture (above all arable land and heterogeneous agricultural areas). The area comprises the entire Hungarian part of the DRB and the detected changes can be interpreted as the consequences of the agricultural transformation because of the political changes since 1989.
- Area V appears to be quite similar to area IV with changes referring to agricultural land use. However, the areas affected are by far smaller than in Hungary and the transformation mainly occurred in the period 1990–2000.

All in all, from a regional point of view the observed changes are concentrated in specific parts of the DRB. The examples above have shown that natural (areas I and II) as well as human processes can be the reason for changes. However, it remains open to which extent errors in the CLC dataset influence the picture. In any case, land use seems to have changed most dynamically in Hungary whereas in Slovenia only very small areas were affected.

3.5 Land Use and Environmental Problems

Land use is a key factor in environmental problems. In this study, we (i) shortly discuss the influence of land use activities on the environmental quality of the DRB and (ii) take an exemplary look into areas protected within the EU framework of Natura 2000. In order to check the impact of human activities on a Central European scale, the study of Nefedova et al. (1992) is a valuable source. The maps provided by them cover the entire DRB (except of its tiny Italian section) and show that—in accordance with our results—the DRB is in its eastern part characterized by high intensity of agricultural land use whereas there are only a few industrial locations with a significant emission potential of pollutants (particularly in the upper Mura catchment, Maribor and Dravsko polje, Pécs, Osijek). To a large extent, this is due to the fact that there are no major urban agglomerations in the DRB—only 5 cities exceed 100,000 inhabitants (with even the largest, Graz, remains below 300,000). Hence, air, water, and soil quality is better than in most other second-order subcatchments of the Danube (Nefedova et al. 1992). However, this does not mean that there are no environmental problems, but they are not as urgent as is many other regions. Considering the Drava River itself, Sommerwerk et al. (2009) address a set of environmental problems ranging from the negative influence of large dams for power production (their number of 49 is the highest one of all subcatchments of the Danube, see Chap. 9 in this volume) to the sewage input from cities in the transformation countries.

3.6 Land Use and Nature Conservation

Our results show a high percentage of land use classes which can be considered natural and seminatural (classes 6–11 in Table 3.2 together covered 60% of the DRB in 2012!). This means that nearly no pollutants are emitted from them. Thus, it is to be expected that protected areas have a large extent in the DRB (including Natura 2000 areas—Fig. 3.7). Natura 2000 is a key instrument to protect biodiversity in the EU. It is an ecological network of protected areas, set up to ensure the survival of the most valuable species and habitats in Europe, based on the 1979 Birds Directive and the 1992 Habitats Directive (European Environmental Agency 2017b).

Natura 2000 areas cover 23.2% of the DRB (Table 3.3). The protected areas show concentrations (i) in the high mountains of the west and north of the alpine

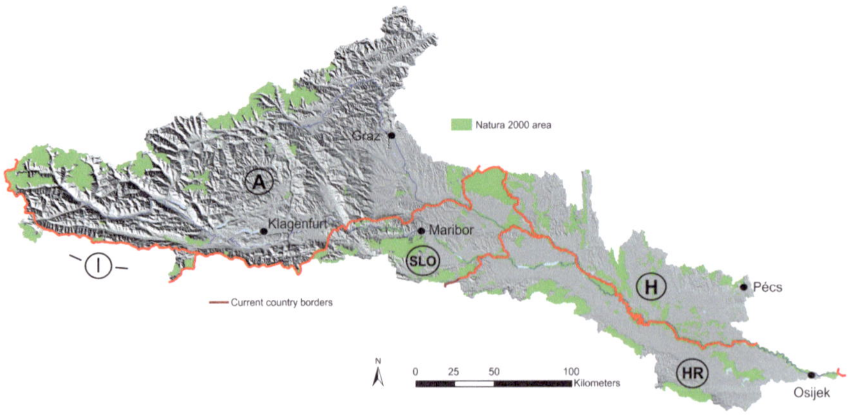

Fig. 3.7 Natura 2000 areas in the DRB. *Data source* European Environmental Agency (2017b)

Table 3.3 Natura 2000 areas in the DRB by countries and areas of the countries within the DRB

Country	Natura 2000 area		Country area belonging to DRB	
	km^2	%	km^2	%
Austria	2,614.37	28.3	22,163.11	55.8
Croatia	2,558.07	27.7	6,645.86	16.7
Hungary	1,683.02	18.2	5,903.17	14.8
Italy	182.33	2.0	363.05	0.9
Slovenia	2,205.15	23.8	4,680.19	11.8
Total	9,242.94	100.0	39,755.38	100.0

Data source European Environmental Agency (2017b)

region and (ii) in the pre-alpine region. In the high mountains, the widespread areas with little or no human impact are of course the basis for the protection status (e.g. High Tauern National Park, Austria). This also applies to the floodplain forests of the pre-alpine region (e.g. Duna-Drava National Park, Hungary), forming a corridor of valuable ecosystems of European importance along the Mura and Drava (Schneider-Jacoby 1996, 15 and Chap. 20). However, one can find protected areas also in agricultural regions as far as the intensity of land use is moderate (e.g. Krajinski park Goričko, Slovenia)—here the predominance of the land use class heterogeneous agricultural areas gives place to a broad variety of ecosystems with high biodiversity. However, the distribution of protected areas in Fig. 3.7 does not so much reflect land use patterns, but rather the environmental policies of the single states (and federal countries in Austria)—comparing the percentage of the country and the protected areas (Table 3.3) gives a hint on the ambitions of environmental policy in the individual countries.

3.7 Conclusions

From a methodical point of view, our study has some limitations because we used a single dataset. For the comparability of the data between countries, however, this was inevitable. Thus, the statements derived from our data are as good as the data are. In an overall perspective data quality is sufficient at least at the scale of the entire DRB. But even at this scale, some artifacts, classification mistakes or simply misinterpretations are evident. For instance, in the Prekmurje region of Slovenia built-up areas are shown in Fig. 3.4 where they definitely do not exist—this area is part of the Goričko Natura 2000 site (Sect. 3.5)! These errors could be corrected in a regional analysis whereas in the present study this was not possible.

Based on a multitemporal analysis of CLC data a good overview of land use pattern and its recent change in the DRB could be given. Thereby the presented data are in good agreement with those from Sommerwerk (2009), discussed in Sect. 3.1. Land use in the DRB is dominated by forests (2012: 46%), due both to natural conditions (especially in the alpine region) and to the historical development (especially in the pre-alpine region). Regionalizing the DRB shows the most pronounced differences between the west and the east (with Slovenia having an intermediate position):

- The western part is naturally characterized by alpine conditions and historically by being part of Austria. In terms of recent land use (changes), this means dynamic economy and high environmental pressures, but also an early establishment of sustainable land use directives and high environmental standards.
- The eastern part belongs to the pre-alpine region and shared its historical development with Hungary. In terms of land use (changes) this means an economic concentration on agriculture, under the communist regime a retarded economic development with low environmental standards and a currently still ongoing transformation process.

Looking at the DRB in its entirety, it can be claimed that still large areas have remained in a status with only moderate human impact proved by a high percentage of high-grade protected areas (Sect. 3.5). However, environmental protection as part of sustainable development is a highly challenging task for the future. Concerning the river itself, the floodplains are still endangered by hydropower projects, gravel extraction and navigation (20% of the river length is navigable) (Sommerwerk et al. 2009) although there are several sustainable initiatives like Drava Life (2017) or a projected biosphere reserve (for details see other contributions in this volume). Concerning the entire DRB catchment, there are some non-negligible problems the solution of which also impacts the entire Danube River Basin (Sommerwerk et al. 2010). However, improving the current environmental status of the DRB seems to be much easier than in many other European river catchments.

References

Alpine Convention (2017) Alpine Convention. http://www.alpconv.org. Accessed 15 Sept 2017

Bruckmüller E, Hartmann PC (eds) (2011) Putzger Historischer Weltatlas, 104th edn. Cornelsen, Berlin, p 312

Bürgi M (1999) A case study of forest change in the Swiss lowlands. Landscape Ecol 14:567–575. https://doi.org/10.1023/A:1008168209725

Copernicus programme (2017a) Copernicus. Europe's eyes on earth. http://www.copernicus.eu. Accessed 15 Sept 2017

Copernicus programme (2017b) CORINE land cover. http://land.copernicus.eu/pan-european/corine-land-cover. Accessed 15 Sept 2017

Drava Life (2017) Drava-Life. Integrated River Management. http://www.drava-life.hr/en/home/. Accessed 15 Sept 2017

European Environmental Agency (EEA) (1999) CORINE Landcover – Technical Guide. https://www.eea.europa.eu/publications/COR0-landcover. Accessed 15 Sept 2017

European Environmental Agency (EEA) (2016) Biogeographical regions in Europe, 2016. https://www.eea.europa.eu/data-and-maps/figures/biogeographical-regions-in-europe-2/map_2-1_biogeographical-regions/image_large. Accessed 15 Sept 2017

European Environmental Agency (EEA) (2017a) Land use. https://www.eea.europa.eu/themes/landuse. Accessed 15 Sept 2017

European Environmental Agency (EEA) (2017b) Data and maps. https://www.eea.europa.eu/data-and-maps. Accessed 15 Sept 2017

FAO (2017) Land Cover Classification System (LCCS). Food and Agriculture Organization of the United Nations, Rome. http://www.fao.org/docrep/003/x0596e/x0596e01e.htm. Accessed 15 Sept 2017

Mazúr E, Drdoš J, Tarábek K, Schappelwein K (1985) Natural Landscape Types. In: Atlas of the Danubian Countries. Thematic map 1:2.000,000 and explanations. Österreichisches Ost- und Südosteuropa-Institut, Vienna

Nefedova T, Bassa L, Plit J, Runova T, Deică P, Dončev D, Mikulík O, Špes M, Tomasi E, Zăvoianu I (1992) Use of the environment and resultant problems in central and Eastern Europe. In: Atlas of Eastern and Southeastern Europe. Österreichisches Ost- und Südosteuropa-Institut, Wien. Thematic map (sheet 1.2-G.4) 1:3,000,000 and explanation volume, 30 pp

Reid RS, Kruska RL, Muthui N, Taye A, Wotton S, Wilson CJ et al (2000) Land-use and land-cover dynamics in response to changes in climatic, biological and socio-political forces: the case of Southwestern Ethiopia. Landscape Ecol 15:339–355. https://doi.org/10.1023/A:1008177712995

Schneider-Jacoby M (1996) Drau und Mur. Naturerbe Verlag Jürgen Resch, Überlingen, Leben durch Flußdynamik, 149 pp

Sommerwerk N, Hein T, Schneider-Jakoby M, Baumgartner C, Ostojić A, Paunović M, Bloesch J, Siber R, Tockner K (2009) The Danube River Basin. In: Tockner K, Uehlinger U, Robinson CT (eds) Rivers of Europe. Elsevier, Amsterdam, pp 59–112

Sommerwerk N, Bloesch J, Paunović M, Baumgartner C, Venohr M, Schneider-Jacoby M, Hein T, Tockner K (2010) Managing the world's most international river: the Danube River Basin. Mar Freshw Res 61:736–748. https://doi.org/10.1071/MF09229

Somogyi S, Rátóti B, Keresztesi Z (1983) Hydrogeography. In: Atlas of the Danubian Countries. Thematic map 1:2,000,000 and explanations. Österreichisches Ost- und Südosteuropa-Institut, Wien

Weber JL (2009) Land cover classification for land cover accounting. In: 14th Meeting of the London Group on Environmental Accounting, Canberra, 27–30 April 2009. https://unstats.un.org/unsd/envaccounting/londongroup/meeting14/LG14_9a.pdf. Accessed 15 Sept 2017

Chapter 4
Climate and Climate Change in the Drava-Mura Catchment

Dénes Lóczy

Abstract Climate is markedly different in the upland and lowland catchments of the Drava River. The Illyric climate of the Austrian section of the Drava Valley is characterised by Mediterranean influences (high number of sunshine hours, late autumn rainfalls, moderately cold winters, but abundant rainfall in the summer), while in the Upper Mura catchment Atlantic weather (frequent rainfalls in all seasons, deeper snow cover, high humidity) prevails for most of the year. The overall result is a relatively mild climate, which, however, shows a mosaical pattern. The orographic and wind shelter effects on precipitation are highly variable in the different parts of the upland catchment. Spatial variations in the predominance of climatic influences (oceanic, continental and Mediterranean air masses) are revealed through a graphic analysis of temperature, precipitation amounts and wind direction. The distribution of precipitation as well as the depth, duration and melting date of snow cover determine (along with ice melt) the river regime of the Drava and its tributaries. To the southeast, in the lowland section there is a gradual transition towards a more continental Pannonian climate, which only has a subordinate contribution to river discharge of the Drava. Climate change trends are expected to further increase contrasts between climate zones.

Keywords Illyric climate · Atlantic influence · Pannonian climate
Inversion · Orographic effect · Precipitation · Snow cover · Winds

4.1 Introduction

The Alpine section of the Drava-Mura catchment extends over ca. 12,000 km^2 out of the total 40,120 km^2. However, this part is responsible for two-thirds (ca 450 m^3 s^{-1}) of the discharge of the Drava at its confluence with the Danube (670 m^3 s^{-1}).

D. Lóczy (✉)
Department of Physical and Environmental Geography, Institute of Geography and Earth Sciences, Faculty of Sciences, University of Pécs, Ifjúság útja 6, Pécs 7624, Hungary
e-mail: loczyd@gamma.ttk.pte.hu

D. Lóczy (ed.), *The Drava River*, Springer Geography,
https://doi.org/10.1007/978-3-319-92816-6_4

Therefore, the climatic conditions of the upland section (in Italy and mostly in Austria) decisively control the water regime of the Drava and Mura rivers—at present and also in the future (Baumgartner et al. 1983; Matulla et al. 2002; Blöschl et al. 2011; Goler et al. 2016).

In the southern ranges of the Eastern Alps and their valleys and basins, a range of mesoclimates developed and further diversified by topography (altitude, valley alignment, slope exposure). Based on observations between 1961 and 2005 (Šraj et al. 2007), precipitation values are somewhat higher in the Drava basin (average: 1,150 mm; maximum: 3,055 mm in the Groβglockner region; minimum: 747 mm in Heiligenblut, Carinthia) than in the Mura basin (average: 921 mm; maximum: 1,987 mm, measured in Niederalpl, Styria, in 2005; minimum: 422 mm in Zalaapáti, Hungary, in 2000).

The spatially highly variable manifestation of the orographic effect on precipitation has been studied by several authors. In his monograph on the Danube-Mura river system, the Hungarian hydrogeographer Lovász (1972) analysed meteorological data for 85 stations in the upland Drava and Mura catchments (mostly in Austria). He identified winter, summer types of vertical temperature distribution, and estimated the intensity of orographic control on precipitation distribution. In some closed basins (e.g. the Isel, Upper Möll, Upper Lieser, northern Gurk, southeastern Mürz catchments) the orographic effect is weak and in wind shelter positions annual precipitation at higher elevations is much below average. In topographically open catchments, however, Mediterranean air is able to flow in through the gaps of mountain ranges.

Steinhauser et al. (1960) investigated the relationships between the number of winter and frost days and altitude above sea level over the time span of one hundred years. He claimed that winter days become increasingly numerous above 1,000 m, while the number of freeze–thaw alternations decreases above ca. 900 m altitude. He explained this observation with the prevalence of free air movement (Atlantic influence) in the upper atmosphere at higher elevations.

In their hydrological monograph of the entire Mura drainage basin, Šraj et al. (2007) described the precipitation conditions of from data for 99 meteorological stations (83 stations in Austria, 2 in Slovenia, 3 in Croatia and 11 in Hungary). Using monthly precipitation data available for all stations for the period 1971–2000 and for 57 stations for 1961–2005, they found positive precipitation trends in the northeast and characteristically negative in southern part of the basin, while temperature was rising throughout the basin.

Modelling approaches have also been employed to the prediction of future climate conditions in the Drava-Mura drainage basin (Gobiet 2010; Strauss et al. 2013). An obvious option is the downscaling of regional climate models, such as the CLM (forced by the Global Climate Model ECHAM5) based on the A1B scenario (Kromp-Kolb and Schwarzl 2009; Smiatek et al. 2009; Schöner and Böhm 2011). The prediction refers to the period 2021–2050 and employs 1976–2007 as reference period. Mean air temperature in Austria will increase by about 1 °C compared to the reference period, with more pronounced warming in summer.

4.2 Climate Zones

The so-called *Illyric climate* of the Upper Drava Valley is due to the inflow of Mediterranean air masses through the gates of the South Alpine ranges, the movement of Atlantic air from northwestern direction and the protective effect of the High Tauern massif (3,798 m) (Auer et al. 2001). Transitional among Mediterranean, Alpine and the drier Pannonian climates, this zone extends from southern Austria to northeastern Slovenia (Fig. 4.1). A large number of sunshine hours (more than 2,000 h above 2,000 m elevations) and summer days (10–15 days with above 20 °C daily mean temperature at elevations between 400 and 500 m) are typical of the Upper Drava Valley (ZAMG 2016). One of the highest absolute maximum temperatures in Austria was recorded in Dellach (Carinthian Drava valley) on 3 August 2013 (39.9 °C). Along the Upper Drava and Gail Rivers Mediterranean influence can be detected in early spring warming (starting in February) and also in the late autumn (November) secondary precipitation maximum (mostly rainfall, 9–11% of annual total). (In the Lesach Valley of the Gail River even the primary maximum is in November with 160 mm—ZAMG 2016.) However, high summer rainfall amounts (30–35% share of annual precipitation, i.e. much higher than under true Mediterranean climate) point to the common occurrence of oceanic cyclones along the Upper Drava. The vegetation at 400–700 m reflects this moderately warm and dry climate: sweet chestnut and fluffy oak forests with periwickle (*Cotinus coggygria*) and manna ash (*Fraxinus ornus*) are found in the Upper Drava Valley. The frequent thermal inversions, however, make the (Isel, Möll, Lieser, Gail and Gurk) valleys and basins cold in winter.

Fig. 4.1 Climate zones of the Drava-Mura catchment (based on ÖBV—Freytag and Berndt 2015; Zaninović 2008; OMSz 2003). 1, Illyric climate; 2, Pannonian climate; 3, Atlantic climate; 4, high-mountain (Alpine) climate. The locations of meteorological stations are shown

The *Pannonian (continental) climate* zone replaces the Mediterranean climate in the southeast of the drainage basin, downstream the Drava from Slovenia to Croatia and Hungary (Zaninović 2008—Fig. 4.1). The cyclones brought by westerly winds and Mediterranean inflows equally weaken here and anticyclonal weather situations are more common. Therefore, annual precipitation amounts fall below 850 mm and total summer rainfall below 200 mm (ZAMG 2016; Zaninović 2008). There are usually less than 110 days with frost, but, because of regular Atlantic air inflows in winter, the number of days with (deeper than 1 cm) snow cover only slightly reduces along the Drava valley floor (from 110 to 80 days—ZAMG 2016). The moderate relief, lower altitude and openness to easterly winds also contribute to the hot and dry summer and sometimes extremely cold winter weather. The Pannonian basin, however, is not exempt from oceanic and Mediterranean climatic influences either. Intense summer rains can cause flood waves on the lowland tributaries of the Drava, often followed by dry spells (Cindrić 2006). The share of November precipitation in the annual amount remains relatively high (8–9%) in the Pannonian Basin, too.

The *Atlantic climate* zone is more extensive in the Upper Mura catchment (Fig. 4.1). Even in January occasional inflow of oceanic air—from northwestern direction, primarily through the Präbichl Pass (1,232 m)—raises temperatures and brings snowfall to the higher regions of the Mura catchment. In February, Atlantic air flows along the southern margin of the North-European anticyclone reach the subcatchments which are open to the north (Lovász 1972). In May, Atlantic influence is associated with cold spells and in June and July large amounts of rainfall moderates warming, but for the rest of the year it only occasionally predominates in the weather of the catchment. Lovász (1972) claimed that oceanic cyclonic activity had been increasing over the period 1901–1950. Today in valleys of Atlantic climate (in northern Styria and West of Graz) northwestern winds prevail, annual precipitation is above 1,000 mm, rainfall amounts exceed 100 mm in all three summer months, air humidity is high (at 14 h above 55%) and no drought period is observable (ZAMG 2016). The maximum yearly precipitations are around 1,800 mm in the area of the Mura headwaters. However, the Atlantic climate zone is often heavily influenced by the inflow of Mediterranean air, manifested in 10–13 summer days (daily temperature maxima $\geq$ 25.0 °C) and a minor secondary precipitation maximum (8–10 rainy days in October) (ZAMG 2016).

The *high-mountain (Alpine)* environment is located above ca 1,500 m altitude (Fig. 4.1). Here the orographic effect on both temperature and precipitation is more marked. The lowest mean temperatures are measured on peaks like Sonnblick (3,111 m, annual mean for 1961–1990: −3.4 °C; February mean: −10.5 °C—Schöner et al. 2008). Although there are almost 200 rainy days in a year, more than 2,600 mm annual precipitation and high humidity, the total duration of sunshine is still above 1,700 h (in the Carnic Alps even 2,000 h), almost equally distributed throughout the months of the year (ZAMG 2016).

In this chapter, an attempt is made to demonstrate the spatial pattern of dynamic agents which are responsible for the complexity of mesoclimates in the upland catchment of the Drava River and regional variations in the detectable trends of climate change.

4.3 Methods

It is assumed that in the vertical profile the deviations of temperature from the general environmental lapse rate are due to impacts of the different types of air masses, prevalent in the upland portion of the Drava-Mura drainage basin. To delimit the areas where these air masses are major controls on climate, distributions of monthly air temperature and precipitation values were analysed at catchment scale from the ZAMG (2016) database, the Climate Atlas of Croatia (Zaninović 2008) and Hungary (OMSz 2003). The data mostly refer to the periods 1981–2010 or occasionally to 1971–2000 or 1961–1990.

To characterise the distribution of thermal conditions, *temperature* data from 26 meteorological stations in the Drava catchment (upstream the confluence with the Mura at Legrad, Croatia) and 30 stations in the Mura catchment have been selected. The lowland section of the catchment, where variations in temperature are subdued, is underrepresented in the database. It is assumed that each meteorological station portrays the climate in its vicinity of tens or hundreds of square kilometres. (The partial basins of the Drava and Mura River were treated separately for a better visual presentation.) In Figs. 4.2 and 4.3, the thermal regime of each station is demonstrated by a triplet of dots showing

- mean annual temperature (°C);
- mean coldest month (January) temperature (°C);
- mean warmest month (July) temperature (°C).

The triplets were plotted against the altitude above sea level of the meteorological stations.

Temperature data were supplemented with *wind direction* information for the winter and summer half years, shown at the dots of mean coldest and warmest month temperatures, respectively. One or two whiskers indicate prevailing wind direction(s) according to wind rose conventions. It is assumed that, in most of the cases, southern and southwestern winds are an indication of Mediterranean influence, northern, northwestern, and southwestern winds usually bring Atlantic air masses and southeastern and eastern winds are generally associated with Pannonian (continental) climate. The altitudinal ranges of the different influences are established on the basis of both thermal and wind regimes and indicated by brackets.

In this first step, however, only the altitudinal region of the individual climatic influence can be determined. For a further and spatially more explicit delimitation of climate zones *precipitation* data from altogether 66 stations were involved in the investigation (Fig. 4.4). Precipitation data also plotted against altitude above sea level, the positions of the stations in (the Drava trunk river, Gail, Lieser, Gurk, Möll, Lavant, Laßnitz, Mura, Mürz and Kainach) subcatchments were indicated in the chart by symbols. This way the climatic influences cannot only be linked to altitudinal regions, but also to major subcatchments.

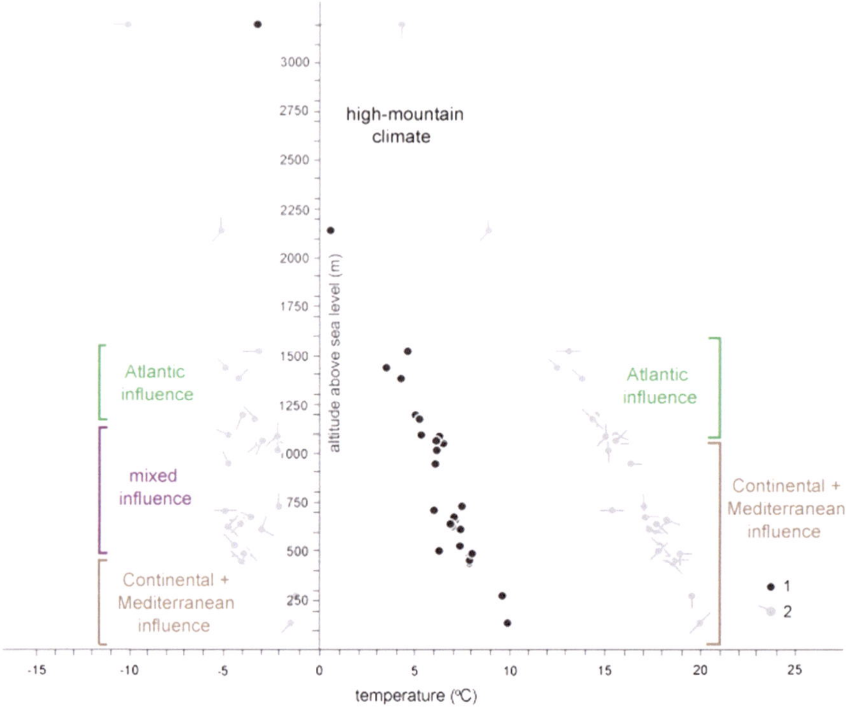

Fig. 4.2 Altitudinal variations in the thermal regime of the Drava catchment upstream the Mura confluence (by D. Lóczy, based on data from Climate-data.org 2016). 1, mean annual temperature (°C); 2, mean monthly temperature in January and July (°C) with prevailing wind directions in the winter and summer half-years, respectively

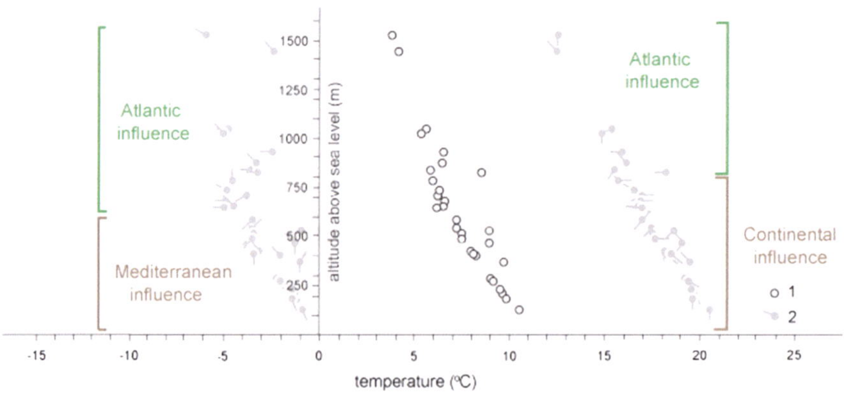

Fig. 4.3 Altitudinal variations in the thermal regime of the Mura catchment (by D. Lóczy, based on data from Climate-data.org 2016). For legend see Fig. 4.2

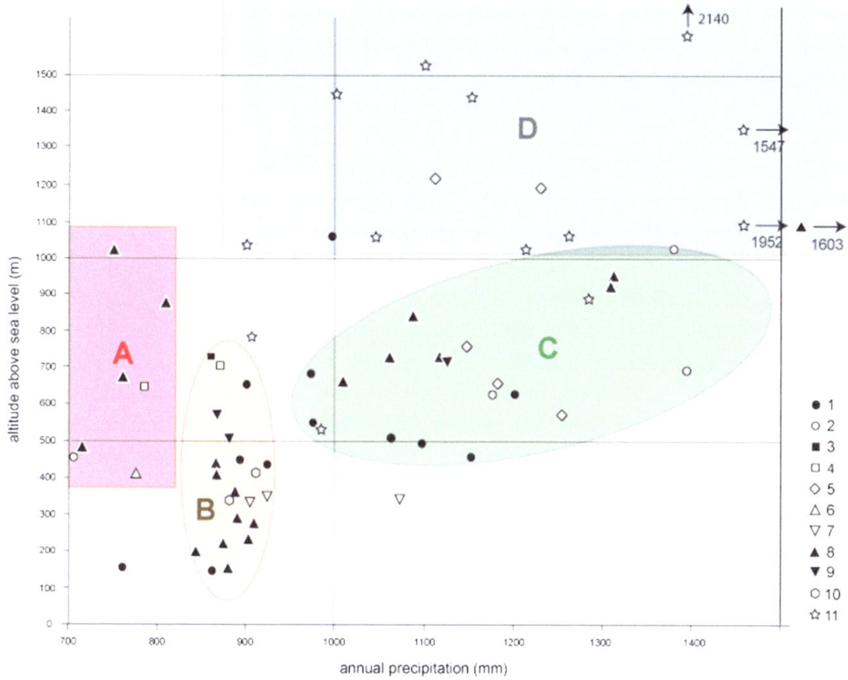

Fig. 4.4 Distribution of precipitation by altitude in the Drava-Mura drainage system and the mesoclimate types identified (by D. Lóczy, based on data from Climate-data.org 2016). A, wind-sheltered basins and valleys; B, region under predominantly continental influence; C, under alternatingly Atlantic and Mediterranean influences; D, mountain climate; 1, Drava valley; 2, Gail valley; 3, Lieser valley; 4, Gurk valley; 5, Möll valley, 6, Lavant valley; 7, Laβnitz valley; 8, Mura valley; 9, Mürz valley; 10, Kainach valley; 11, stations located high above river valleys, close to watersheds

Climate change in the 20th century can be detected through the comparison of temperature and precipitation averages for the periods 1901–1950 and 1971–2000. Example study areas (the headwaters of the Drava represented by the meteorological station Lienz; the Mediterranean valley of the Gail represented by the station Villach, the Mura valley of Atlantic climate represented by Bruck an der Mur and the transition towards Pannonian climate represented by Leibnitz).

4.4 Results: Spatial Pattern of Climatic Influences

Previous research has indicated that the vertical distribution of temperature and precipitation is only loosely adjusted to elevation (Steinhauser et al. 1960; Lovász 1972; Frei and Schär 1998). The graphs where the values of climatic elements are plotted against altitude above sea level more pronouncedly demonstrate the significance of dynamic factors in shaping mesoclimates.

The distribution of *temperature* values over the Upper *Drava* catchment upstream the Mura confluence (Fig. 4.2) shows that in the province from 500 m (8 °C) to 1,500 m (3 °C) above sea level mean annual temperatures decrease with altitude at a rate of ca 5 °C/1,000 m, which is close to the environmental lapse rate. (Negative deviations in some valley floors are due to openness to the west and stronger Atlantic influence, i.e. cooler summers.) In July mean monthly temperature drops from 18 to 13° C. Looking at mean monthly temperatures in January, however, no unequivocal vertical change can be observed: the values range between −2 and −5 °C in the mentioned altitudinal province. Here winter cooling (which is clearly manifested in the values of high-mountain stations) is reduced by Atlantic influence, shown by the prevalence of western and northwestern winds. In summer the same winds, however, reduce atmospheric warming in regions above ca 1,050 m altitude, which can be regarded the lower limit of Atlantic influence in summer. In winter, this limit is more difficult to determine. Seasonally, the inflow of Atlantic air masses alternates with continental impacts above 900 m and with Mediterranean effects below 750 m (Lovász 1972). Therefore, a broad zone of 'mixed influence' is displayed in the graph (Fig. 4.2).

The same graph for the *Mura* catchment is seemingly similar, but the differences are clearly noticeable (Fig. 4.3). Winter temperatures remain relatively high (between −3.5 and −5 °C) and even tend to grow up to 1,000 m altitude. This is due to frequent inflows of Atlantic air masses. The outliers of dots of mean annual temperature and July temperature between 350 and 550 m altitudes, however, are due to continental influence (anticyclonal conditions and intense insolation). In the hilly and lowland sections of the Mura catchment (below ca 400 m), winter cooling is reduced by Mediterranean air and summer warming is enhanced by continental air (Fig. 4.3).

Long-term annual *precipitation* shows a very irregular distribution (Fig. 4.4). It is rather difficult to detect a strong orographic control for any of the subcatchments. The closed Upper *Drava* valley is rather dry with only ca 1,000 mm precipitation at 1,000 m altitude above sea level. The middle valley sections between 450–700 m elevation receive 950–1,200 mm from Mediterranean (and Atlantic) cyclones and show a transition towards continental climate in the Klagenfurt Basin and Slovenia. Among the tributaries, the Gail valley presents Mediterranean, the Möll Atlantic and the Gurk pure continental character. In other areas, Atlantic and Mediterranean climatic influences broadly overlap.

On the floor of the *Mura* valley total precipitation amounts range from 1,600 to 700 mm. There are valley sections in wind shelter position scattered between 500 and 1,000 m elevations, where annual precipitation remains below 800 mm (area A in Fig. 4.4). Atlantic influence is typical of the Upper Mura valley above 650 m elevation, where 1,000–1,300 mm precipitation falls in a year (area C). The Lower

Mura plain is under continental influence with precipitation amounts below average (920 mm) (area B). Among the tributary valleys that of the Mürz, Kainach and Laβnitz are similarly continental (also area B).

Above 1,000 m elevation, mountain climate predominates (area D). It is striking that annual precipitation amounts on middle mountain tops of 1,000–1,100 m have a huge range: from 900 to almost 2,000 mm.

4.5 Climate Change

The Alps were among the first regions of Europe where climate change was observed (Rudloff 1967). The long-term trends of climate change, however, are difficult to detect as the year-to-year oscillations in climatic elements (temperature, precipitation, depth and duration of snow cover) are much greater. Long-term trends are often found non-significant in modelling (Auer et al. 2001; Smiatek et al. 2009). In general, the statement applies that temperature rise since the 1970s can be better documented and predicted for the future than changes in the amount and annual regime of precipitation. In Austria mean summer temperatures have increased by almost 2 °C and winter temperatures by ca. 1° C since the 1970s (Schöner and Böhm 2011) and a further growth of 1 °C until 2050 is predicted. The number of heat days (days with maximum temperatures $\geq$ 30 °C) in the growing season showed a 300% increase since the 1970s. Climate change in the lower Drava catchment can be characterized by observations in Osijek, Croatia (Cindrić et al. 2009). Based on one hundred years of observation, a positive winter temperature trend (+0.06 °C/10 years in Osijek) is typical for the continental region. Out of the ten warmest years four were recorded in the 21st century (with 12.9 °C annual mean temperature in 2000). The decline in annual precipitation along the Drava River derives from dropping spring (Osijek: −4.1% in 10 years) and autumn (Osijek: −3.0% in 10 years) rainfall amounts. The number of rainy days (precipitation $\geq$ 1 mm, −12%/10 years) and the duration of dry spells (−10%/10 years) both significantly decreased. The year 2000 was also the driest in Osijek (316 mm annual precipitation).

The trend of winter precipitation varies by regions: in the southern valleys and basins, the values tend to drop, but occasionally show high extremes. Although the rate of change is slow, in the Alpine regions of the Drava catchment higher and higher precipitation amounts are recorded decade by decade (Schmidli et al. 2002). Eitzinger et al. (2016) found annual precipitation between 2002 and 2014 to be higher than the long-term (1981–2010) average in several Austrian study sites. Seasonal distribution is generally favourable with most precipitation falling in the growing season (April–September), but a shift towards the winter half-year (October–March) is also discernible (Strauss et al. 2013). In 29% of dry years between 2002 and 2014, precipitation totals were both below winter and summer averages (Eitzinger et al. 2016). More fundamental changes in precipitation and evapotranspiration are predicted for the second half of the 21st century (BMLFUW 2010).

A simple comparison of climate data for the selected sites from the early and late 20th century confirms climate change trends (Table 4.1).

In closed mountain basins, winters have become colder (Table 4.1, Lienz). Spring and autumn rainfalls resulted in almost 10% growth in annual precipitation, pointing to increasing Mediterranean cyclone activity. In contrast, in the Illyric basins and valleys (represented by Villach—Table 4.1) the whole winter half-year have become remarkably warmer, but the climate turned to only moderately wetter. Some warming can be detected in the valleys of Atlantic climate (Table 4.1, Bruck an der Mur) with higher spring and summer precipitation (intensifying oceanic influence). Winter temperatures have risen close to the boundary of the Pannonian zone (Table 4.1, Leibnitz), but no change was observed in summer. With the exception of July, an aridification trend can be observed.

At the same time *drought* hazard is increasing, particularly in eastern Austria. Droughts caused severe damage to agricultural crops in the years 2003, 2006, 2010, 2012, 2013 and 2015 (Eitzinger et al. 2016). Even exceeding the damage caused by hail, in 2012 the share of drought (and heat wave) damage was the highest in the 120 million Euros of income loss due to weather extremes. Throughout Central Europe, water deficit and/or heat effects are identified as a primary cause of yield decrease for various crops (Semenov and Shewry 2011). Eitzinger et al. (2016) claim that the periods with low rainfall coincide with critical phenological stages of crops: the tillering of cereals (April), flowering of maize (July) and row closure of sugar beet (July/August).

Climate change profoundly affects the *hydrological* cycle (Schöner and Böhm 2011; Goler et al. 2016). The early summer flood stages of the Drava and its tributaries tend to fall, while in autumn water levels are becoming somewhat higher (Prettenthaler et al. 2007; Holzmann et al. 2010).

High rates of warming are recorded in high-mountain environments, such as the High Tauern (Auer and Böhm 2010). The most spectacular impact is observed in the mass balance of the Pasterze, the largest part of a combined mountain *glacier* system in the Austrian Alps (High Tauern) (WGMS 2015). Descending from 3500 to 2100 m and drained by the Möll River, the Pasterze Glacier of 16.3 km^2 area (2012) is partly covered by a 5–20 cm thick debris layer. In 2011/12, the annual mass balance was −1,298 mm water equivalent, and in 2012/13, it was −600 mm w.e. A mean mass loss of 560 mm between 1969 and 1998 and 1,210 mm between 1998 and 2012 can be calculated. Rapid retreat is also typical of other Austrian glaciers in recent decades. The duration of snow cover also shows a decline in Austria (Schöner et al. 2000).

Climate change also affects other earth systems, including soil moisture (Kromp-Kolb and Schwarzl 2009), geomorphological (Rickenmann 2009) and ecosystem processes and even has serious socio-economic implications (Formayer et al. 2008).

Table 4.1 Comparison of climatic parameters in selected catchments between 1901–1950 and 1971–2000 time series (data from Lovász 1972 and ZAMG 2016). **Bold** numbers indicate significant change

Station	Mean monthly temperature (°C)		Change (°C)	Average monthly precipitation (mm)		Change (mm)
	1901–1950	1971–2000		1901–1950	1971–2000	
Lienz, year	7.5	7.0	−0.5	837	915	+78
January	−4.0	−5.2	**−1.2**	35	42	+7
February	−1.6	−1.9	−0.3	46	35	−11
March	3.3	3.1	−0.2	43	59	**+17**
April	8.1	7.6	−0.5	55	66	+11
May	12.8	12.7	−0.1	73	85	+12
June	16.1	15.9	−0.2	94	99	+5
July	17.6	17.9	+0.3	123	119	−4
August	16.8	17.2	+0.4	105	100	−5
September	13.3	13.0	−0.3	79	88	+9
October	8.1	7.3	−0.8	75	96	**+21**
November	1.7	0.6	**−1.1**	64	76	+12
December	−2.8	−4.2	**−1.4**	44	50	+6
Villach, year	8.0	8.6	+0.6	1189	1230	+41
January	−4.1	−2.2	**+1.9**	55	61	+6
February	−1.5	−0.2	**+1.3**	55	65	**+10**
March	3.8	4.1	+0.3	73	72	−1
April	8.5	8.3	−0.2	98	103	+5
May	13.5	13.2	−0.2	102	107	+5
June	16.7	16.7	0	129	133	+4
July	18.4	18.6	+0.2	127	129	+2
August	17.6	18.0	+0.4	121	129	+8
September	13.8	14.6	**+0.8**	122	126	+4
October	8.4	9.4	**+1.0**	118	107	**−11**
November	2.5	3.7	**+1.2**	111	119	+8
December	−1.9	−0.7	**+1.2**	78	79	+1
Bruck an der Mur, year	7.9	8.1	+0.2	788	887	+99
January	−3.0	−2.3	**+0.7**	36	36	0
February	−0.7	0	**+0.7**	35	41	+6
March	3.8	3.8	0	35	53	**+18**
April	8.0	7.7	−0.3	55	60	+5
May	12.9	12.8	−0.1	81	91	+10
June	15.9	15.9	0	103	121	**+18**
July	17.6	17.8	+0.2	109	126	**+17**
August	16.9	17.5	**+0.6**	97	110	**+13**
September	13.4	13.7	+0.3	76	79	+3

(continued)

Table 4.1 (continued)

Station	Mean monthly temperature (°C)		Change (°C)	Average monthly precipitation (mm)		Change (mm)
	1901–1950	1971–2000		1901–1950	1971–2000	
October	8.3	8.5	+0.2	62	61	−1
November	2.8	2.7	−0.1	53	63	+10
December	−1.3	−1.2	+0.1	46	46	0
Leibnitz, year	8.9	9.2	+0.3	943	915	−28
January	−2.9	−2.1	**+0.8**	43	38	−5
February	−0.7	−0.1	**+0.6**	41	40	−1
March	4.5	4.5	0	48	49	+1
April	9.5	9.4	−0.1	73	62	−11
May	14.4	14.3	−0.1	91	93	+2
June	17.6	17.6	0	115	118	+3
July	19.4	19.4	0	106	119	**+13**
August	18.4	18.6	+0.2	112	109	−3
September	14.7	14.9	+0.2	95	85	**−10**
October	9.3	9.6	+0.3	90	78	**−12**
November	3.6	3.9	+0.3	70	72	+2
December	−0.6	−0.2	+0.4	59	52	−7

4.6 Conclusions

From the literature reviewed in this chapter, the following changes, originally identified for the territory of Austria (Formayer et al. 2001), can be confirmed for the Drava-Mura catchment, too:

- marked rise in annual mean temperature, less pronounced in maximum monthly temperature in mountain environment;
- considerable growth in the number of sunshine hours in winter;
- higher precipitation amounts in mountains, lower in the lowland;
- snow depth increases in the upland section and decreases in the lowland;
- the duration of snow cover has reduced in the lowland (established with high uncertainty);
- the number of sunshine hours has grown in the mountains and in winter;
- relative air humidity has markedly decreased;
- cloud cover has increased in summer, but in winter only increased at higher elevations, reduced in the valleys.

The above changes will probably further enhance the contrasts between the climatic (and land use—see Chap. 3) subdivisions of the Drava-Mura catchment.

Among the direct impacts of climate change in the Alpine regions the following are usually enumerated (Niederer 2013):

- increasing heat pressure in urban agglomerations;
- increasing frequency of summer drought affecting crop cultivation;
- higher flood risk;
- reduced slope stability and more frequent mass movements;
- retreat of the lower limit of snowfall to higher elevations;
- deteriorating water, soil and air quality;
- changes in diversity at the levels of Alpine species, habitats and landscapes;
- spreading of pests and alien species.

Although there are several urban agglomerations of various size in the catchment (Graz, Klagenfurt, Knittelfeld, Lienz, Spittal an der Drau and Villach in Austria, Maribor in Slovenia and Osijek in Croatia), increasing heat pressure has only been described from the Graz agglomeration (population in 2014: 605, 143) (Lazar 1999). All the other impacts of climate change, however, have been identified by researchers, first of all, in the upland catchment of the Drava River. The modifications of climatic elements are expected to transform the entire physical environment.

References

Auer I, Böhm R, Schöner W (2001) Austrian long-term climate 1767–2000. Österreichische Beiträge zu Meteorologie und Geophysik 25. Zentralanstalt für Meteorologie und Geodynamik (ZAMG), Wien

Auer I, Böhm R (2010) Klimawandel in den Hohen Tauern. alpine space – man & environment 11: Zwei Alpentäler im Klimawandel. 5–82. https://www.uibk.ac.at/alpinerraum/publications/vol11/teil_a.pdf

Baumgartner A, Reichel E, Weber G (1983) Der Wasserhaushalt der Alpen: Niederschlag, Verdunstung, Abfluß und Gletscherspende im Gesamtgebiet der Alpen im Jahresdurchschnitt für die Normalperiode 1931–1960. Oldenbourg Verlag, München – Wien

Blöschl G, Viglione A, Merz R, Parajka J, Salinas JL, Schöner W (2011) Auswirkungen des Klimawandels auf Hochwasser und Niederwasser. ÖWAW – Österreichische Wasserwirtschaft 1–2/2011. 21–30

BMLFUW (2010) Anpassungsstrategien an den Klimawandel für Österreichs Wasserwirtschaft – Kurzfassung. Bundesministerium für Land- und Forstwirtschaft, Umwelt und Wasserwirtschaft, Wien. 30 S. file:///C:/Users/Dell/Downloads/Anpassungsstrategien_Kurzfassung.pdf. Accessed 03 Feb 2017

Cindrić K (2006) The statistical analysis of wet and dry spells in Croatia by binary DARMA (1,1) model. Hrvatski meteorološki časopis 41:43–51

Cindrić K, Gajić-Čapka M, Zaninović K (2009) Observed climate changes in Croatia. In: Fifth National Communication of the Republic of Croatia under the United Nation Framework Convention on the Climate Change (UNFCCC). Meteorological and Hydrological Service of Croatia (DHMZ), Zagreb. http://klima.hr/razno/publikacije/climate_change.pdf

Climate-data.org (2016) Climate Data for Cities Worldwide. AM Online Projects, Oedheim, Germany. https://en.climate-data.org

Eitzinger J, Daneu V, Bodner G, Kubu G, Loiskandl W, Macaigne P, Thaler S, Schaumberger A, Wittmann C, Weidle F, Kann A, Murer E, Krammer C, Trnka M, Hayes M (2016) Drought monitoring system for Austrian agriculture – AgroDroughtAustria (Final Scientific Report of project "AgroDroughtAustria" of the Austrian Climate Change Research Program). BOKU-Met Report 25. https://meteo.boku.ac.at/report/
Formayer H, Weber A, Volk F, Eckhardt S, Kromp-Kolb H, Boxberger J (2001) Ermittlung der verfügbaren Feldarbeitstage für die Landwirtschaft in Österreich. Endbericht des Forschungsprojektes Nr. 1086 des BMLF
Formayer H, Clementschitsch L, Kromp-Kolb H (2008) Regionale Klimaänderungen in Österreich. Universität für Bodenkultur (BOKU), Wien. 103 S. http://www.wau.boku.ac.at/met.html. Accessed 03 Feb 2017
Frei C, Schär C (1998) A precipitation climatology of the Alps from high-resolution rain-gauge observations. Int J Climatol 18:873–900
Gobiet A (2010) Klimamodelle und Klimaszenarien für Österreich. Österreichischer Wasser- und Abfallwirtschaftsverband (Hrsg.). Auswirkungen des Klimawandels auf Hydrologie und Wasserwirtschaft in Österreich – Präsentation aktueller Studien. Wien. 11–24
Goler RA, Frey S, Formayer H, Holzmann H (2016) Influence of climate change on river discharge in Austria. Meteorol Z. https://doi.org/10.1127/metz/2016/0562
Holzmann H, Lehmann T, Formayer H, Haas P (2010) Auswirkungen möglicher Klimaänderungen auf Hochwasser und Wasserhaushaltskomponenten ausgewählter Einzugsgebiete in Österreich. Österreichische Wasser-Abfallwirtschaft 62:7–14
Kromp-Kolb H, Schwarzl I (Hrsg.) (2009) Anpassung an den Klimawandel in Österreich. Endbericht StartClim2008. Umweltbundesamt, Wien. 59 S
Lazar P (1999) Untersuchungen zum Stadtklima in Graz. In: Helbig A, Baumüller J, Kerschgens MJ (eds) Stadtklima und Luftreinhaltung. München, C.H, Beck, pp 403–413
Lovász Gy (1972) A Dráva-Mura vízrendszer vízjárási és lefolyási viszonyai (River regime and runoff in the Drava-Mura water system). Akadémiai Kiadó, Budapest. 158 pp (in Hungarian)
Matulla C, Groll N, Kromp-Kolb H, Scheifinger H, Lexer MJ, Widmann M (2002) Climate change scenarios at Austrian National Forest Inventory sites. Climate Res 22:161–173
Niederer P (Hrsg.) (2013) Anpassung an den Klimawandel im Berggebiet. Schweizerische Arbeitsgemeinschaft für die Berggebiete (SAB), Bern. 9 S
OMSz (2003) Magyarország éghajlati atlasza/Climate Atlas of Hungary. Hungarian Meteorological Service, Budapest
ÖBV – Freytag & Berndt (2015) Österreich. Klima. In: Schulatlas. Interaktiver Atlas Österreichs. http://www.schulatlas.com/Oesterreich/Interaktiver_Atlas
Prettenthaler F, Dalla-Via A (Hrsg.) (2007) Wasser & Wirtschaft im Klimawandel. Konkrete Ergebnisse am Beispiel der sensiblen Region Oststeiermark. VÖAW Verlag. Institut für Stadt- und Regionalforschung - Forschungsberichte, Band: 31. Studien zum Klimawandel in Österreich. 190 S
Rickenmann D (Hrsg.) (2009) Extremereignisse: Ereignisbezogene Dokumentation-Prozesse Bergstürze, Hochwasser, Muren, Rutschungen und Lawinen. Institute für Alpine Naturgefahren und Forstliches Ingeneurwesen, Universität für Bodenkultur (BOKU), Wien. 41 S
von Rudloff H (1967) Die Schwankungen und Pendelungen des Klimas in Europa seit dem Beginn des regelmässigen Instrumenten-Beobachtungen (1670). Springer Fachmedien, Wiesbaden. 371 S
Schöner W, Auer I, Böhm R (2000) Climate variability and glacier reaction in the Austrian Eastern Alps. Ann Glaciol 31:1–37
Schöner W, Auer I, Böhm R (2008) Long term trend of snow depth at Sonnblick (Austrian Alps) and its relation to climate change. Hydrol Process 23:1052–1063
Schöner W, Böhm R, Haslinger K (2011) Klimaänderung in Österreich – hydrologisch relevante Klimaelemente. Oesterr Wasser Abfallwirtsch 1–2:11–20
Schmidli J, Schmutz C, Frei C, Wanner C, Schär C (2002) Mesoscale precipitation in the Alps during the 20th century. Int J Climatol 22:1049–1074

Semenov MA, Shewry PR (2011) Modelling predicts that heat stress, not drought, will increase vulnerability of wheat in Europe. Nature Scientific Reports 1

Smiatek G, Kunstmann H, Knoche R, Marx A (2009) Precipitation and temperature statistics in high-resolution regional climate models: evaluation for the European Alps. J Geophys Res 114: D19107. https://doi.org/10.1029/2008JD011353

Šraj M, Horvat A, Koprivšek M, Vidmar A, Brilly M (2007) Analysis of water regime of the Mura River by homogenized discharge data. University of Ljubljana, Ljubljana. ftp://152.66.121.2/Floodrisk/_DC/docs/3_09_brilly.pdf

Steinhauser F, Eckel O, Lauscher F (1960) Klimatographie von Österreich. Bd.3, Lieferung 2. Springer, Wien. 137–380

Strauss F, Formayer H, Schmid E (2013) High resolution climate data for Austria in the period 2008–2040 from a statistical climate change model. Int J Climatol 33(2):430–443

WGMS (2015) Global Glacier Change Bulletin No. 1 (2012–2013). Zemp M, Gärtner-Roer I, Nussbaumer SU, Hüsler F, Machguth H, Mölg N, Paul F, Hoelzle M (Eds), ICSU(WDS)/IUGG(IACS)/UNEP/UNESCO/WMO, World Glacier Monitoring Service, Zurich, Switzerland. 230 pp. https://doi.org/10.5904/wgms-fog-2015-11

ZAMG (2016) Klimadaten von Österreich. Zentralanstalt für Meteorologie und Geodynamik. http://www.zamg.ac.at/fix/klima/oe71-00/klima2000/klimadaten_oesterreich_1971_frame1.htm. Accessed 25 Jan 2017

Zaninović K (ed) (2008) Klimatski atlas Hrvatske/Climate atlas of Croatia 1961–1990, 1971–2000. Državni hidrometerološki zavod, Zagreb, 200 pp

Chapter 5
Hydromorphology of the Lower Drava

Ulrich Schwarz

Abstract Together with the Sava, the Drava is the most important tributary of the Danube flowing from west to east. In contrast to the Sava, however, the hydrological and sedimentological regime of the Drava is mainly driven by the Alps. The headwaters of the upper Drava and the most important tributary, the upper Mura, show glacial regimes and their sediment load mainly originates from the Central Alps and partly from the Southern Calcareous Alps. The Drava follows in broad valleys with braided channel pattern and narrow gorges in Austria and Slovenia before it enters the lowland. Downstream the Mura confluence, the river tends to present anabranching characteristics, but with decreasing slope turns to the meandering type. Regarding the sediment in the middle course, all kinds of gravel fractions occur whereas on the lower course sandy fractions prevail. Lateral channel shift is also typical along the lower course. The upper course and parts of the middle course in Austria, Slovenia, and Croatia are used for hydropower generation and affect the sediment balance as well as the long-term hydromorphological regime of the entire lower course. Hydromorphological inventories of the middle and lower Drava and the lower Mura since the late 1990s and hydromorphological assessments since 2005 indicate still rich riparian ecosystems under changing conditions. The need to observe and evaluate the long-term effects of hydropower dams and interrupted sediment continuum as well as the changing hydrographical conditions call for the consideration of new options for river management and river and floodplain restoration in the coming decades.

Keywords Hydromorphology of drava · River section types
Hydromorphological alterations and assessment · Loss of floodplain
Overall restoration potential

U. Schwarz (✉)
FLUVIUS, Floodplain Ecology and River Basin Management,
Hetzgasse 22/7, 1030 Vienna, Austria
e-mail: Ulrich.Schwarz@fluvius.com

D. Lóczy (ed.), *The Drava River*, Springer Geography,
https://doi.org/10.1007/978-3-319-92816-6_5

5.1 Introduction

The Drava-Mura river system with a drainage basin of 40,150 km^2 (Fig. 5.1) and an average discharge of about 550 $m^3 s^{-1}$ is the third largest in the Danube Basin. It is predominantly influenced by Alpine conditions in the headwaters namely by discharge and sediment budget. Huge gravel deposits dominate in the basins of Graz and Klagenfurt and again downstream the Drava breakthrough in Slovenia accumulated along the foothills expanding downstream to the Mura confluence. Here the river has a strong mountainous (Alpine) character. The relief of the Lower Drava consists mostly of lowlands with Quaternary sediments and loess-mantled terrace systems. The course partly follows geological lineaments (e.g. high banks in Hungary). Also the lower course is dominated by sands and silts and the banks have much lower cohesion than those in the neighboring lower Sava system. Therefore, the lower Drava used to be characterized by strong lateral shifting and meandering under natural conditions.

The area of the Drava confluence with the Danube, the Kopački Rit, is still the largest riparian wetland (roughly 25,000 ha area) of the entire middle Danube (i.e. from Bratislava on the Slovakian-Hungarian border to the Iron Gate/Djerdap on the Serbian-Romanian border). This triangular wetland of international importance (Ramsar site) lies in a geomorphological depression, and subsidence is compensated by accumulation of fine material over millennia. The north-to-south Hungarian Danube section ends at a resistant loess steep bank (Erdutski breg), where the river turns 90° and then again nearly 180° following the steep right bank towards Vukovar in eastern direction. The confluence situation and the strongly

Fig. 5.1 The Drava River Basin covers territories in Italy, Austria, Slovenia, Hungary and Croatia

accumulating sandy lowland Drava River, reduces the slope of the Danube channel to 0.00001. As a consequence, floods are lasting long, up to three months in the lowest-lying areas. (This is one of the reasons why the area was not drained and ameliorated.) This situation is not typical for the still alpine Danube in this section. Downstream the Tisa and Sava confluences, just within the next 100 river km, the Danube changes its entire hydromorphological regime. The Drava has been of great importance for the survival and rejuvenation of the Kopački Rit area.

Regarding the drainage network of Drava and Mura rivers, the catchment is clearly structured in the larger Alpine tributaries coming mainly from the north (the Isel, Möll, Lieser) and by the only major right-bank tributary, the Gail. From the north the only and by far largest tributary, the Mura, enters the Drava just upstream of its lower course at about rkm 235. (The contemporary length development of the river was observed for the last about 40 years and indicate discrepancies of several rkm due to shifts and cutoffs.) All typical channel pattern types, from straight over braided to meandering, are present in a broad range. Downstream the Mura confluence, the Drava is of transitional type from anabranching towards meandering.

The hydrological regime is determined by the alpine region. The highest discharge occurs between May and July. The Upper Drava has still a glacial regime (climatic change with fast melting glaciers) whilst the Mura, its most important tributary, has a nival regime (peak already in May). The higher discharge in autumn is due to the more Mediterranean precipitation pattern in the middle-southern (including parts of the Southern Alps) and lower course of the river, but is not predominant as along the Sava River, which is strongly influenced by its Balkan tributaries. The natural water level fluctuation is 5–6 m near Botovo (Croatia, downstream of the Mura confluence), the mean long-term annual average of discharge ranges between 237 m^3 s^{-1} (low water; absolute minimum around 70 m^3 s^{-1}), 526 m^3 s^{-1} (mean water) and 850 m^3 s^{-1} (high water). The discharge for the 10-year flood is about 2,100 m^3 s^{-1} and the 100-year flood about 3,200 m^3 s^{-1}.

5.2 River Section Types and Reference Conditions for the Lower Drava

River section types are meant to describe river stretches with prevailing hydromorphological conditions based on so-called *reference conditions* without human intervention and were developed for the lower Drava and Mura rivers in earlier studies (Schwarz 2007) using historical map analysis, as a basic framework for hydromorphological assessments and for restoration options (Schwarz 2013).

The Drava hosts several stretches with at least near-natural features—here illustrated by two photographs (Figs. 5.2 and 5.3).

The following presentation and hydromorphological assessment focus on the lower Drava reach from the Mura confluence (rkm 235) to the mouth into Danube. The two main river section types for lower Drava are

Fig. 5.2 Two examples for reaches serving as potential reference sites. Just 10 rkm downstream of the Mura confluence the anabranching river builds gravel bars and islands (photo by Arno Mohl)

Fig. 5.3 Two examples for reaches serving as potential reference sites. One of the most dynamic meander reaches is at about rkm 180, highlighting the driving force, the lateral shift of river channels: erosion of land on the one side, rejuvenation of habitats on the other side (photo by Darko Grlica)

- transitional anabranching from the higly dynamic partially braided Drava upstream the Mura confluence to the meandering downstream (D-II, rkm 235–185) and
- meandering large lowland river with sand bed and extensive floodplain (D-III, rkm 185) (Figs. 5.2, 5.3, and 5.4).

The morphological description of the reference conditions offers a comprehensive way to compare the current situation with the reference state of a river section and indicate changes in the fluvio-morphological processes. Assessments for the entire riverine landscape provide valuable information on the targets of long-term restoration.

The detailed description of the reference conditions of section types should comprise available data on the position in the river continuum, morphological river type, channel width, valley floor shape, slope and characteristics of the valley, channel planform and pattern, rate of lateral shift, longitudinal profile (channel slope and structure, flow characteristics and variation), river bed structure and substrate, cross section characteristics and with/depth variance, bank structure, basic flood indicators as well as properties of the riparian landscape (substrates, forms and vegetation) (Schwarz 2007) (Fig. 5.5).

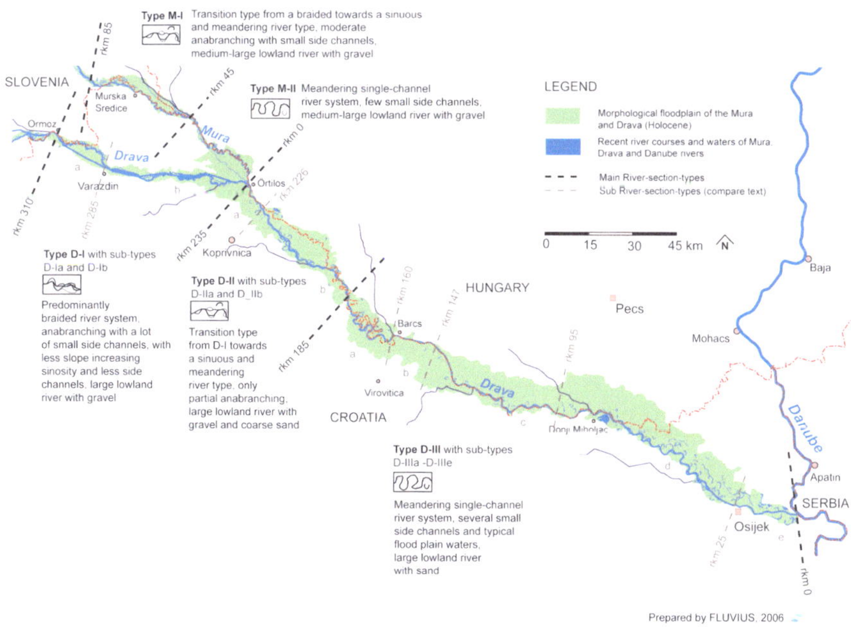

Fig. 5.4 River section types of the lower Drava and Mura system

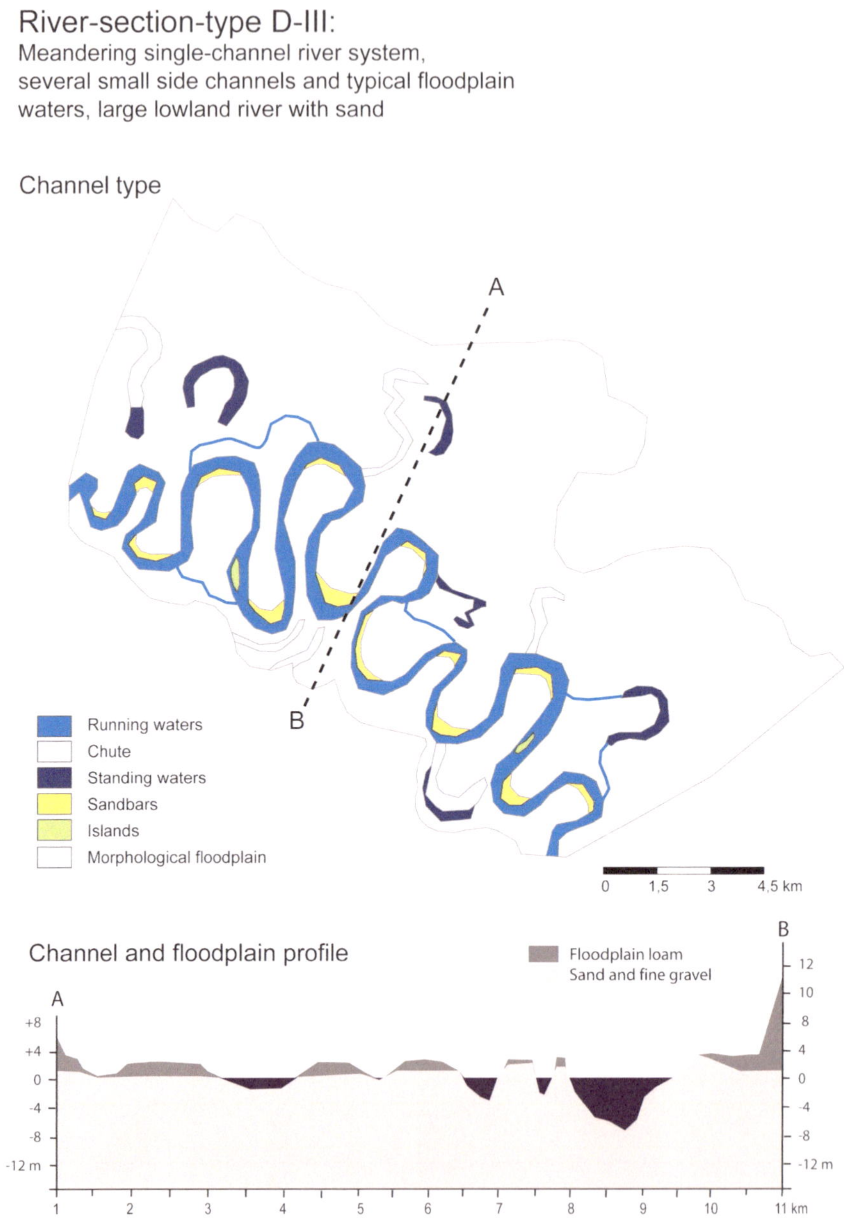

Fig. 5.5 Visualisation of the main river section type for lower Drava (Schwarz 2007)

5.3 Hydromorphological Assessment

For the International Association of Danube Research (IAD) a first pilot survey and physical habitat assessment based on the CEN Standard from 2004 (CEN 2004) was conducted between 2005 and 2007 (Schwarz 2007). For this assessment, the Lower Drava downstream the Mura confluence was surveyed by boat and over land. A GIS database was developed to allow an easy and fast analysis of about 400 longitudinal assessment stretches of individual length, for which channel, each left and right bank and floodplain was assessed (Table 5.1).

Table 5.1 Main hydromorphological parameter groups and subparameters to be assessed in a five-class scoring system. The first class (near-natural) serves as reference condition

Main parameter groups	Main parameter	Subparameters, description
Channel	Planform and cross-section (width and depth)	(a) Bankfull width (b) Entrenchment depth (to bankfull) (c) Average stream width (c) Mean depth of water body (d) Maximum depth of waterbody
	Average velocity (littoral, channel)	Flow classes: no flow (stagnant), low flow (just visible—approx. 0.3 m s^{-1}), medium flow (0.35–0.65 m s^{-1}), high flow (>0.7 m s^{-1})
	Channel type	Single thread, parallel channels, braided/meandering, braided, sinuous, constrained (natural/artificial)
	Navigation channel	No navigation channel, <1/3 of the bottom area, 1/3–2/3 regarding width and depth, 2/3–3/3 with strong impact (waves, ship propellers)
	Riverbed features	Bars, islands, riffles; accretion between groynes; large woody debris
	Channel substrates	Undisturbed, dredging, groynes/rip-rap, bed reinforcement, navigation
	Grain size of sediment (littoral, channel)	Inorganic: bedrock, boulder, cobble, gravel, sand, silt, clay, concrete and other artificial material
	Composition of channel substrates	No artificial changes, no changes in >70% of the evaluated section, reduction of grain size due to backwaters, backwaters with mostly changed flow velocity and grain size, totally impounded sections
	Channel stabilisation	Artificial material and extent
	Migration barriers (longitudinal)	Type of barrier such as dam or weir
	Longitudinal continuity	(a) Height of structures (b) Channel substrates (c) Migration barriers (for the migration capacity of biota and sediment)

(continued)

Table 5.1 (continued)

Main parameter groups	Main parameter	Subparameters, description
	Impoundment	Length
	Lateral connectivity	Whole floodplain area is connected, >50% of the floodplains are connected, 50–75% disrupted, 75–90% disrupted, <10% are connected
	Hydropeaking	To be defined for large rivers, it seems to be important to record daily changes above 15 cm
Banks/ Riparian Zone	Bank profile	Type, extent natural, remaining, bank structure: fine substrate/flat (to medium) slope, versus (very) steep slope
	Extent of natural vegetation	Percentage of assessment reach
	River engineering on banks (rip-rap)	Natural > 75%, 50–75%, 20–50%, < 20%, 0%
Floodplain	Width	Active floodplain, loss of floodplain (percentage)
	Land use	Artificial, agricultural, forest/near-natural areas, wetlands, water bodies
	Oxbows/side channels, tributaries	Connection type: main channel, tributary, small side-arm, open-end oxbow (lower end open), oxbow, near-separated (no permanent plant growth on the connecting zone, gravel), large secondary channel, floodplain lake, reservoir

The second CEN Standard on the scoring (evaluation) was published after the study in 2010, but basically the parameters were assessed in a five-class system using arithmetic means, individual assessments for channel, banks and floodplains as well as an overall assessment based on the arithmetic mean of all parameters.

The CEN Standard from 2004 is currently under substantial revision moving from the pure physical habitat description towards process-based approaches (sediment, morphology, hydrological conditions considering river scaling approaches from catchment down to river reach) following the outcomes of the REFORM project (http://www.reformrivers.eu/—Rinaldi et al. 2017).

5.4 Results

5.4.1 Basic Fluvio-Morphological Parameters

In a first step, basic fluvio-morphological parameters and the assessment of riverbanks should illustrate the current conditions before summarizing the hydrological assessment by the CEN method (see Table 5.2 and Fig. 5.6). In comparison with

Table 5.2 Major fluvio-morphological parameters in comparison with reference conditions derived from historical maps for the Lower Drava River

Parameter	Drava D-II (reference state/current situation)	Drava D-III (reference state/current situation)
Reach length (km)	68/50	295/185
Channel width (m)	100–1500/80–450	200–400/120–300
Meander wave length (km)	4/6.2	3.8/5.3
Meander amplitude	3.1/1.1	4.5/2.2
Sinuosity	1.5/1.2	2.2/1.5
Number of islands	90/15	45/6
5 meander development stages (in percent of the reach length, compare Fig. 5.6)	II (20%)/(70%)	II (15%)/(50%)
	III (60%)/(30%)	III (45%)/(50%)
	IV (20%)/(0%)	IV (35%)/(0%)
		V (5%)/(0%)

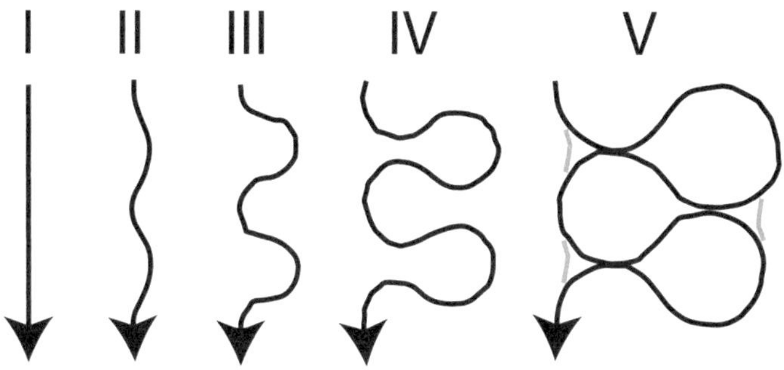

Fig. 5.6 Different stages of meander development used for the morphological characterization after Lászlóffy (in: Bognar 1990)

reference length, the reduction of the lower meandering Drava (D-III) is considerable, and reaches nearly 40%. Average channel width was reduced to up to one-third of original size for D-II and lost most of its variability in channel width.

The sinuosity (the ratio between channel length and valley length) and meander parameters, including the five stages of meander development (Fig. 5.6), clearly indicate the considerable reduction of meander activity for all sections. Only selected subsections such as D-IIIa and D-IIId (cf. Fig. 5.4) still host typical and vital meander sequences. The detailed evaluation of the distribution of meander development stages indicates mostly initial stages of meanders and very few reaches in the fifth stage (with completed cutoffs). This is the expression of the main 19th-century regulation and slow re-formation of meanders within reaches of low maintenance after World War I (declining navigation from Osijek upstream to Barcs, part of the iron curtain border between Hungary and the former Republic of Yugoslavia).

Table 5.3 Assessment of banks

Highly dynamic banks (steep banks with erosion)	41 (7%)
Shallow banks (associated with point bars)	29 (5%)
Others (mostly nearly natural banks)	266 (43%)
Old structures (collapsed rip-rap and groynes)	34 (5%)
Major bank revetments and structures (rip-rap, groynes, side-arm closures)	247 (40%)

A total length of 617 km banks of the Drava include along the studied 235 rkm stretch both riverbanks plus all major side channels and hydraulic structures such as groynes (Table 5.3). Nearly 12% of all banks are highly dynamic steep banks and shallow point bars. Together with the group of invariant banks, showing no clear trend to erosion or accumulation, over 50% of all banks are not systematically stabilized by rip-rap, which is of high value compared to western European rivers of similar size. However, the group of old collapsing structures (rip-rap and groynes) can be counted along with the major bank revetments (together 45%) as they stabilize the river over decades.

The morphological floodplain in the respective section spreads originally over 240,981 ha and only 53,693 ha has been retained as active floodplain, which is a loss of 78% (in general, length reduction in Hungary is about 82% and only 70% in Croatia).

This hydromorphological overview may reflect a too poor overall hydromorphological situation along the Lower Drava. But looking at the given river continuum (no dam within the reach) and on reach scale considering all in-channel features, banks and floodplain vegetation of the shrunken but still continuous active floodplain, a better situation can be presented.

5.4.2 *CEN-Based Hydromorphologcial Assessment*

As described above, the river was subdivided into about 400 reaches, allowing a precise description and assessment of all in-channel features, banks, and floodplain on both banks in a five-class assessment scheme.

5.4.2.1 Channel Assessment

At least a few scattered reaches of the lower Drava (together about 45 km out of 235 km) could be attributed to the best (blue) quality class, showing all features of a near-natural channel in comparison to the reference condition. On the other hand, only about 7% are extensively and severely modified (orange and red classes, see Fig. 5.7). About 38% belong to the quality class 2 (green), indicating a still high potential to provide most of the hydromorphological functions of reference reaches.

Fig. 5.7 Overall evaluation of the lower Drava (left) and for all rivers in the Drava and Mura river system (right) (Schwarz 2007)

The remaining 34% are rated as third (yellow) class, mostly regulated with stabilized banks. The two most significant human pressures are flood protection and hydropower generation (upstream from the studied section) (see Chaps. 9 and 10 in this volume). But for the Lower Drava in particular, the conditions maintained for the nearly non-existing navigation (mean and low water regulations) and the commercial gravel and sand extraction are significant pressures (the latter was considerably reduced since 2010).

5.4.2.2 Assessment of Banks and the Riparian Zone

Less than 20% of the surveyed banks reflect natural conditions such as steep and very shallow banks at point bars (indicating lateral erosion, accumulation and channel shifting). From the remaining 80% about the half are neutral banks without continuous bank protection, but the other half is reinforced by rip-rap. Compared to reference conditions, the potential total length of erodible steep banks is estimated 2–3 times higher compared to the current situation. The riparian vegetation shows whether the banks are intact or not, but neophytes are spreading along regulated and less frequently flooded banks.

5.4.2.3 Floodplain Assessment

Today about 78% of the morphological floodplain is cut off or excluded from the regularly flooded area. The remaining area still hosts most of the typical softwood and hardwood habitats in particular along the Lower Mura and Drava. However, flood dynamics along strongly regulated reaches are affected by incision of the main channel and aggradation of fine sediments on the floodplain (increasing disconnectivity).

The overall assessment results obtained in 2005 for the lower Drava (arithmetic mean of the channel, banks and floodplain assessments) indicate that more than the

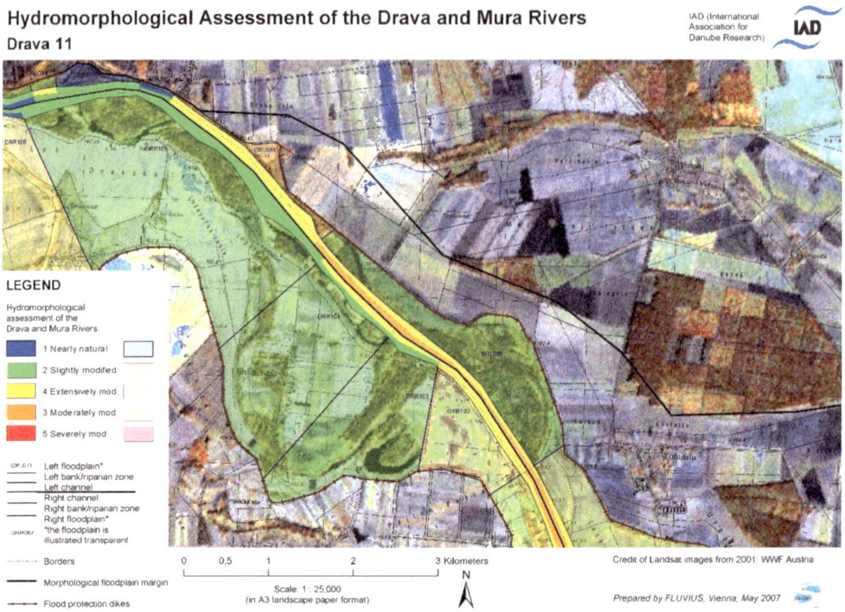

Fig. 5.8 Map example indicating the individual assessment of channel banks and floodplain as colour ribbon and segment visualisation

half of the river belongs to the first and second class and one third is in the moderately modified class three (see the left pie chart in Fig. 5.7 and the map in Fig. 5.8). This result, excellent in comparison with many other European rivers of similar size, however, should be critically viewed. The long-term processes driven by the chain of hydropower plants upstream alter the sediment balance and hydrological behavior at least for the minor but ecologically important flood events. The impact of continuous sediment deficit (channel incision) in regulated reaches, lowering of groundwater on the one side and disconnection of the active floodplain areas should lead to degradation on the long term. Looking at the catchment scale (right pie chart in Fig. 5.7), the significant alterations due chains of hydropower plants in Austria, Slovenia and Croatia are evident in the 37% ratio in the two poorest classes.

Regarding the official national classifications for the 2015 River Basin Management Plans for the Lower Drava (Hungary and Croatia), the results are not comparable for different reasons. First of all, the Water Framework Directive is based on water bodies which comprise much longer assessment units (in the case of the Drava over dozens of km for one water body) and different methods (in the case of Croatia only based on pressure related risk assessments and not on a particular hydromorphological survey and assessment). *Hungary*[1] established its own

[1]https://www.vizugy.hu/index.php?module=vizstrat&programelemid=149.

methodology fitting the CEN standard, which refers the entire Lower Drava to the second (good) class. No significant quantitative hydrological alterations were designated (same results are stated regarding ecological status). For the 2015 report, *Croatia*[2] only uses a single pressure—impact risk assessment applying the "one out–all out" principle—as applied for biological quality elements. In other words, the worse parameter defines the final assessment. The Lower Drava was assessed to be in the third class and only the lowermost 20 km downstream Osijek (utilized by navigation) was assessed in the fourth class (being another heavily modified water body). Ecological status was mainly in class 3, hydropower reach class 4 and downstream of a most polluted tributary on Lower Drava west of Osijek to the mouth class 5. (This was the situation before 2009.)

5.5 What Happened Since 2005?

Hydromorphological inventories are time consuming and the six-year monitoring cycle (as proposed in the WFD) is very ambitious. Moreover, there are no systematic approaches how to update the extensive inventories. However, it is extremely important to observe and record all changes and developments in the following periods, even to upgrade methodologies with the new CEN standard as intended, from the pure physical habitat descriptions towards process-based approaches which also consider sediment and hydrological behaviour. For lack of space, only the most important changes and trends are presented below.

In *Austria* about 20 km of the river length of at least class 3 (yellow) (Fig. 5.7, right pie chart) were affected by hydropower impoundments and turned into the red class (two plants built downstream of Graz on the Mura). On the Upper Drava and Mura, however, several river restoration projects were carried out, not only creating continuity by fish passes, but also improving the morphological situation (e.g. between Lienz and Spittal) and locally on the Upper Mura in Styria. But the length of new impoundments exceed the length of restored river reaches.

As of particular interest in the upper catchment, climate change is most visible in the mountain ridges of the upper headwaters. At a first glance, those reaches are still in close-to-natural conditions (the glacial headwaters of the Isel, Möll and Mura). However, the retreat of glaciers and melting processes will make their discharge regimes alter considerably in the future, when glaciers will disappear or retain a minimum extension. Therefore, such river section types are under high pressure from the climate change. Since all geomorphic processes are driven by the freezing and melting of water, their discharge and sediment regime controls all processes downstream.

[2]http://www.voda.hr/sites/default/files/plan_upravljanja_vodnim_podrucjima_2016._-_2021_0.pdf.

Although in *Slovenia* there is a new initiative to construct further hydropower plants on the Mura, no significant change has taken place yet. (Mota is under planning, six other dams have been designed, e.g. on the transboundary reach with Austria.) Local restoration projects (Life projects on the Mura, mainly side channel connections and bank improvements and along the residual Drava branch downstream of hydropower plant in Maribor), outreach additional and new bank stabilization efforts on other sections.

In *Croatia,* too, further hydropower plants were intended but are not physically in the planning stage. Water management agencies currently work on the first large-scale Life project of the country concerning river restoration. However, some additional bank stabilization works for the protection of infrastructure (bridges and flood-control dykes) and housing enlarge the length of modified banks. In particular, the most downstream reach from the Danube confluence to Osijek and further upstream to Belišće (rkm 53) was improved for navigation, by introducing low-water correction, but mainly stabilization of erodible steep banks, leading in total to a reduction of class 2 (and even some class 1) stretches. Also on the Lower Drava two new bridges north of Osijek were built (the highway is now close to completion) deteriorating the Lower Drava corridor. At the same time, commercial dredging has been significantly reduced. The general trend, strategic planning (for instance, for the reconnection of oxbows) as well as the management of flood-control dykes and stakeholder involvement are promising.

In *Hungary,* the Danube-Drava National Park stretches all along the river course. Commercial dredging was reduced within the past decade and several small improvements as side-arm reconnection and some bank removal are evident and concepts how to reconnect oxbows are under implementation (see Chaps. 19–21). A detailed mapping of the current channel centerline of the river compared to the old official rkm from the early 1980s shows discrepancies of the overall river length by meander cutoffs on the one hand and natural processes (lateral shift) on the other (see Chap. 11). Finally, the loss of about 10 rkm since the beginning of river regulation has been almost fully compensated by lateral shifting (Schwarz 2017).

In summary, further slight degradation can be observed for the Drava and Mura in general (new hydropower plants in Austria, additional bank reinforcements/regulation in Croatia) and the problem of ongoing channel incision and disconnection of active floodplain by aggradation is still unsolved. But keeping in mind the very special situation of the Lower Drava, for long decades part of the iron curtain and now dividing Hungary and Croatia, EU countries facing complicated transition processes, the establishment of the Transboundary Biosphere Reserve Mura-Drava-Danube can be seen as a positive step to preserve and restore the river corridors and their functions for the benefit of next generations.

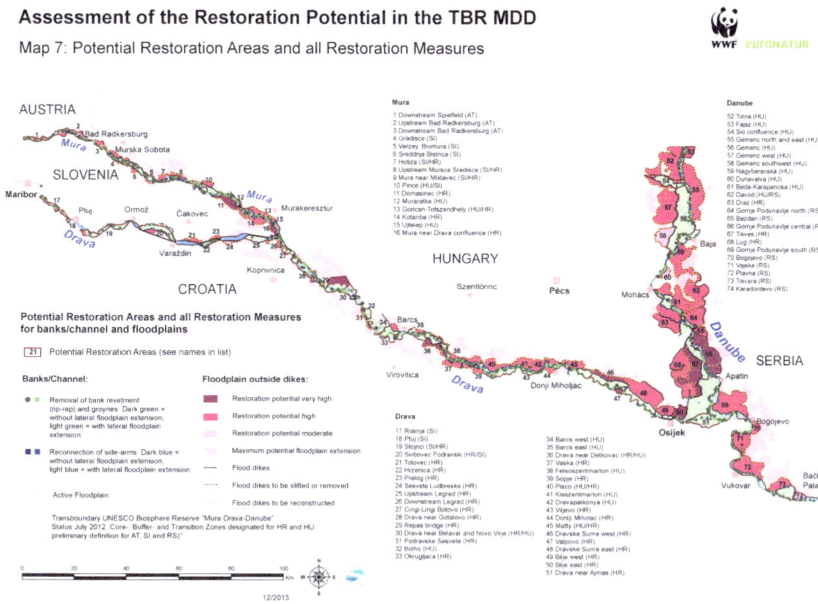

Fig. 5.9 River and floodplain restoration potential along the entire river corridors

5.6 Proposals for Restoration Since 2012

Regarding the Lower Drava stretch (rkm 0–235), a long-term (maximum) river and floodplain restoration potential was identified (Schwarz 2013), also confirmed by Croatian Waters regarding potential side-arm and oxbow reconnections and studies for several Hungarian projects, mainly within the national park. EU Life projects in both countries are under implementation (DravaLife in Croatia[3] and an oxbow reconnection project in Hungary,[4] a transboundary project).

For the respective reach 205 bank sections with a total length of 201.6 km (each 980 m long on the average) could be subjected to rip-rap and groyne removal and 46 major side-arms with a total length of 146.8 km (average 3.2 km) could be reconnected. Together with the 48 proposed areas (in total 207,557 ha) for floodplain restoration (147,027 ha to be reconnected outside the flood-control dykes), these impressive figures clearly confirm both the need and opportunity to improve the hydromorphological conditions by river and floodplain restoration (Fig. 5.9).

---

[3]http://www.drava-life.hr/en/home/.

[4]http://www.olddrava.com/.

5.7 Conclusions

Since the mid-1990s the lower Drava (from the Mura confluence to the Danube confluence, 235 rkm) was investigated repeatedly and 2007 the first hydromorphological assessment for a large river in the middle and lower Danube basin, based on the first CEN standard, was published (Schwarz 2007) and delivered important findings for the approach to the mapping of the entire Danube (Schwarz 2015; Schwarz et al. 2015).

The Lower Drava is influenced and altered by upstream hydropower plants in Austria, Slovenia and Croatia, the various river regulation works (bank stabilization, some groynes and former meander cutoffs) as well as the loss of floodplains by constructing dykes. However, it still retains nearly all features of the original riparian landscape along several reaches and the whole course was designated a Biosphere Reserve containing two Ramsar areas and one riparian National Park, which underline the ecological importance of the river corridor.

However, the initial investigations and the hydromorphological pilot study should be viewed as a substantial baseline work only to follow up the middle and long-term processes and alterations along the Lower Drava. Sediment flux, the prevailing reaches with incision and also the short accumulational reaches and their associated channel features such as gravel and sand bars have to be surveyed and monitored jointly and data exchanged. Hydrographical analysis should underlie the investigation of flood dynamics (amplitude, magnitude, duration) considering all effects of the hydropower chain upstream (discharge distribution, role of reservoirs during low to annual flood discharges, hydropeaking which is observable at least 50 km downstream of the last barrage).

Finally, both the pressures and responses (impacts) should be evaluated at different scales providing the basis for understanding the cause-effect linkages that govern the observed river system variation. The developments within the last 10–15 years partly indicate further deterioration along some river reaches, but also initial restoration activities reflecting a slow paradigm change in river management as well as a better transboundary cooperation.

References

Bognar A (1990) Geomorfologija Baranje. Znanstvena monografija (Geomorphology of Baranja. A scientific monograph). Zagreb. 312 pp (in Croatian)

CEN (2004) EN 14614:2004. Water quality—guidance standard for assessing the hydromorphological features of rivers. German Institute for Standards, Berlin. 22 pp

Rinaldi M, Belletti B, Bussettini M, Comiti F, Golfieri B, Lastoria B, Marches E, Nardi L, Surian N (2017) New tools for the hydromorphological assessment and monitoring of European streams. J Environ Manage 202(2):363–378. https://doi.org/10.1016/j.jenvman.2016.11.036

Schwarz U (2007) Pilot study: hydromorphological survey and mapping of the Drava and Mura Rivers. IAD-Report, FLUVIUS, Floodplain Ecology and River Basin Management, Vienna, p 140

Schwarz U (2013) Assessment of the restoration potential in the Transboundary UNESCO biosphere reserve "Mura-Drava-Danube". For WWF Austria, Vienna, p 167

Schwarz U (2015) Hydromorphology of the Danube. In: Liska E (ed) The Danube River Basin. Springer Verlag, Berlin – Heidelberg. 469–480 (The Handbook of Environmental Chemistry 39)

Schwarz U (2017) LIFE Project "DRAVA LIFE – Integrated River Management" Action A.2 LIFE Drava Map. Technical Report, Wien. 116 p

Schwarz U, Holubova K, Cuban R, Matok P, Busovsky J (2015) JDS 3 Hydromorphological survey. In: ICPDR Joint Danube Survey 3. A comprehensive analysis of Danube water quality. Vienna. 40–71

Chapter 6
Hydrological Characteristics of the Drava River in Croatia

Lidija Tadić and Tamara Brleković

Abstract The Drava River is one of the biggest right-bank tributary of the Danube and one of the most important rivers in continental Croatia. Its great environmental value is in conflict with its economic potential in hydroelectricity, navigation, and irrigation. During the history, many hydraulic structures were constructed on the river and caused changes in its hydrological regime. In addition, in the early 1980s, climate change manifested in less precipitation and higher air temperature became more evident. A hydrological analysis of the Croatian section of the Drava River pointed out the consequences of both trends. There are three different sections of the Drava River:

- the upstream section dependent on dam constructions;
- the middle part affected by the tributaries and
- the lower section under the influence of the Danube River backwater.

Analyses of water level and discharge time series between 1960 and 2015 were made for five hydrological gauging stations in Croatia. All characteristic minimum, average, and maximum values of these two parameters have shown a decreasing trend with two exceptions in the most upstream part of the river where three dams were constructed. Extreme hydrological events, floods and droughts were also analyzed. High water levels, which were a great threat in the past, are better controlled by dykes today, but they still endanger the area along the confluence of the Drava and Mura and the lowermost section due to the backwater effect of the Danube River. Droughts or periods of low discharges are becoming a more and more serious problem in the last 15 years, but, according the standardized streamflow index for detected drought severity, there is no clear increasing trend in drought frequency.

L. Tadić (✉) · T. Brleković
Faculty of Civil Engineering, University of Osijek, Drinska 16a, 31 000 Osijek, Croatia
e-mail: ltadic@gfos.hr

T. Brleković
e-mail: tamaradadic@gfos.hr

D. Lóczy (ed.), *The Drava River*, Springer Geography,
https://doi.org/10.1007/978-3-319-92816-6_6

Keywords Water level · Discharge · Flood · Low water · Human impact
Backwater effect

6.1 Introduction

Out of its total length (749 km), the middle section of the river in Croatia is 322.8 km long to Terezino Polje (152.350 rkm). Here the Drava is mostly a lowland river with a small gradient and numerous meanders, to its confluence with Danube River (1,382.300 rkm). The area of its catchment in Croatia is 7,015 km^2, less than 20% of the total catchment area of 41,238 km^2. According to its hydrological regime, the Drava River is a typical glacial river with highest water levels in late spring and summer (May, June, and July) and lowest water levels during winter period (January and February). Recently, however, important shifts in these characteristic periods occur. In 1993 and 1998, high water levels were observed in October and in 2012 in November. The hydrological characteristics of the Drava River, like most European rivers, have been changing over time caused by human impact, including construction works carried out at various dates in history. It is obvious that river length has been significantly shortened and accumulation has happened at Varaždin. In the past, the riverbed with meanders and side-arms in wide alluvial lowland had a completely different hydrological regime than nowadays (Fig. 6.1). More recently, river regime modifications are partly attributed to climate change.

In addition to the overall morphological division of the Drava River to upper, middle and lower sections, the Croatian section can be divided into three units by river regime (Fig. 6.2). On the upstream section (A), three multipurpose dams and reservoirs were established (hydropower plants Varaždin in year 1975, Čakovec in 1982 and Dubrava in 1989). This section ends at 255.050 river km, downstream

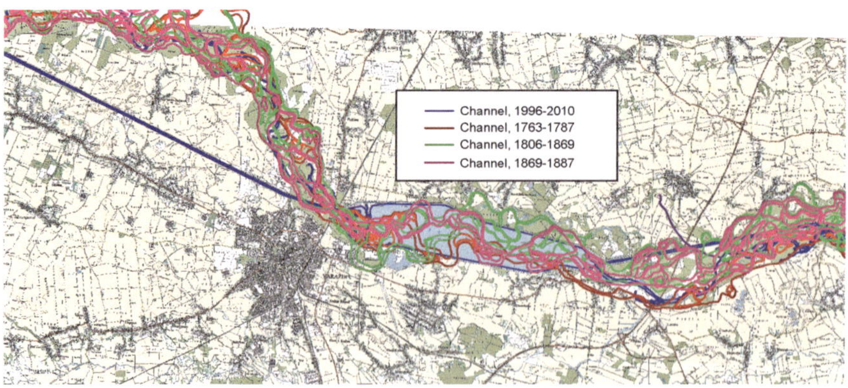

Fig. 6.1 Changes in the Drava riverbed on the Varaždin section from the 19th to 21st century

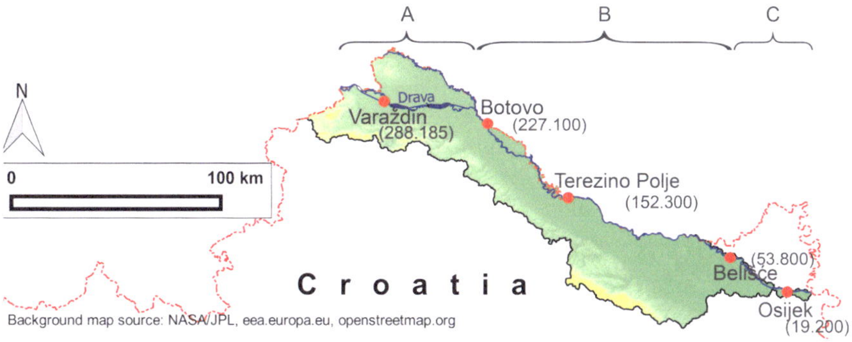

Fig. 6.2 Drava River in Croatia with gauging stations and the sections identified

from the last dam and it is significantly influenced by accumulations. The representative gauging station for analysis is Varaždin (288.145 rkm). Second section (B) is influenced by tributary catchments and ends at 53.800 rkm. The time series from three stations were analyzed for this section: Botovo (226.500 rkm), downstream from confluence of the Mura River, Terezino Polje (52.350 rkm) and Belišće (53.800 rkm). The third section (C) stretches from the gauge Belišće to the confluence with the Danube River. The representative gauging station for this section is Osijek (19.100 rkm) (Fig. 6.2). This entire section is influenced by the Danube River backwater, which has a significant impact on high water levels. Therefore, this section is endangered by floods from both the Drava and Danube rivers. The impact of high water levels from the Drava River is much weaker after dam constructions in the upstream catchment.

In the last decade, many papers have been published on river hydrology, including the hydrological properties of the Drava River. Different authors were motivated by the observed changes in river regimes and they sought to find the cause of these changes. Water levels and discharges were analyzed regarding climate change and water balance of the catchment.

An overview of most significant research is presented in chronological order. In the lower section of the Drava River, the period between 1900 and 1995 was characterized by an increase of potential evapotranspiration and reduction of surface runoff and soil moisture content (Zaninović and Gajić-Čapka 2000). It is also observed that a negative deviation of the Drava River water levels completely follows the positive air temperature deviation (Bonacci and Gereš 2001). The discharge analysis of the Drava River showed a negative trend, similar to the discharge trends for other Croatian rivers (Pandžić et al. 2008). The comparison of the number of precipitation days which are significantly reduced in the cold period and slightly rose in the warmer period with the Drava River discharges, also showed a negative discharge trend. In this case, the coefficient of variation of mean annual discharges decreases and the variation coefficient of the mean discharges of the colder period increases (Gajić-Čapka and Cesarec 2010).

There are also papers in which authors cite anthropogenic influences as well as climate change as the main cause of hydrological changes. After construction of the three hydropower plants on the Drava River, alterations in discharges, water levels, and sediment transport can be detected. Minimum and average discharges and all characteristic water levels of the lower Drava River are being reduced (Bonacci and Oskoruš 2010). Besides dams and hydropower plants, numerous barriers on the upper section in Austria and Slovenia have been built on the Drava River.

Hydrological observations on the Drava River started at the beginning of the 20th century, in year 1926, but not at all stations. For this reason, the hydrological properties of the Croatian section of the Drava River will be analyzed based on the time series of discharges and water levels observed at five gauging stations between 1960 and 2015.

6.2 Water Levels

Minimum, average, and maximum water levels observed at five hydrological gauging stations (Varaždin, Botovo, Terezino Polje, Belišće and Osijek) in the period between 1960 and 2015 were studied. Declining linear trends of water levels were found for all marked water levels and all gauging stations except for Varaždin. Minimum and average water levels for this gauging station showed a positive trend and maximum water levels negative, but least emphasized in relation to other stations. The impact of dams is evident in this case. The intensity of water levels reduction is clear from the linear regression equations listed in Table 6.1. Other stations showed no variation in the slope of the linear regression lines, but the declining trend of the minimum water levels at the Terezino Polje station is remarkable.

Monthly water levels were tested according to Standard Normal Homogeneity Test (SNHT) in order to define homogeneity of the time series or presence of any significant differences in the occurrence of the minimum, average, and maximum water levels. Figure 6.3 shows sub-periods of characteristic water levels obtained by Standard Normal Homogeneity Test.

Table 6.1 Linear regression equations for all characteristic water levels

Hydrological gauging station	Water level (m a.s.l.)		
	Minimum	Average	Maximum
Varaždin	$y = 1 \times 10^{-5} + 166.4$	$y = 8 \times 10^{-6} + 167.32$	$y = -6 \times 10^{-6} + 168.14$
Botovo	$y = -4 \times 10^{-5} + 123.35$	$y = -2 \times 10^{-5} + 123.86$	$y = -2 \times 10^{-5} + 124.83$
Terezino Polje	$\mathbf{y = -0.001x + 101.98}$	$y = -9 \times 10^{-5} + 102.3$	$y = -8 \times 10^{-5} + 102.94$
Belišće	$y = -3 \times 10^{-5} + 86.46$	$y = -3 \times 10^{-5} + 87.121$	$y = -4 \times 10^{-5} + 88.18$
Osijek	$y = -5 \times 10^{-5} + 83.2$	$y = -5 \times 10^{-5} + 83.72$	$y = -4 \times 10^{-5} + 84.39$

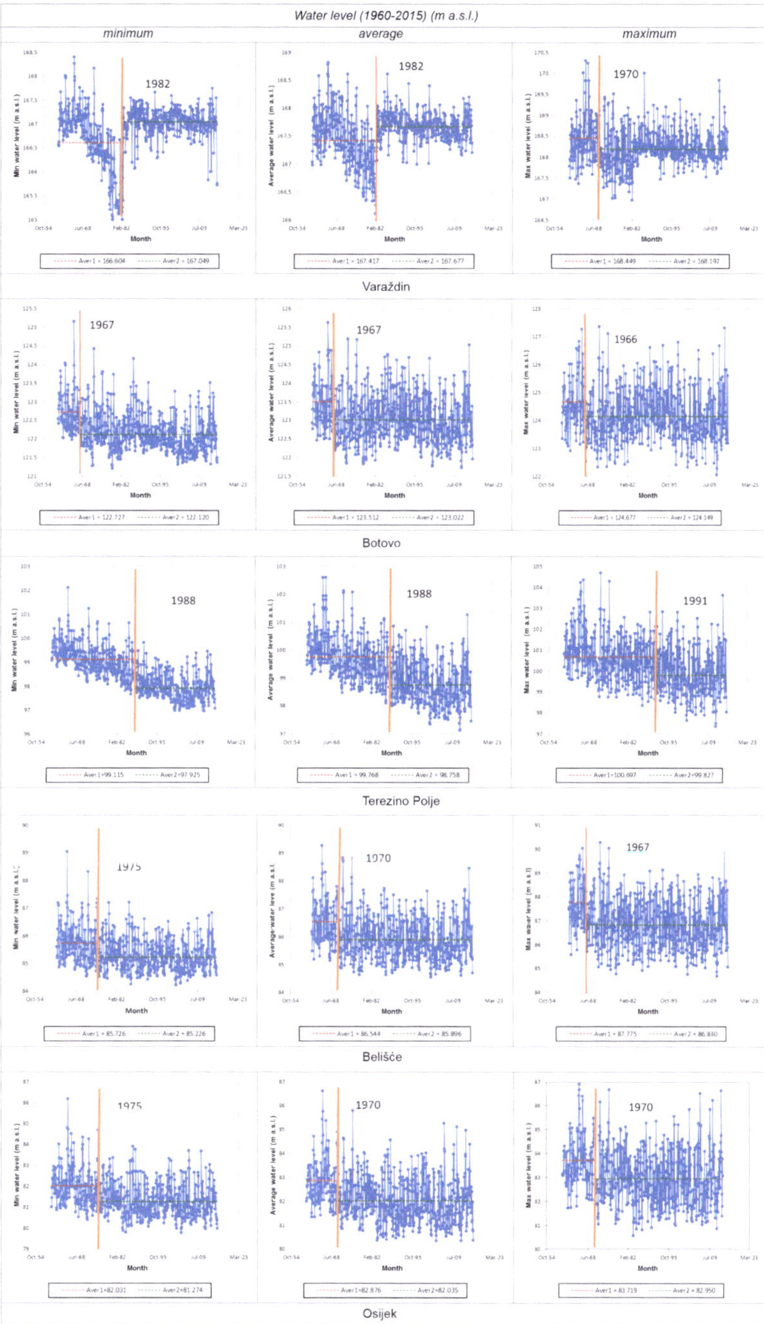

Fig. 6.3 Sub-periods of characteristic water levels obtained by Standard Normal Homogeneity Test for all gauging stations

For all gauging stations and for all characteristic water levels, the time series were split up into subperiods. In the most cases, the breaking year between the subperiods is in the late 1960s and early 1970s. At the hydrological station Varaždin, there is a more pronounced impact of dam construction and downstream controlled water regime. The second subperiod starts in 1982 and it is characterized by higher minimum and average water levels by 26 cm and 44 cm, respectively. Besides, maximum water levels are lower in the second sub-period by 26 cm, and the breaking year corresponds with the construction of the first dam in 1970. In the second subperiod average, water levels are lower from 40 to 120 cm at the stations Botovo, Terezino Polje and Belišće. The largest-scale lowering of minimum water levels during second subperiod (1988–2015) is detected at the gauging station Terezino Polje with the magnitude of 120 cm in the first subperiod. The lowering of average water levels between 76 and 84 cm is characteristic for second subperiod for downstream station Osijek which is under influence of the Danube River backwater.

To better define observed changes in water levels, coefficients of variation were analyzed. Major oscillations of water levels can have severe negative consequences for the ecological system, water management, water use and, particularly, water quality. In Fig. 6.4, coefficients of variation for subperiods are shown.

According to Fig. 6.4, the range of monthly water level oscillation decreased for most of the gauging stations. The reduction is the most remarkable for the most upstream station Varaždin because of dam construction. The reduction in the variability of water levels is negligible for the middle and lower sections, where Osijek is situated, where the coefficients of variation for maximum water levels increased. The main reason for this is the cumulative influence of Drava and Danube high water levels.

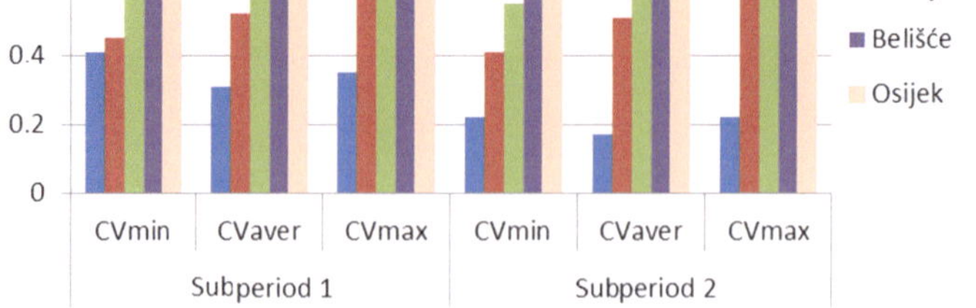

Fig. 6.4 Coefficients of variation of characteristic water levels for subperiods

6.3 Discharges

Based on the analyzed water levels, it can be concluded that the hydrological regime of the Drava River is dominated by anthropogenic influence since most of the changes in precipitation and air temperature have occurred in 1980s (Gajić-Čapka and Cesarec 2010; Tadić et al. 2016a, b). Discharge analysis was carried out for the time series of the period 1960–2015 for three hydrological stations: Botovo, Terezino Polje, and Belišće. For the upstream gauging station Varaždin, since accumulations were built, discharges are no longer measured and for Osijek, because of the backwater influence of the Danube, discharges have never been measured. In Fig. 6.5, results of homogeneity test are shown. Subperiods have been detected with breaking years between 1967 and 1980. Comparing to the first subperiod to the second, discharges have been reducing. The greatest difference is between average and maximum discharges in the first and second subperiods at Belišće station. (It should be mentioned that between 1994 and 2002 discharges were not recorded.)

The breaking years between the first and second subperiods for water levels and for discharges correspond. Thus, it can be concluded that significant riverbed

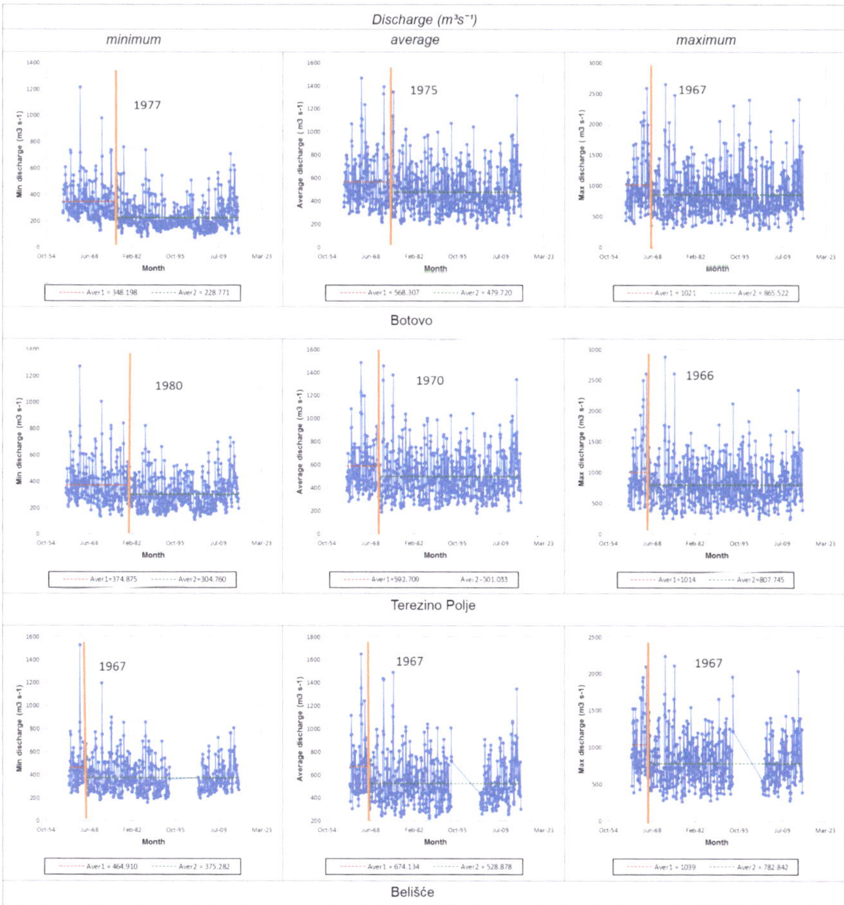

Fig. 6.5 Homogeneity test of the discharge series for three gauging stations

deepening occurred. Considering the changes of precipitation in the 1980s (Tadić et al. 2016a, b), it can be claimed that climate change has been the main cause of reduction in discharge. The impact of the snow and its contribution to total discharge should be also mentioned. Since the early 1970s, the number of days with snow cover in continental Croatia has been reduced (Bonacci et al. 2012). Changes in land cover also caused changes in the surface drainage in smaller subcatchments of the Drava River in Croatia and Hungary (see Chap. 5 in this volume). However, numerous dams were built on the Drava and its tributaries in Croatia and in countries upstream. Therefore, anthropogenic influence on river regime can be considered a predominant factor.

6.4 Maximum Water Levels (Floods)

Organized flood protection activities in the Croatian part of the Drava River basin started in the 19th century. In the last one hundred years, several disastrous floods occurred (in 1926, 1965, 1966, and 1972). After each flood, flood protection system was reconstructed and improved. In total, 435 km of dykes were built along the Drava River. The most important turning point in flood control was the major flood in 1972. Since then has been developed modern approach to the flood protection including technical and non-technical measures. Construction of dyke system and three dams significantly reduced flood frequency and damage. Table 6.2 presents five the most extreme water levels in the period 1960–2015. On only the most downstream section (C), represented by gauging station Osijek, three of five floods have occurred in the 21st century. All of them were caused not by the Drava River, but high water levels on the Danube.

Table 6.2 shows that from the flood protection point of view, the lowermost section of the Drava River (to ca 20 km upstream the confluence) is highly endangered since it is under the influence of the Danube. The hydrological analysis of the occurrence of maximum water levels on this section showed that until 1970s the majority of floods were caused by high water levels on the Drava River and after that date all floods (except in 2014) were caused by the Danube River backwaters. Coincidence of high water levels of both rivers would afflict enormous damage on the surrounding area. Fortunately, it has never happened yet, but this event cannot be excluded (Tadic et al. 2016a, b—Fig. 6.6).

There were 1,000 random pairs of water levels in Osijek and Bezdan generated (blue dots in Fig. 6.6) using the Normal copula function. Horizontal and vertical bars represent water levels at the Osijek gauging station on the Drava River and the Bezdan gauging station on the Danube River, which correspond with the first and second stage of flood warning. There is 0.7% probability of simultaneously occurring water levels that correspond to the 2nd stage or higher in Osijek and in Bezdan and 34.7% probability of simultaneous occurrence of the water level that corresponds to the 1st stage or higher at Osijek and at Bezdan (Tadić et al. 2016a, b).

Table 6.2 Five year with the most extreme maximum water levels (in bold: years of the 21st century)

Varaždin	Botovo	Terezino Polje	Belišće	Osijek
Maximum water levels, 1960–2015				
1965	1972	1972	1972	1965
1966	**2014**	1966	1966	1975
1964	1966	1975	1975	**2013**
1986	1975	1965	**2014**	**2006**
1975	1965	**2014**	1965	**2010**

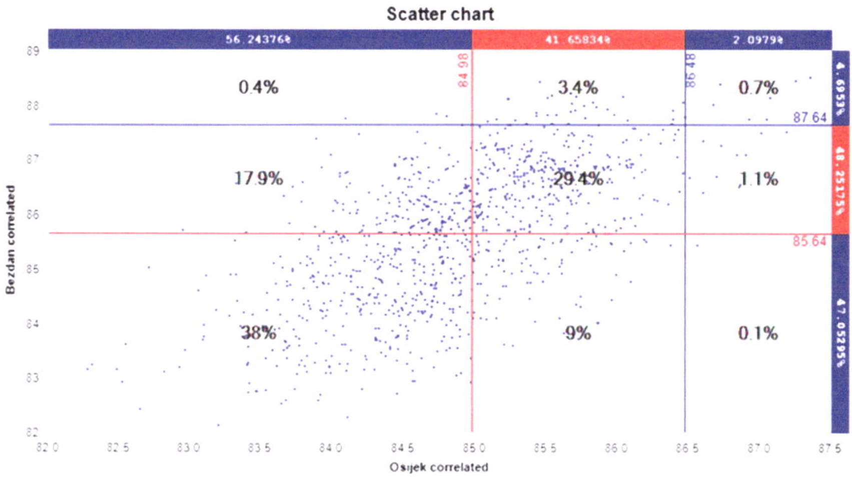

Fig. 6.6 The probability of coincidence of maximum water levels of the Drava and Danube rivers (Tadić et al. 2016a, b)

6.5 Minimum Water Levels (Droughts)

Increasing flood frequency is only one of the climate change impacts. Hydrological droughts or extreme low water levels are also becoming more and more common. In order to study this phenomenon, five extreme minimum water levels on representative gauging stations have been analyzed (Table 6.3). At the upstream gauging stations, Varaždin and Botovo, a clear influence of constructed reservoirs can be detected. Besides, the most downstream section (C) represented by the gauging station Osijek is under the Danube backwater influence. The section between 152 and 53 rkm is the most endangered by hydrological drought. This section depends on the Drava River tributaries and the runoff from their catchments. The early 21st century is characterized by several extremely dry years (2001, 2003, and 2011) in this area (Tadić et al. 2015).

Especially on the middle Drava section, we can expect more pronounced drought periods (Table 6.3), which may have enormous impact on water utilization (irrigation, navigation) and the environment in general. All the five years with the

Table 6.3 Five years with the most extreme minimum water levels (in bold: years of the 21st century)

Varaždin	Botovo	Terezino Polje	Belišće	Osijek
Minimum water levels (1960–2015)				
1982	1978	**2012**	**2002**	**2003**
1979	1971	**2002**	1987	1985
1978	1979	**2003**	**2003**	1983
1981	1972	**2011**	**2006**	**2011**
1982	**2001**	**2001**	**2001**	**2002**

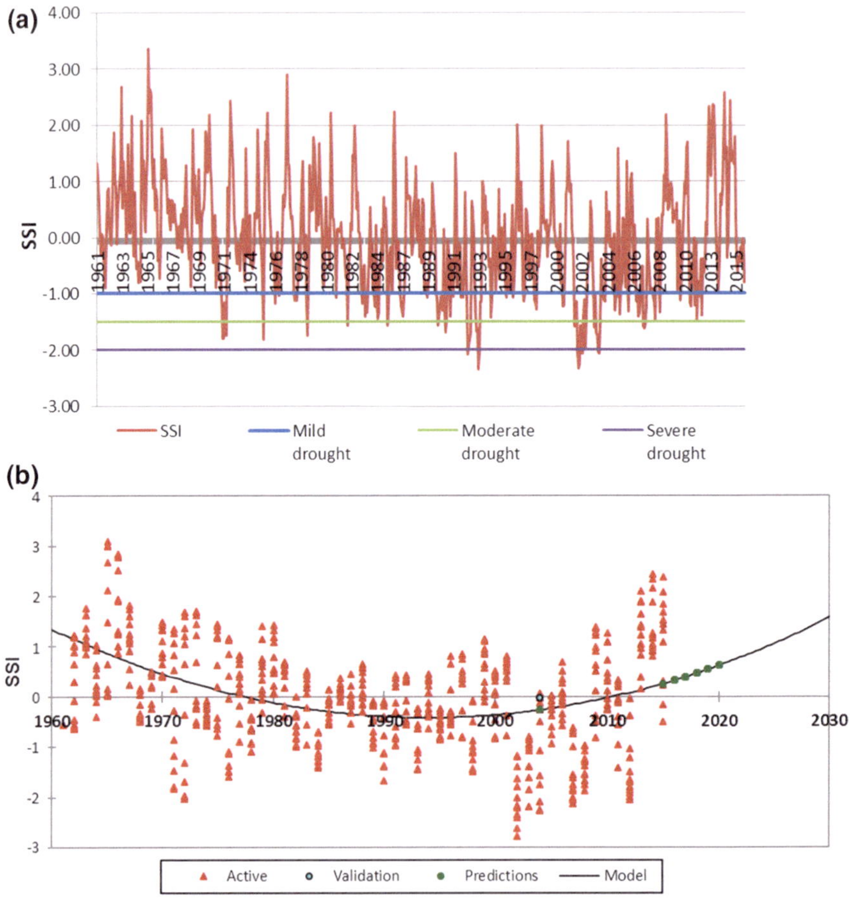

Fig. 6.7 Standardized streamflow index (SSI) calculated for the Terezino Polje gauging station (**a**). Nonlinear regression analysis for the same gauging station with prediction period of five years (**b**)

lowest water levels obtained from the gauge of Terezino Polje occurred in the 21st century. To inspect this problem more deeply, standardized streamflow index (SSI) was calculated using average monthly discharge time series (1961–2015).

According to this drought index, values between 1 and −1 represent normal years (Ljubenkov and Cindrić-Kalin 2016). SSI values below −1 represent mild, moderate, or severe drought (Fig. 6.7a). The application of nonlinear regression analysis on calculated SSI values with a prediction period of five years (Fig. 6.7b) shows that a lower frequency of extreme low water levels and hydrological droughts can be expected.

6.6 Conclusions

The hydrological analysis of the Drava River in Croatia is presented in the chapter based on water level and discharge measurements at five hydrological gauging stations which operated in the period between 1960 and 2015. Characteristic water levels (minimum, average and maximum) show a decreasing trend with the exception of the gauging station Varaždin which is under the influence of a hydropower plant on the upstream river section. The homogeneity test (SNHT) applied on the water level time series indicates two sub-periods with a breaking year in the late 1960s and early 1970s. Characteristic discharges also decrease and time series are also non-homogeneous. The breaking years between two subperiods appear between 1967 and 1980.

The analysis of extreme hydrological events, floods and droughts, points out that floods endanger the lowermost section (designated as C). Besides, hydrological droughts are more significant in the middle section (B) which is under the influence of small tributaries. On the uppermost Croatian section (A) there are no extreme events in recent times due to the influence of dams, constructed in the period 1975–1989. Because of last-century anthropogenic impact, the first of all dam constructions, the Drava River regime has been changed and controlled. These activities are reducing negative effects of the extreme water levels, but, at the same time, have caused major and lasting deterioration of the environment.

References

Bonacci O, Gereš D (2001) Impact Assessment and Adaptation to Climate Change: Hydrology and Water Resources // The First National Communication of the Republic of Croatia to the UN Framework Convention on Climate Change (UNFCCC) / Jelavić V (ed) Ministry of environmental protection and physical planning, Zagreb 175–189

Bonacci O, Oskoruš D (2010) The changes of the lower Drava River water level, discharge and suspended sediment regime. Environ Earth Sci 59(8):1661–1670. https://doi.org/10.1007/s12665-009-0148-8

Bonacci O, Dadić Tadić L (2012) Hidrološki aspekti pojave snijega u Hrvatskoj. Hrvatske vode 81:117–130 (in Croatian)

Gajić-Čapka M, Cesarec K (2010) Trend and variability in discharge and climate variables. Hrvatske vode 71:19–30

Ljubenkov I, Cindrić-Kalin K (2016) Ocjena suše primjenom standardiziranog indeksa oborine i protoka te njihov odnos na primjeru Sinjskog polja. Građevinar 68(2):135–143 (in Croatian). https://doi.org/10.14256/JCE.1337.2015

Pandžić K, Trninić D, Likso T, Bošnjak T (2008) Long-term variations in water balance components for Croatia. Theoret Appl Climatol 95(1–2):39–51. https://doi.org/10.1007/s00704-007-0366-5

Tadić L, Dadić T, Bosak M (2015) Usporedba različitih metoda za ocjenu suše na području kontinentalne Hrvatske. Građevinar 67(1):11–22 (in Croatian). https://doi.org/10.14256/JCE.1088.2014

Tadić L, Dadić T, Leko-Kos M (2016a) Variability of hydrological parameters and water balance components in small catchment in Croatia. Adv Meteorol. https://doi.org/10.1155/2016/1393241

Tadić L, Bonacci O, Dadić T (2016) Analysis of the Drava and Danube rivers floods in Osijek (Croatia) and possibility of their coincidence. Environ Earth Sci 75(18). https://doi.org/10.1007/s12665-016-6052-0

Zaninović K, Gajić-Čapka M (2000) Changes in components of the water balance in the Croatian lowlands. Theoret Appl Climatol 65(1–2):111–117. https://doi.org/10.1007/s007040050008

Chapter 7
Sediment Transport of the Drava River

Enikő Anna Tamás

Abstract Sediment transport data collection on the Croatian-Hungarian section of the Drava River started in 1961 at the Barcs and Drávaszabolcs (Hungary) gauging stations, in 1980 at Donji Miholjac and Botovo (Croatia) stations, in 1991 at Terezino Polje (Croatia), and in 1998 at Bélavár (Hungary). In the present study, recent data measured after the construction of the last hydropower plant on the river have been used and analyzed. Sampling and analysis methodology is described, and an overview of the sediment transport processes along the river is given. The conclusion is that typical river morphological processes like armouring on the upstream reach and riverbed erosion downstream of reservoirs can be identified. For a detailed analysis and reliable prediction, a more systematic approach and a regular and well-designed field data collection and monitoring activity would be needed.

Keywords Drava river · Sediment sampling · Suspended load
Bedload · Bed material

7.1 Introduction

The most important parameters describing fluvial sediment transport are sediment load, Q_s, meaning the amount of sediment (volume or mass) passing through a given cross-section during a specified time; sediment yield, G_s, which is the mass of sediment passing by during a specified period of time; and, for suspended sediments, sediment concentration, c_s, which is the ratio of the mass of sediment and the volume of the water in which it is contained.

The measurement of fluvial sediment is based on sampling procedures, as a result of which, relying on protocols, sediment load and concentration can be calculated. Based on regular sediment measurement, it is essential to estimate the

E. A. Tamás (✉)
Institute for Hydraulic Engineering and Water Management, Faculty of Water Sciences, National University of Public Service, Bajcsy-Zs. u. 14, 6500 Baja, Hungary
e-mail: tamas.eniko.anna@uni-nke.hu

D. Lóczy (ed.), *The Drava River*, Springer Geography,
https://doi.org/10.1007/978-3-319-92816-6_7

correlations between flow characteristics and sediment parameters, for different water regime conditions. The best way to achieve this is to draw up sediment rating curves (Graf 1984).

Sediment transport data collection on the Drava River started in 1961 at Barcs and Drávaszabolcs stations (Hungary), in 1980 at Donji Miholjac and Botovo stations (Croatia), in 1991 at Terezino Polje (Croatia) and in 1998 at Bélavár (Hungary) (VITUKI 2003). In the present study, recent data measured after the construction of the last hydropower plant on the river have been used and analyzed.

Sediment investigation of the Drava River in recent times included different sampling campaigns as well as regular monitoring activities of both Hungarian and Croatian water management bodies. In frame of the Hungary-Croatia IPA Crossborder Co-operation Programme 2007–2013 (HU-HR) project number HUHR/1001/1.1.2/0009, entitled "Drava morphological monitoring" there was a possibility to examine the morphological conditions of the river, and the repeated measurement of hydraulic characteristics. The project featured a co-operation between the South-Transdanubian Water Management Directorate and the Croatian Water Management Organization (Hrvatske Vode), and was selected for support in 2011. The work was strongly related to the requirements of the EU Water Framework Directive, which implies that during preservation work and action in order to maintain the good status of the waterbodies, the regular monitoring of certain Drava reaches has to be carried out, with the purpose to follow the changes in the status of the river. The present study highly relies on the results of this project (Rákóczi et al. 2012). While the above-mentioned project included comprehensive analyses of sediments and bed material, reference is also made to the suspended sediment investigations of Croatian authorities, which were analyzed by Gilja et al. (2009).

It has already been reported in recent literature that the amount of suspended sediment along the Drava River has been greatly reduced, which can cause serious consequences (Bonacci and Oskoruš 2010 and Chap. 9 in this volume; Kiss et al. 2011). This, according to several authors, can be attributed to the massive construction on the Drava River since 1975, in the form of numerous hydrotechnical works, including various hydroelectric power plants.

7.2 Methods

In frame of the project of 2012, measurements were carried out at five cross-sections along the Hungarian-Croatian common reach of the Drava river: Botovo, Bélavár, Barcs, Drávaszabolcs, and Belišće. The measurements comprised three discharge measurements, suspended and bed-load sampling, and bed material sampling and were carried out in May, June and August, respectively. In situ measurements were performed by the South-Transdanubian Water Directorate (Pécs, Hungary), while the laboratory analysis of the sediment and bed material samples was carried out by Eötvös József College (now the Faculty of Water Sciences, National University of Public Service, in Baja, Hungary).

7.2.1 Sampling

In the following, the methods of sediment sampling in Hungary are described, as the majority of data and analyses on which the present study is based, refer to Hungary. However, we have to mention that in Croatia, sampling methods are not always comparable, and thus the data obtained from the different measurement methodologies must be very carefully used because of possible inhomogeneity issues. In Croatian methodology, for example, it is a common practice to determine suspended load concentrations from a single, near-surface sample (Gilja et al. 2009).

When sampling suspended load in the whole cross-section, in Croatian practice six verticals and three samples per vertical are used, while in Hungary it is usually three verticals and two samples per vertical. Before analyses, the samples taken from different depths are joined in both countries, thus the results in the grain size distribution give a vertical-average only instead of a 2D distribution in the cross-section.

7.2.1.1 Suspended Load

The most effective way of sampling suspended load is with a pump. An advantage is that it is not needed to regain the sampler onboard between the points. Thus, this method is the fastest, which is an issue, particularly at high velocities and when sampling is done in the navigation route. During sampling, it is very important to ensure that the sampling nozzle faces the flow, the pipe is not bent and to let enough time before taking samples to flush the pipe.

Sampling needs to be carried out with care to adjust the revolutions per minute value (RPM) or the discharge of the pump for the velocity through the nozzle V_{in} should not differ much from the velocity of the flow v at the given point:

$$0.8\,\text{v} \leq V_{\text{in}} \leq 1.5\,\text{v}$$

In case the velocities are outside this range, the RPM of the pump should be accordingly adjusted, or a tap should be installed at the end of the pipe to ensure that intake velocities match. In order to determine intake velocity, the discharge of the pump (q_{p}) has to be divided by the cross-section area of the nozzle (f_{n}):

$$v_{in} = q_p / f_n$$

In practice, we perform sampling with a constant pumping discharge, assigning a fixed intake velocity to different velocity ranges of the flow, keeping the hydraulic coefficient between the values 0.8 and 2.0. This ensures a maximum 20% difference in concentrations, which is acceptable.

7.2.1.2 Bedload

The bedload of rivers is moving intermittently over the surface of the riverbed. As bedload samplers disturb the current, they have an effect on the transported bedload. When choosing the appropriate sampler, we have to minimalize disturbance. Sampling of the bedload usually happens with the Helley-Smith sampler (in fine sediment, e.g. sand) or the Károlyi sampler (in coarser sediment). Both samplers have different sizes and gaps for different river types, depending on the grain size and the mass flow of the sediment.

The samplers are lowered to the bottom of the river, and, depending on the sampling time (usually 10–15 min), based on the mass of the sample taken, bedload transport can be calculated:

$$q_b = \frac{G}{b \cdot T \cdot \rho_s},$$

where

G dry mass of the sample;
b width of the gap of the sampler;
T sampling time and
ρ_s density of the bedload sample

In our experience, it is very useful to equip these samplers with an underwater camera, in order to be able to see the clogging of the sampler or anything blocking it; furthermore, to determine the exact sampling time needed to collect a reasonable amount of sediment in the apparatus.

The size of the Károlyi sampler used was: length 1,620 mm, height: 310 mm, wingspan: 200 mm, gap size: 314 × 122 mm, weight: 35 kg.

On the reaches where bedload is mainly sand, the Helley-Smith sampler was also used, fitted with a camera as well. Depending on the expected coarseness of bed material, two different size Helley-Smith samplers were used on the Drava river. For fine sand, a smaller sampler (length: 960 mm, height: 200 mm, wingspan: 400 mm, gap size: 152 × 152 mm, weight: 17 kg), and for coarse sand a bigger (length: 1,550 mm, height: 240 mm, wingspan: 510 mm, gap size: 150 × 150 mm, weight: 35 kg).

7.2.1.3 Bed Material

The sampling of the bed material is usually achieved with a bucket-sampler. Because armouring of the riverbed is to be expected on the river Drava, the edge of the sampler was sharpened to facilitate its penetration into the riverbed (length: 535 mm, diameter: 180 mm, weight: 9 kg).

7.2.2 Processing of Samples

The grain size distributions of bedload and bed material were determined using Taylor sieves—separating fractions (0.063; 0.125; 0.25; 1.0; 2.0; 4.0; 8.0; 12.0; 16.0; 24.0; 32.0; 48.0; 63.0; 96.0; 125.0 mm) or by settling velocity method—separating fractions (>0.10 mm; 0.05–0.10 mm; 0.02–0.05 mm; 0.01–0.02 mm; 0.005–0.01 mm; <0.005 mm). Drying was carried out at 105 °C. After that, dry matter content measurement (analytical precision) was done. In a case when the ratio of fractions with diameters less than 0.15 (0.1) mm is higher than 10%, hydrometry method must be used to establish the grain size distribution.

For suspended load, which is usually contained in about 5L water samples, the first important step is to measure the exact amount of the sample, to know from how much water we will measure sediment concentrations. The samples are then left to settle. When the sediment settles in the bottom of the containers, the excess water is carefully sucked from the containers and approximately 0.5L is left. The amount of clean water removed is precisely measured and recorded in the protocol. The samples are dried in electronic oven for 24 h at 105 °C temperature. Then, we measure the weight of each sample and its dry matter content on a precision scale. Concentration is calculated as

$$C_{ss} = \frac{m_d}{v_s},$$

where

m_d dry matter weight and
v_s total volume of the sample

In the case of suspended load, too small a grain size for screening, the grain size determination is done with a special Atterberg-type settling device, which is operating on the principle of the Stokes equation. The main part of the settling velocity meter is a cylindrical glass tube with an inner diameter of 35–40 mm. There are six level markings on the tube. On the bottom of the tube, there is a stricture and a tap. It ends in a ca 4 mm diameter rider. On the top of the tube, there is a funnel with a throttle. With the throttle open, the tube is filled up with distilled water, and the sample is also poured into the tube through it. There is a vent in the axis of the throttle, connected with a 0.1–0.2 mm diameter nozzle, to secure that outflow velocity does not exceed 0.2–1 cm s^{-1}. The tube has to be mounted on a stand with its axis vertical.

Before filling the samples in the tubes, we leave them dissolve in aqueous solution of sodium metasilicate to avoid coagulation. Then we fill the tubes with sodium metasilicate solution and we pour the samples into it. The grain size fractions are <0.10 mm; 0.10–0.05 mm; 0.05–0.02 mm; 0.02–0.01 mm; 0.01–0.005 mm and >0.005 mm. The grain size distributions of the different samples can be drawn up as percentages from the data as a distribution curve. We read the values of d_m, d_{eff}, d_g, d_{10}, and d_{60} from the diagram and we calculate U unevenness factor ($U = d_{60}/d_{10}$), according to the Hungarian measurement standard.

7.2.3 Analyses

Concentrations of suspended load and grain size distribution curves of suspended sediment, bedload and bed material were determined from each sample. Total yields were estimated if a simultaneous discharge measurement was available.

For a better description of sediment transport processes, we determined the correlation between discharge and suspended sediment concentration. In this, all available sampling results were used (2004–2016) and the Barcs station, with the longest time series, was selected as an example.

Altogether the data of five cross-sections were used, out of which in the upstream four the data of some formerly executed repeated discharge measurements and sediment samplings from the period 1998–2002 were available (VITUKI 2003). In this case, comparative analyses to detect eventual changes were also carried out.

7.3 Results

7.3.1 Suspended Load

Suspended load in the river is highly dependent on the water regime, and is highly changeable. At Barcs, suspended sediment yield ranged from 697 to 35,5016 g s^{-1} (n = 70), while at Drávaszabolcs from 1,298 to 124,481 g s^{-1} ($n = 70$) in the period from 2004 to 2016. At Botovo, the measured values fell between 1,029 and 137,095 g s^{-1} (n = 45), According to the findings of the detailed analyses in 2012, the suspended load (or concentration) of the river was increasing from Botovo downstream. The increase was even more prominent from Barcs to Drávaszabolcs, then, until Belišće (54 fkm) it decreased a little, and almost equalled to the values measured at Barcs. The growth of the suspended sediment load can probably be attributed to the nearby location of the Croatian Hydropower Plant at Donja Dubrava, the reservoir of which entraps the majority of the arriving suspended sediment, thus forwarding a relatively clear, sediment-free flow downstream the dam. Because of the relatively high deficit in the sediment transport capacity, the suspended sediment is entrained from the bed. Upstream of Barcs, however, bed material is mainly gravel and the process is limited by the low ratio of fine fractions in the bed material. Downstream Barcs bed material is finer and a larger amount of sand can be taken into suspension, increasing the concentration rapidly. A little more downstream the sand riverbed starts to widen, which results in sediment deposition and the decrease of the concentration of suspended load.

In 2003, the average of the grain size diameter of the suspended sediment was 0.058 mm at Botovo (227.5 river km), and practically the same: 0.06 mm at Bélavár (198.5 rkm) in 1999, when sampling was performed on the recession of a floodwave. At Barcs (154.1 rkm) 0.105 mm and at Drávaszabolcs (78 rkm) 0.15 mm was recorded, which means that the grain size of the suspended load was indeed increasing downstream. A possible explanation for this phenomenon can be that on the two upstream stations the sediment supply is source-limited for two reasons. The same reasons described above explaining the changes in concentrations can be attributed for the changes in grain size parameters as well: the settling of suspended load in the reservoirs, and the relatively large grain size of bed material, from where the river cannot bring fine fractions into suspension, even at high velocities (VITUKI 2003) (Fig. 7.1).

Based on the measurements carried out in 2012, we can conclude that the grain size distribution of the suspended load on the investigated Drava reach became rather homogeneous and did not change much along the river (Fig. 7.2).

The average diameter of the suspended load seemed to have decreased since 2003, but, considering the relatively low discharges at the time of the 2012 sampling (between 356 and 706 $m^3\ s^{-1}$), and the grain size distribution of the later measurements until 2016, the change in the diameter is not significant. In the latest samplings the average diameter ranged from 0.1 to 0.17 mm at Barcs and

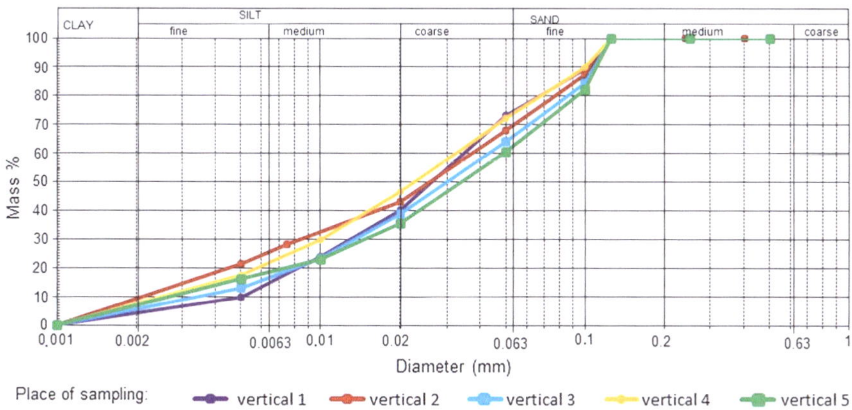

Fig. 7.1 Typical grain size distribution of suspended load of the Drava river at Barcs (29.08.2012)

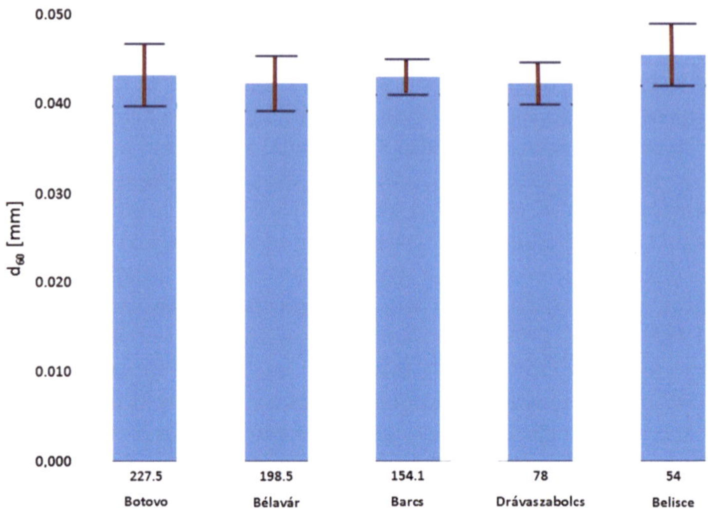

Fig. 7.2 Average grain size of the suspended sediment along the Drava river in 2012 (with standard deviation)

0.12–0.17 mm at Drávaszabolcs during a relatively high discharge, which correspond to the values determined in 2003 (Fig. 7.1).

7.3.2 Bedload and Bed Material

The bedload of the Drava river downstream of Botovo is seemingly very changeable, but this phenomenon can be attributed to the periodic armouring of the riverbed, which can make it very stable. Depending on the water regime, the process lasts from a few days to some weeks. The armour layer resists the flow and prevents particles from being washed away from the surface of the riverbed. At Barcs, bed material is finer and the armouring process is no more prevalent. Thus, here bedload transport is higher, as well as at Drávaszabolcs. The widening of the riverbed on the lower slope reaches until Belišće causes a decrease in bedload transport similarly to suspended load concentrations.

The variation in average grain sizes of bedload samples does not allow well-based conclusions. As bedload sampling was only performed on three occasions in both sampling periods, there are insufficient data available for detailed analyses. At the two upstream sections, the grain size distributions of bed material are very similar. Grain sizes are much larger than at the lower three which are on the sand-bed reach of the river. Bedload and bed material grain sizes are very similar.

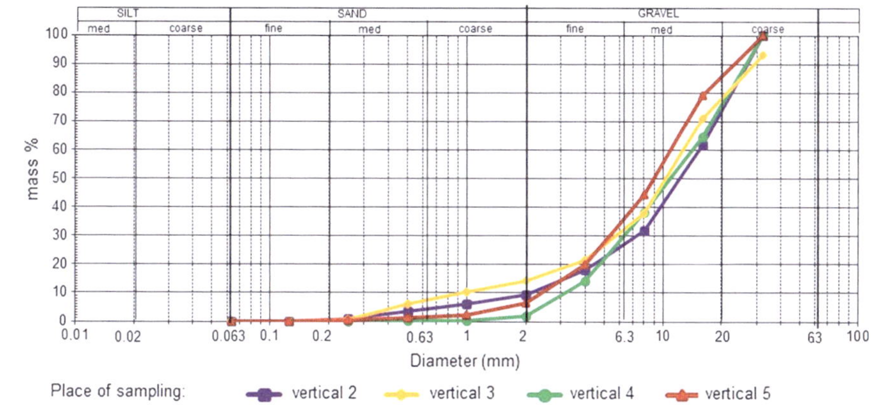

Fig. 7.3 Typical grain size distribution of bed material at Botovo (08.05.2012)

The shapes of the grain size distribution curves at Botovo indicate that there is a constant riverbed erosion process going on and armouring is predominant (Rákóczi et al. 2012) (Fig. 7.3).

At Bélavár, armouring is less developed, and can only be observed near the right bank (vertical 5). In all the other verticals no significant erosion or deposition is shown by the bed material. Dynamic equilibrium is assumed (Fig. 7.4).

More downstream, where the slope of the river considerably decreases, sand becomes dominant in bed material. At Barcs, depending on the water regime, fine and medium gravel can also be observed, but amounting to less than the half of the samples, usually below 20%. At Drávaszabolcs and Belišće medium sand prevails in the composition of bed material, and gravel can be seldom observed. These two latter sections show a near-equilibrium riverbed (Rákóczi et al. 2012) (Fig. 7.5).

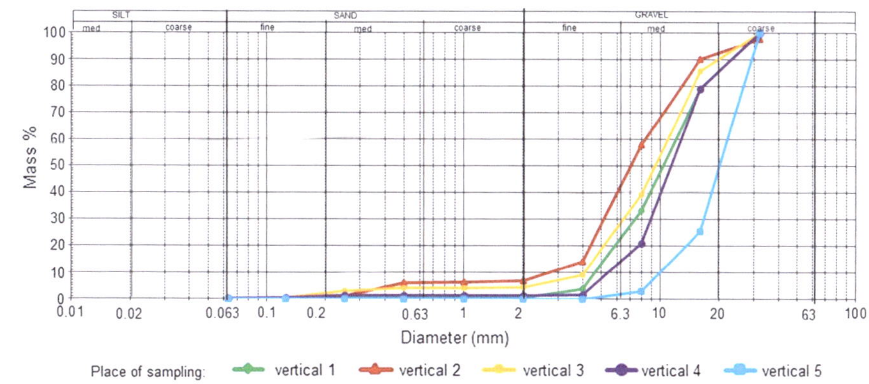

Fig. 7.4 Typical grain size distribution of bed material at Bélavár (09.05.2012)

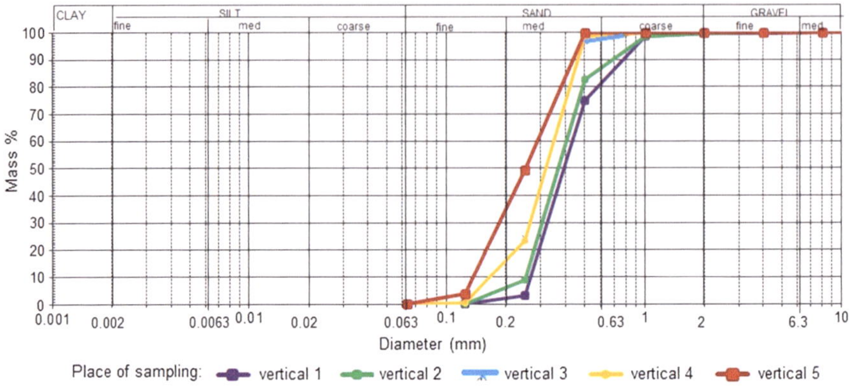

Fig. 7.5 Typical grain size distribution of bed material on the lower reaches (30.08.2012)

The grain size distributions of bedload and bed material are similar. At the upstream two sections, where gravel is predominant, average grain sizes are between 9.5 and 12.8 mm, at Barcs, between 2 and 3 mm, at Drávaszabolcs between 0.4 and 0.5 mm and at Belišće around 0.3 mm (Fig. 7.6).

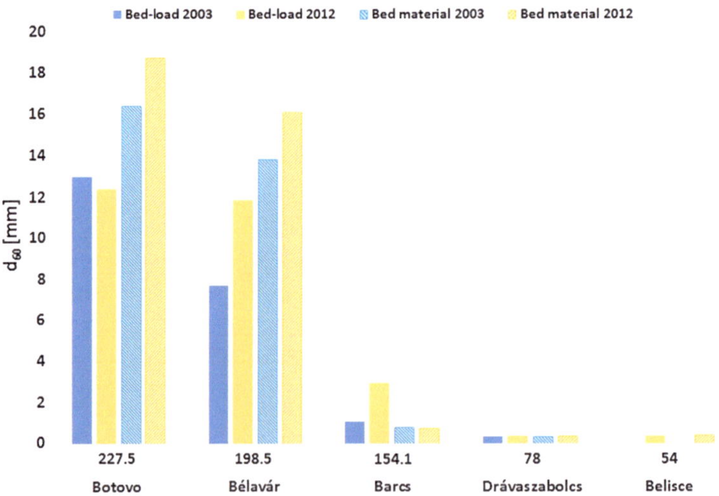

Fig. 7.6 Bedload and bed material grain sizes along the Drava river in 2003 and in 2012

7.3.3 Connections Between Discharge and Sediment Transport

Establishing mathematical correlation between discharge and sediment load makes it possible to assess the change in sediment regime if discharge changes (Gilja et al. 2009). However, as sampling requires human presence and is a rather resource-demanding procedure, data are scarce and connections are not very well correlated, especially at high stages/discharges (Fig. 7.7).

The number of sediment samplings over the investigated period is relatively low and does not cover the high waterlevel range sufficiently to establish a reliable correlation between sediment concentration (or sediment load) and discharge. Sediment sampling during high discharges cannot only be difficult because of the lack of resources. As floodwaves are usually short, the time interval between flood forecast and the occurrence of the peak waterlevels is sometimes too short to organize these complex measurements. It is also to be noted that navigating and anchoring a measurement boat on the Drava river during high waterlevel can also be dangerous or even impossible because of the high velocities, especially on the upstream reach (Rákóczi et al. 2012).

Despite the low number of data, the division of the time series into shorter time periods suggests a certain increase in the suspended load concentrations (quantities), especially in the higher discharge ranges. However, the trend cannot be proven significant; in order to establish it a reliable database of sediment measurements would be needed in the future (Fig. 7.8).

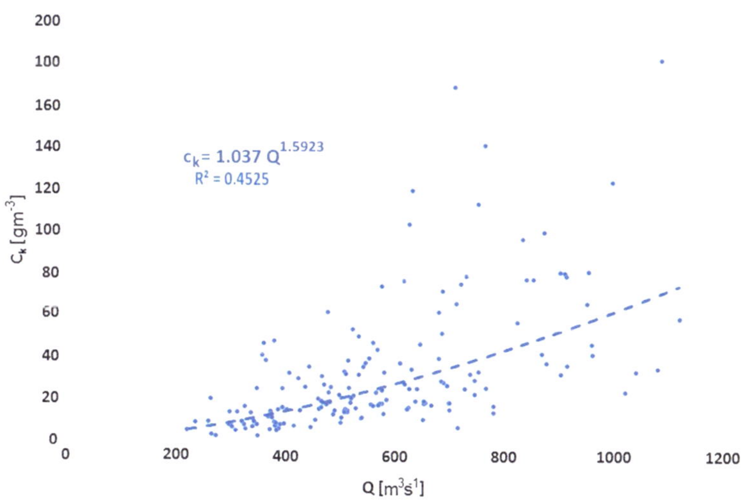

Fig. 7.7 Suspended load concentration plotted against discharge at Barcs, 1991–2016 (n = 164)

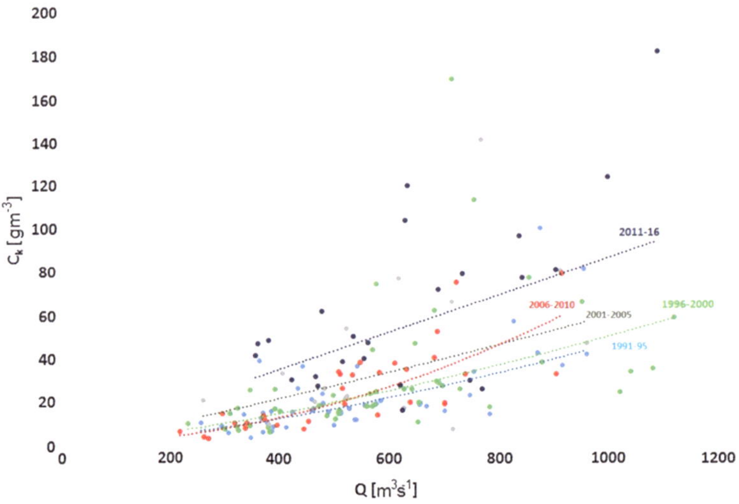

Fig. 7.8 Suspended load concentration plotted against discharge at Barcs, 1991–2016 (n = 164) divided into time periods

7.4 Conclusions

According to several authors, sediment transport of the Drava is supply-limited downstream of reservoirs, but a more thorough investigation is needed to describe the phenomenon correctly and to predict its future consequences. However, we found evidence of riverbed erosion on the lower reach, supporting the statement that the construction and operation of three Croatian dams and their reservoirs are mainly responsible for the downstream deficit of suspended sediment yield. On the upstream reach, intensive armouring of the riverbed is observed, which also needs a more detailed future investigation, with particular emphasis on the collection of a more detailed set of field data (e.g. thickness of the sand layer over the armoured gravel bed) in order to be able to evaluate the ongoing processes using the newest methodological findings (e.g. Kuhnle et al. 2017). Climate change and/or variability as well as other anthropogenic influences (material excavation) could be additional reasons for these phenomena (Bonacci and Oskoruš 2010; Rákóczi et al. 2012).

Today the goal of the sediment sampling is not only to describe sediment transport in the flow, but further to provide calibration and validation data for numeric modelling. Sediment measurements are different in the different countries in Europe (Schwarz et al. 2008). Methodologies and samplers vary, both for field and laboratory analyses. Even in Hungary, sampling and laboratory techniques have been modified several times in the past. Also, sediment sampling was never really systematic, and the sampling campaigns did not follow the hydrological processes. That is why sediment data can hardly be compared. The data series are inhomogeneous and cannot be statistically analyzed. Sampling has to be carried out as to be able to obtain a true picture about the changes of sediment transport across the flow, along the flow and with respect to variability with depth. The sampling points have to be determined based on morphological and flow conditions. Discharge measurement has to be executed in parallel to sediment sampling. For a few years, water autorities in both Hungary and Croatia have been using Acoustic Doppler Current Profilers (ADCP) for the measurement of the discharge. This opens up new possibilities for future analyses. Despite this fact, we still have to emphasize that the availability of hydromorphological data is extremely important for assessments under the Water Framework Directive, also to support ecological status evaluation. However, the lack of information on some large rivers, including the Drava, is evident. The changes in the hydrological and sediment regime of river systems induced by hydromorphological alterations are not well understood, so in the near future there is an urgent need for a harmonised database (Schwarz 2008 and Chap. 5). To this end, the first step is to intensify and reorganize hydromorphological monitoring, including sediment sampling and data management.

References

Bonacci O, Oskoruš D (2010) The changes in the lower Drava River water level, discharge and suspended sediment regime. Environ Earth Sci 59:1661. https://doi.org/10.1007/s12665-009-0148-8

Graf WH (1984) Hydraulics of sediment transport. McGraw-Hill, New York

Gilja G, Bekić D, Oskoruš D (2009) Processing of suspended sediment concentration measurements on Drava River. In: International Symposium on Water Management and Hydraulic Engineering, Ohrid/Macedonia, 1–5 September 2009. Paper: A67

Kiss T, Andrási G, Hernesz P (2011) Morphological alteration of the Drava as the result of human impact. AGD Landscape & Environ 5(2):58–75

Kuhnle RA, Langendoen EJ, Wren DG (2017) Prediction of sand transport over immobile gravel from supply-limited to capacity conditions. J Hydraul Eng. https://doi.org/10.1061/(asce)HY.1943-7900.0001292

Rákóczi L, Szlávik L, Sziebert J, Tamás EA, Süveggyártó AM, Koch G (2012) Sediment analysis study under the project of "Dráva morphological monitoring" HUHR/1001/1.1.2/0009. Research report (in Hungarian with English summary)

Schwarz U (2008) Hydromorphological inventory and map of the Mura and Drava Rivers 2005. (IAD Pilot study). Large Rivers 18(1–2):45–59

Schwarz U, Babić-Mladenović M, Bondar C, Gergov G, Holubova K, Modev S, Rákóczi L, Rast G, Steindl J, Sorin T, Tamás EA (2008) Assessment of the balance and management of sediments of the Danube waterway: current status, problems and recommendations for action. Working paper for the Danube River Basin. 1–60

VITUKI (2003) A Dráva hordalékjárásának vizsgálata a legfrissebb adatok figyelembevételével (Investigation of the sediment regime of the Drava based on the newest data). Research report 714/1/6128-01, Responsible person: Szekeres J, Budapest (in Hungarian)

Chapter 8
Flood History and River Regulation

Hrvoje Petrić, Enikő Anna Tamás and Dénes Lóczy

Abstract Historical sources on the indirect consequences of Drava floods are available from medieval times and in more detail since the 16th century. The first documented flood on the Upper Drava occurred on 15 June 1553 and on the Lower Drava on 2 June 1770. The approximate sequence of flood events can be deduced from the summary table of historical floods. Several case studies describe the relocation of settlements following devastation by major floods. Flooding has ever been both curse and blessing for the population of neighbouring areas: it not only presented a threat to lives and properties, but also fertilized lands with nutrient-rich alluvial deposits. In the course of history people have learnt to fight floods and regulate streams employing various hydrotechnical measures, which brought about both beneficial and detrimental changes in the hydrography and ecology of riverine environments.

Keywords Floods · Channel shift · Settlement relocation · Flood control River training

H. Petrić
History Department, Faculty of Humanities and Social Sciences, University of Zagreb, Ivana Lucica 3, HR-10000 Zagreb, Croatia

E. A. Tamás (✉)
Institute for Hydraulic Engineering and Water Management, Faculty of Water Sciences, National University of Public Service, Bajcsy-Zsilinszky u. 14, Baja 6500, Hungary
e-mail: tamas.eniko.anna@uni-nke.hu

D. Lóczy
Department of Physical and Environmental Geography, Institute of Geography and Earth Sciences, Faculty of Sciences, University of Pécs, Ifjúság útja 6, Pécs 7624, Hungary
e-mail: loczyd@gamma.ttk.pte.hu

D. Lóczy (ed.), *The Drava River*, Springer Geography,
https://doi.org/10.1007/978-3-319-92816-6_8

8.1 Introduction

The hydromorphology of most rivers has been altered through a long history of human interventions (Grabowski et al. 2014). Water management, including flood control, reservoir construction and river training, manipulates the natural regime of rivers. Many rivers are channelized to improve navigation and flood control. Countless dams have been built for water abstraction and the generation of hydroelectric power (Giller 2005). More than 70% of the large rivers of Europe, North America and the former Soviet Union are strongly regulated, and there are more than 800,000 dams worldwide affecting approximately two-thirds of fresh water inflow into the oceans (Dynesius and Nilsson 1994; Rosenberg et al. 2000). River regulation provides or ensures important services for humanity, as navigation possibilities, providing irrigation supply, drinking water and electric power generation, but at the same time it often degrades ecosystems and limits natural resources.

Historical sources report on the floods of the Drava since medieval times. For Carinthia the first mention of flood destruction is from 792 (Peturnig 1969). On the lowermost section of the Drava, at Osijek, the first documented flood occurred on 2 June 1770 (Petrić and Obadić 2007). (The cause, however, is unknown.) Table 8.1 summarizes the major events in the history of floods along the Drava River.

Naturally, the inconsistency and large gaps in the data series do not allow the estimation of flood frequency for the Drava, but some floods which virtually affected the entire length of the river can be identified. (For flood frequency along the Lower Drava see also Sect. 11.5.3).

Floods often had a fundamental impact on economic and settlement development. Along the Lower Drava meander evolution and channel wandering accelerated after major floods (Lóczy et al. 2016). A good example is Strmec village in Petrijanec parish, Varaždin county, Croatia, which used to be located next to the river. Due to flooding and meandering, the village center was moved away from the river (Zorković 1995, 73). In the 16th century this led to a dispute between Varaždin citizens and Ivan Ungnad, the owner of Varaždin old castle. During a devastating flood water penetrated into the city of Varaždin and formed the Isle of Zrepičar between the old and new water flow. The old river bed remained in the north, so the Isle of Zrepičar fell under jurisdiction of the city of Varaždin. Citizens used to graze their cattle on the isle from spring to autumn. Beside shepherd cots, the isle had no inhabitants. When in 1543 Baron Ivan Ungnad became the owner of the city castle, he decided to usurp this isle and turn it into his own property. His governor of the castle, castellan Juraj von Bother started inhabiting the isle with refugees from regions under Turkish invasion, without prior consent of the city magistrate. This is how north of Drava a new settlement, Novo selo village, emerged which still exists (in Međimurje county). All the way, up to the mid-20th century, it belonged to Varaždin area, despite the fact that it was built on the opposite river bank. In 1549, Varaždin people complained of trespassing and protested before Zagreb Chapter (Horvat 1993, 79–80).

Table 8.1 Dates of major floods of the Drava

Time interval	Carinthia	Slovenia	Croatia	Hungary
Before 1000	792			
11–15th centuries	1118, 1195/96, 1236, 1280, 1293, 1316, **1342**, 1480	**1342**	**1342,** 1346	1147, 1245, 1294, 15 September **1342**, May 1346, 1480
16th century	15 June 1553, 28 October **1567**			
17th century	October 1615, **1632**, 1660, 1680		1698	
18th century	1755, 1767		1718 3 June and 4 November 1770	1780
19th century	16 August and 16 October 1810, October 1820, November **1851**, 13 September 1864, 18 September **1882**, 1885, 1886, 1888, October 1889, August 1891, 1896	1851, 1874, 1885, 1893	1804, 1807, **1814**, 1827, 1833, 1836, 1840, 1843	October 1814, June **1827**[a], 1830, 1836, 1853, 1879, 1891
20th century	September 1903, May 1907, July 1910, May 1914, 7 September 1916, 20 July 1920, 25 July, 21 October, 1 November 1926, April 1928, 1931, 7 October and 18 November 1935, 28 September 1942, 10 August 1945, 8 July 1946, 18 June and 28 October 1953, 30 July and 11 December 1954, 22 March and 15 June 1956, August 1965, 1 September **1965**, 19 August and 4 November **1966**[a], 13 June 1972, 2 October 1973, 1975; September 1983; August 1987; June and July 1991	1901, 1910, 1925, 1926, 1933, 9 September 1963, 25 October 1964, July–September **1965** (at Ptuj: 2,600 $m^3\ s^{-1}$), 20 August 1966 (1,961 $m^3\ s^{-1}$ at Dravograd), 6 November 1966, 13 June 1972, 2 October 1973, 7 April 1975, 23 May 1978, 4 October 1978, 10 June 1980, 18 October 1980, 21 May 1984, 9 June 1985, 29 August 1986, 5 July 1989, 18 June 1991, 9 October 1993, 15 September 1994 (1,562 $m^3\ s^{-1}$ at Dravograd), 18 November 1996 (1,361 $m^3\ s^{-1}$ at Dravograd), 28 June	7 July 1926, 28 October 1930, 5 July 1935, 27 May 1936, 4 June 1937, 27 September 1937 (1,646 $m^3\ s^{-1}$ at Botovo), 26 May 1938, 13 July 1946, 12 July 1948,10 June 1951 (1,884 $m^3\ s^{-1}$ at Botovo), May–June 1954, 5 June 1956, 3 July 1959, 4 June 1962, 14 March 1963, 27 October 1964 (2,035 $m^3\ s^{-1}$ at Botovo), 4 August 1965 (2,196 $m^3\ s^{-1}$ at Botovo), 22 August 1966 (2,587 $m^3\ s^{-1}$ at Botovo), 2 June 1967, July–August 1970, 19 July **1972**[a] (2,833 $m^3\ s^{-1}$ at Terezino Polje),	17 May 1924, 16 June 1926, 3 April 1940, July 1954, 1963, 1964, 25 June 1965 (3,190 $m^3\ s^{-1}$ at Barcs), 17 August 1966, 18 July 1970, 15 June **1972**[a] (3,040 $m^3\ s^{-1}$ at Barcs), May 1975 (2,570 $m^3\ s^{-1}$ at Barcs), June 1976, October 1979, October 1980, May 1984, May 1985, July 1989 (1,730 $m^3\ s^{-1}$ at Barcs), October 1993 (1,770 $m^3\ s^{-1}$ at Barcs), October 1998 (1,781 $m^3\ s^{-1}$ at Barcs)

(continued)

Table 8.1 (continued)

Time interval	Carinthia	Slovenia	Croatia	Hungary
		1997 (1,330 m^3 s^{-1} at Dravograd), 7 October 1998 (1,672 m^3 s^{-1} at Dravograd)	28 November 1973 (2,032 m^3 s^{-1} at Botovo), 3 July 1975 (2,473 m^3 s^{-1} at Botovo), June–July, October 1979, 21 July 1981, 10 May 1985, 8 May 1987, 6 July 1989 (2,055 m^3 s^{-1} at Botovo), 3 November 1990, 29 June 1991, 7 December 1992, 25 October 1993 (2,302 m^3 s^{-1} at Botovo), 6 June 1995, 7 April 1996, October–November 1996, 10 October 1998 (2,398 m^3 s^{-1} at Botovo), 22 May 1999	
21th century	8 November 2000, June 2004, July 2005, September 2008, August 2009, July 2010, July 2012, November 2012 (2,500 m^3 s^{-1} at Lavamünd), November **2012** (2,700 m^3 s^{-1} at Lavamünd), February and November 2014 (1,600 m^3 s^{-1} at Lavamünd)	7 November 2000 (1,169 m^3 s^{-1} at Dravograd), 19 November 2002 (1,256 m^3 s^{-1} at Dravograd), 4 October 2005 (1,205 m^3 s^{-1} at Dravograd), 1 October 2008, 19 September 2011, 5 November **2012** (at Ptuj: 3,200 m^3 s^{-1}), 13 August 2014	19 November 2000, 28 November 2002, 27 June 2004, 7 October 2005, 31 May 2006, 28 June 2009, 7 November **2012** (2,071 m^3 s^{-1} at Botovo), 11 June 2013, 19 September 2014	January 2001, May 2002, 10 October 2005, 4 April 2006, 15 February 2009, May–June 2010 (1,590 m^3 s^{-1} at Drávaszabolcs), November 2012 (1,570 m^3 s^{-1} at Drávaszabolcs), 15 June 2013, 5 February 2014, 19 September 2014, 17 June **2016**

Sources Peturnig (1969), ES (1995), Rohner et al. (2004), Mikoš et al. (2004), Petrić and Obadić (2007), Moser (2014), Tadić et al. (2016), Kiss (2018) and time series of gauging stations. The most destructive floods are shown in **bold.** Days of highest discharge are indicated

[a]Floods with probably the largest discharge or inundated area in the given time interval

In the early 1740s the first associations for the construction of embankments on the left bank of the Drava were built between Drávaszabolcs and Dárda (now in Croatia) (Borzavári no date). Around 1770, after a major flood, this was extended up to the border of Somogy County, where the defence lines had been completed. During the 1827 flood 18 dyke breaches happened and, together with the floods of tributary streams, inundated the entire lower floodplain level. The floodwater found its way back to the river channel at Dárda. A continuous, 140-km-long embankment was not ready before the 1880s. The floods of 1972 and, to a lesser extent, of 1975 demanded strained defence work on the embankments, but no breach occurred. The defence lines along meander cutoffs were particularly vulnerable to flood damage (Buchberger 1975; Remenyik 2004, 2005).

8.2 River Regulation

8.2.1 Goals and Methods of River Regulation

River regulation or training refers to the structural measures which are taken to improve a river and its banks. River training covers high-water (i.e. flood management with technical measures), mid-water and low-flow training (Zorkóczy and Károlyi 1993). In practice, the term most commonly denotes mid-water training, i.e. structural improvements of the actual river channel to convey water flow, sediments and eventual ice. River regulation often serves multiple purposes (Al-Khafif 1970), also including agricultural water management (provision of irrigation water), or hydropower generation.

Among the types of river training, closure of river branches, short-cutting of river bends, flow guiding structures, embankments, bank and bed protection and dredging can be mentioned (van Duivendijk and te Slaa 2004). The selection and design of the most appropriate structure depends largely on site conditions and hydraulic characteristics of the river reach. River training, apart from various types of earthworks, is basically achieved by structures: sills, weirs, spur-dikes (groynes), jetties, longitudinal dikes (guide banks or guide bunds) and bank protection (revetments, hard points) (Al-Khafif 1970; Zorkóczy and Károlyi 1993).

A main goal of river regulation is to meet navigation requirements through forming a continuous river channel with a thalweg of minimum specified depth, width and radius for all possible river discharges. In a meandering river low-water regulation can only be achieved by means of a system of bank protection works, spur and longitudinal dykes (van Duivendijk and te Slaa 2004).

8.2.2 The Drava River Valley Before Regulation

Before river regulation started, the Drava river was constantly changing its course and its left-bank floodplain was characterised by the web of its old channels. During floods, only higher areas were not inundated, a wide belt (ca 115,000 ha) along the river was waterlogged. In a document dated in 1294, the regular flooding in the vicinity of the settlements Siklós, Nagyfalu and Egyházasharaszti was emphasized. In the Ormánság region too several extensive areas were regularly flooded in the past (Ihrig 1973).

The Drava River used to flow in two branches of almost identical dimension in the 12th–13th century. The Drava and the Mura river often flooded the fields around Csáktornya (the present-day Čakovec, Croatia). During floods, floating debris and island formation commonly changed the course of the river on the relatively flat lowland. As the valley was inhabited already in the ancient times, with a growing population until the 18th century, the need for the regulation of the watercourses and the protection of the settlements arose. The investigation of opportunities was ordered first by Queen Maria Theresia in 1753 (Ihrig 1973).

8.2.3 Beginnings of River Regulation

Water regulation and flood protection works on the Drava began at very early dates. In a document from 1609 a possible transport of large blocks of stone to Drnje village, Koprivnica-Križevci county, is mentioned to help build defence (SAV; Androić 1968). There is a possibility that the blocks were actually used as firing material, transported on river boats all the way from Varaždin. This information shows that the Drava River was navigable in the Koprivnica region up to Drnje and attests to river regulation in the Koprivnica area (Petrić 2000, 76–77).

There are also historical documents witnessing the regulation of the Drava in the Varaždin area. In the parliament session held on 5–6 October 1633 there is a mention of opening the riverbed of the Dravica near Varaždin and on 6 November 1634, on the opening of a Drava backwater channel near Varaždin (Prothocolla generalium …; Vujasinović 2000, 36). It seems that in 1649 major water regulation and flood control works took place on the Drava river (PG, no. 1, 148). On 4–5 November 1654 the Varaždin parliament session appointed a panel to create a new bed for the Dravica near Varaždin (PG, no. 1, 191–192). In 1678 measures against flooding were introduced the river port of Zavrč and Varaždin as well as part of Podravina region (Podrauinas) (PG, no. 1, 382). Public works on the embankment of river in Podravina, and near Varaždin were mentioned from 1692 (PG, no. 1, 548).

At the turn of 17th century, river regulation took place in 1696 and 1698 (PG, no. 2, 87, 173). Floods on the rivers Drava, Sava and Kupa in 1698 brought peril to the landowners and estates, committees and panels were established to fight flooding. They had power and authority to engage in river embankment works. The same year it was decided that embankments were to be built along the rivers Sava and Drava (PG, no. 2, 177–178, 184).

Together with the Mura, the Drava had a tremendous influence on the market town of Legrad, at the confluence of these rivers. Previously the Drava had been a part of Međimurje (southern part of Zala County), but due to the relocation of riverbed became part of Podravina. The Drava made access to its functional region difficult, from the early 18th century towards south (less obvious), and later on towards north (more obvious). The Northern gravitation area was more severely affected by the loss of Legrad's role as a central settlement in the region. The 'relocation' of Legrad from Međimurje to Podravina was discussed by church historian Josip Bedeković, who believes that it happened around 1710–1712, during a major flood (Bedekovich, Natale, chapter LIX, p. 3). Another data confirming that the Drava indeed had run south of Legrad dates back to 21 February 1691 in Legrad's parish records. It indicates that in those days certain parish people from Legrad had been scheduled for marriage; it had been recommended to hurry up to Legrad, while there is still thick ice on the river (The RC Parish Office Legrad, Liber memorabilium parochiae Legrad). As a consequence of the 'relocation' of Legrad from Međimurje to Podravina and the disruption of functional ties with its gravitation zone, in about 70 years (1698–1771), the population of Legrad decreased by more than a 1,000, i.e. over 40%. Around 1800, population grew, but in 1808 there was another significant decline. In the second half of the 19th century population stabilized between 2,150 and 2,350 (Petrić 2005, 41).

In 1777, more detailed plans of water regulation of Drava and some of its tributaries (nearby villages of Đelekovec, Torčec and Botovo) existed (CAN box 10); saved documents from 1804 witness of Drava riverbed regulation in Podravina region (CNA box 21, no. 11/285). On 17 August, 1779 captain Balaško reported from village Drnje to Koprivnica town magistrate, that his general command had ordered him and his troops to clean the Koprivnica Stream (SAV box 6, no. 232, 234/779). In 1810 it was ordered that terrain between Drava river and the embankment had to be reinforced by planting willow and poplar trees.

8.2.4 Case Studies of Floods in the 18th and 19th Centuries

Some of the worst floods of the Drava were recorded on 3 June 1770. People could enter into Drnje church only by boat. All corn, wheat and hay was destroyed. Mass celebration was held by parish priest Martin Korolija and chaplains Baltazar Zrinski and Mirko Sabo. Parish priest Korolija believed that miracle happened and Mother of God, Our Lady of Drnje, saved the church from flooding. Another flood, but a less destructive one, was recorded on 5 November of the same year (The RC [Roman Catholic] Parish Office Drnje [Croatia], Liber memorabilium parochiae Drnje, 74).

The suburbs of Varaždin were frequently flooded. After the 1718 flood, the Croatian Diet decided that the Drava banks should be reinforced. In the early 19th century floods occurred in the years 1804, 1807, 1814 and 1827 (Težak 2004, 40). In 1807, the Drava flooded the entire Varaždin suburbia, with water in the town itself, flooding all lower-lying roads, basements and alehouses (Horvat 1993, 280). In spring 1814, the Drava inundation brought mud to meadows and washed away field crops, affected the northern and northeastern suburbs, damaged gardens and fences, homes and barns. Water flow was so strong that it broke a few supporting beams on a wooden bridge (Horvat 1993, 285).

In the 18th century the Drava changed its course again. Maps from those years show villages like Žabnik and Štefanec further to the north than today. In 1778 an old road ('*via regalis*') led from Varaždin to Ludbreg through these villages. The Drava, however, was gradually shifting to the south, flooding the northern parts of both villages, a manor in Štefanec and even the king's road. A map from 1800 shows a newly built road further south from both villages and the first homes alongside. By the turn of 19th century, the parish had been in Žabnik, then moved to Bartolovec (Kanonske vizitacije Zagrebačke (nad)biskupije, 189), where a new village was built around St. Bartholomew's chapel. Today, there are only smaller remains of Žabnik and Štefanec.

The influence of the river on population can be established by studying demographic movement related to floods. Although in the late 18th and early 19th centuries no significant difference in demographic trends was found for the settlements along the Drava compared to those in the broader region (Obadić 2007) if we regard a longer period of time, we can establish that the Drava flood could have influenced demographic change, the resettlement of population, since it was more difficult to transport food into devastated areas if floods destroyed crops.

The Drava floods influenced the location choice of the new church of St. Martin near Ludbreg (between Varaždin and Koprivnica). This was so described by Katarina Horvat-Levaj: "Such church building continued its building history of their own landlords—counts Patačić of Zajezda—established in the Martijanec estates in 1746. Moreover, due to certain architectural designs, the majority of these medieval elements would have remained untouched by this very day, had there been no Drava. Losing its role of natural protection, this unregulated river, due to snow melting in the Alps and its lower parts, flooded neighbouring areas and started to endanger St. Martin parish church. Catastrophic flooding in the mid-18th century, as recorded in the parish records, in 1767 caused that a new, stone-built church was erected on a safer location. In sync with Baroque-styled spaces, and copying a Baroque-styled Patačić castle in Martijanec, the builders had chosen an elevated location south of the road; although Baroque never had this rule, the church and its shrines were facing east, leaving its representative western front to distant vision and sight from main incoming roads from Varaždin." (Horvat-Levaj 2006, 83).

In 1814 the crop yields were poor and the Drava repeatedly flooded, causing great damage (Table 8.1). In early September, it flooded again, this time so badly that the best arable land was under water, many houses and farm building were destroyed, the roads were damaged and livestock drowned and got killed.

The bridge of Barcs was so heavily damaged that it was impossible to import food from Hungary. The 1814 floods are also reported in the memorial of the Franciscan monastery in Virovitica, recording that "the flooding river Drava destroyed the crops in a large area, from the Drava watermills to the new church in Bušetina. In addition to flooding, considerable damage was also inflicted by rain and frost, which is why the animal feed (hay) was scarce and the local vineyards had little grapes." (Franciscan monastery Virovitica, HDCV)

The floods of 1814 caused great harm in Valpovci district, too. In Virovitica, Voćin and Valpovo (Virovitica county) the loss totalled 470,000 forints. Due to the flooding of the Drava and Karašica, the harvest in 1816 failed, partly because the winter crops failed because of the damp soil. The floods damaged not only the settlements situated near big rivers, but also those next to smaller rivers and streams, even backwaters (Gavrilović 1977, 59–88).

Flow regulation on the lower Drava began earlier. A series of dykes formed lines of flood defence. In the period 1730–1740, the landowners in Hungary joined together in an association for flood protection, building their first protective structures on the left river bank, between the villages Szabolcs and Dárda. In 1770–1773, the Baranya county authorities extended and fortified this mound. Where the bank was sufficiently high, embankments were not built. Different approaches to flood control were taken by the Virovitica and Baranya county authorities. The latter was interested in maintaining a dirt road used for transport alongside the Drava upstream. This road, first mentioned in archives in 1739, ran on the left bank. The Baranya county authorities wanted the river regulation works to start from the confluence with the Danube, thus cutting the water route, and partly accelerating river flow to help carry away all driftwood, thus making barge traffic easier. Since the right bank was higher, Virovitica county was better protected and their representatives advocated the reverse priority control, i.e. to start regulation in the upper course of the river (Kiss 1996, 98–100).

A devastating Drava flood occurred in September 1812 (The RC Parish Office Sigetec, Liber memorabilium parochiae Sigetec) and again in 1814. For *Drnje*, however, the most devastating was the 1827 flood: it ruined a bridge across the Drava; entire part of the village Drnje was literally ripped away. The inhabitants, however, managed to escape. Military Border authorities awarded them with grazing lands near Sigetec and Peteranec. The Drnje people built new homes there (today's Pemija street). Flooding had such effects that in 1828 state elementary school and military company headquarters were moved to Peteranec, to an elevated terrain safer from flooding) (The RC Parish Office Drnje, Liber memorabilium parochiae Drnje; The RC Parish Office Peteranec, Liber memorabilium parochiae Peteranec). In 1830, more buildings in Drnje had to be relocated (CNA, Generalkomanda, box. 45, no. R 39–18) (Fig. 8.1). For Drnje, this meant losing a number of central economic and administrative functions. It fell into an economic crisis, stagnation, deterioration and decline reflected in population decrease. Between 1826 and 1839, the total population of Drnje fell by almost a third and that of Botovo by a seventh, while in Torčec there was 42%, in Peteranec a tenth and in Hlebine a third population growth. The population of Sigetec and Đelekovec

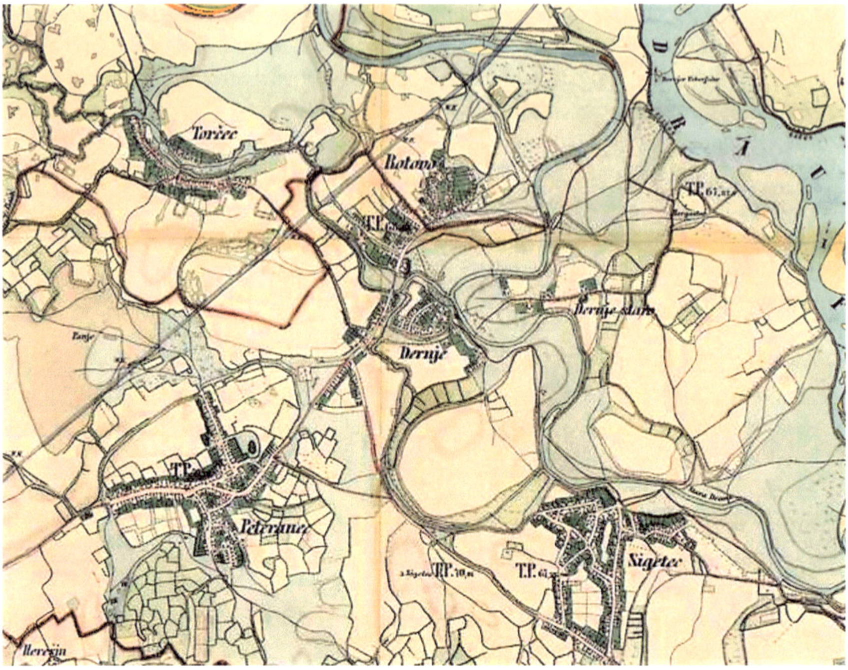

Fig. 8.1 The environs of Drnje (Dernje, Dörnye) on the map sheet of the Second Military Survey, ca 1850. *Source* www.arcanum.hu

remained approximately the same. In Prekodravlje, Gola witnessed a 130% increase in population. It is interesting that in the period 1826–1839, except for Drnje and Botovo, no other settlements in neighbourhood experienced such a decrease in overall population.

Brod is another village the very existence of which depended on Drava. The name Brod itself (meaning boat) refers to a crossing place on the Drava, on roads leading from the Military Border towards the Hungarian county Somogy, into the villages of Vízvár and Bobovec (Babócsa) lying on the left bank (Horvat 1933, 46; Horvat 1941, 18). From the beginnings, Brod was considered an easy, favoured crossing location over Drava, but, threatened by frequent floods, unsuitable for living. The worst obstacle to further settlement growth and development was the river itself. After each flood, the settlement would be fretted away, so people would ask for help from military authorities, for the regulation of the Drava and, more frequently, for resettling the population. In 1777 the authorities passed the Varaždin military order to take steps towards preventing the flooding of the Drava River (CNA, Generalkomande Karlovac-Varaždin, no. 21).

A 1780 map depicts the planned Drava regulation in the Brod area and shows that Brod is located right in the way of direct flood wave of Drava. Thus, the Drava regulation would not spare Brod from floods. Saving Brod was directly linked to the protection of village Heresznye, across Drava, on the Hungarian side, just the opposite to Brod (Slukan Altić 2002, 138).

For Brod, the destruction of the entire village called for the relocation of population to the same bank of the Drava, but to a new, elevated position, more removed from the Drava. A new settlement, Ferdinandovac, took shape around the year 1844. In the mid-19th century Brod was found southeast of the spot, where Ždalica stream (Ždala) meets the Drava, on the right bank. On maps and charts from 1845, it is obvious that Brod village had three north-to-south parallel (main) streets, following the direction of the Drava and two smaller east-to-west streets, perpendicular to the main streets. On the left (Hungarian) bank of Drava river, facing Brod (Slukan Altić 2002, 146) on the opposite side, there was Hrasinja village (today's Heresznye). The same structure is on the map dating back to 1780 (Slukan Altić 2001, 19; Slukan Altić 2002, 138).

It seems that there were twin settlements on both banks of the river (Slukan Altić 2002, map on pp. 138, 146). The last remains of Brod village were completely abandoned as clearly visible on the map from 1847 (Fig. 8.2). Instead of the former village only the name 'Stara Brod' (old Brod) remained. Mirela Slukan Altić (2002, map on the p. 147) is absolutely right to consider Brod one of the best examples of the influence of the Drava on settlements. She also believes that since the riverbed shifted, it brought destruction from the new Drava flow (Slukan Altić 2002, 135).

According to research by Paškal Cvekan, the resettlement of the entire population of Brod in 1844 was the direct consequence from river erosion as the village had excavated canals. In the same year a new settlement Ferdinandovac was built

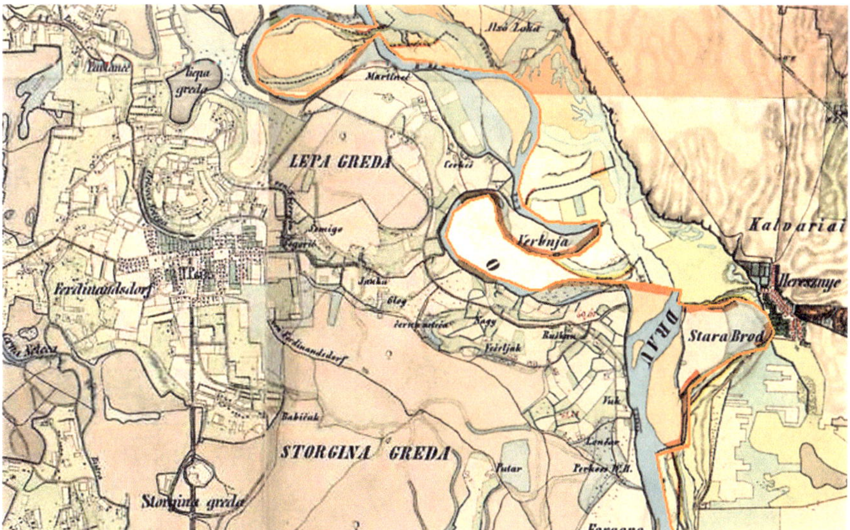

Fig. 8.2 The Ferdinandovac (Ferdinandsdorf, Décseszentpál) and Old Brod area on the map sheet of the 2nd Military Survey, ca 1850. *Source* www.arcanum.hu

(The RC Parish Office Ferdinandovec, Liber memorabilium parochiae Ferdinandovec) only a few kilometers north of Brod. Military Border authorities organized relocation to Ferdinandovac of all former Brod inhabitants, "eighty-three families with *830 people from Brod*" (Cvekan 1996, 41, 60).

In 1814 a new parish was established in Brod and later relocated to Ferdinandovac (The RC Parish Office Ferdinandovec, Liber memorabilium parochiae Ferdinandovec). Life was still difficult for the inhabitants, as recurrent floods in 1821, 1827, 1833, 1836, 1840 and 1843 presented a hazard. The most horrific flood happened in 1827, when the Drava inundated homes and barns, and even in church there was two ells (more than one metre) deep water. Flooding destroyed all crops, so people from Brod had to resettle in the surrounding villages. Pavao Vuk, who in the period 1829–1843 served as the parish priest in Brod, wrote to military authorities claiming that floods endangered the whole village and asked that the Drava river should be regulated and canals dug. He suggested that the entire village should be relocated to a more appropriate place (Cvekan 1996, 40–41).

Although from the above we could conclude that life was relatively unattractive in the flooded Brod with negative effects of the river on the number of inhabitants, other data actually contradict this view. Population of Brod nearly doubled in the period 1800–1844 (Krivošić 1983, 163; Cvekan 1974, 14; Cvekan 1996, 60).

On the map from 1780 (the first survey of the Drava) and in the plan of the regulation project (from 1797) there was a proposed meander intersection, just west of Brod, which was never executed. According to research by Mirela Slukan Altić, the key to saving Brod was regulation on the opposite, Hungarian, bank at Heresznye and this restricted the opportunities (Slukan Altić 2002, 135). Mirela Slukan Altić added some facts to previous research by Cvekan (1974, 1996) on the catastrophic flood of 1847/48, when the river breached the embankment south of Brod and turned a former village into an islet. The panel established that in front of the topographically lowest Vizvarske jame there was a meander, and its tip was eroded, breached and the water continued to flow into the Vizvarski jarak, just south of Brod (Slukan Altić 2002, 135). As it was impossible to regulate the Drava in the vicinity of Brod and Heresznye, nature itself took care of this problem: the river channel changed and Brod, which used to lie on the right (southern) bank, was found on the left (northern) riverbank. At the same time, the river moved its flow further away from Heresznye village, establishing its present course. The inhabitants of Brod finally decided to settle south of the new Drava channel, cut off from the Military Frontier's part of Podravina, where they always belonged. The site of old Brod is now in the Prekodravlje, Republic of Croatia (Veliki atlas Hrvatske, 16).

8.2.5 Regulation of the Drava River in Modern Times

The first river regulation interventions on the Drava River included cutoffs: between 1784 and 1848 64 cutoffs were made downstream of the Mura confluence, which made the river course almost 40% shorter. The first mapping of the Drava took place between 1842 and 1846, including both cross-sections and a longitudinal profile.

In the Varaždin and Koprivnica areas, the most important hydrotechnical works were performed in the second half of 19th century. In 1859, the gazette *Narodne novine* vol. 256, 257, 264, 265 have articles about huge flooding in Varaždin, bringing great damage. This is the reason why the town council ordered the reinforcement of river banks. Riverbank reinforcement, embankment construction and the drainage of neighbouring fields enabled more intensive use of arable land.

The intersection of meanders, continued until 1885, induced a riverbed incision and bank erosion. The cutoffs in the first regulation plans turned out to be insufficient to stabilize the riverbed. Later on the so-called 'German' regulation technique was applied, which means longitudinal bank stabilization against lateral erosion.

The second survey of the river was made between 1886 and 1887, from Zákány to the confluence with the Danube. On this basis, new regulation methods were designed and applied: stone and hurdle structures were built to stabilize the riverbed first between Osijek and the confluence in 1887.

Regulation planning continued until 1904, based on which the works continued on the reaches near Barcs, between Tótújfalu and Budakovac, as well as between Zaláta and Kisszentmárton until 1915. World War I halted regulation works and, as the river became the borderline between Hungary and Yugoslavia, regular navigation also ceased and the riverbed degraded (Ihrig 1973).

Later a new regulation plan was elaborated by the Hungarian River Engineering Office of Osijek in 1904, deviating from the 'German' regulation technique and applying elements of the 'French' principles, in the form of groynes and spur dykes, primarily in the 1931 plan to regulate the reach between the country border and Barcs, developed by the River Engineering Office of Nagykanizsa. It used bends in the development of the new course. However, because of the economic crisis, this plan was not realized.

In 1957, an agreement was concluded between the two riparian countries and the Hungarian-Yugoslavian Water Management Committee was established in order to co-ordinate works of common interest (Ihrig 1973).

In 1958, the bathymetry of the Drava reach between 65 and 75 rkm was completed and a new regulation plan proposed for river regulation carried out until 1965. Later, between 1966 and 1968, the bathymetry between the Danube confluence and Őrtilos (236 rkm) was prepared and in 1970 the first Drava Atlas was published. The two riparian countries jointly developed the General Regulation Plan for the Drava River in 1974, ratified in 1975. Since then, several regulation structures have been built by both of the riparian countries to prevent lateral erosion and to stabilize the riverbed. The last cutoffs on the Drava River have been executed in 1982 at Vízvár and in 1991 at Zaláta (György 2017).

8.2.6 *Navigation*

As the water regime of the Drava River is nival-pluvial, navigation has been well developed for centuries. Being an integral part of the Danube-Main-Rhine waterway system, an increase of regular commercial navigation activities had been expected on the Drava (VITUKI-VIZITERV 1977), which, however, did not materialize. The navigation route parameters along with maximal and minimal waterlevels for navigability have been set and influenced regulation activities. The elaboration of waterway parameters has been completed for the reach between the confluence with the Danube to Vízvár in 1977 (VITUKI-VIZITERV 1977).

Waterway classes are identified by Roman numbers from I to VII. Classes I to III denote waterways of regional or national importance. Waterways of class IV or higher are of economic importance to international freight transport. The class of an inland waterway is determined by the maximum dimensions of the vessels which are able to operate on this waterway. Decisive factors in this respect are the width and length of inland vessels and convoys, as they constitute fixed reference parameters. Restrictions regarding the minimum draught load of vessels, which is set at 2.50 m for an international waterway, as well as the minimum height under bridges (5.25 m in relation to the highest navigable water level) can be made only as an exception for existing waterways. The Drava River is officially navigable from its confluence to 198.6 rkm, waterway class is I–IV, only by small boats in the upper reaches and by larger vessels downstream Donji Miholjac, Croatia, to the confluence (about 90 km length). Maximum vessel sizes for waterway class IV fit in the 'Johann Wesel' type (length L = 85 m, width B = 9.5 m, tonnage T = 1,000–1,500 t). This means a pushed convoy of a single barge can pass this fareway. The single significant port on the Drava is at Osijek (Hartl and Hofbauer 2013).

8.2.7 *Hydropower Plants on the Drava River*

River regulation for hydroelectric power generation and/or irrigation is a provisional service of growing importance (Giller et al. 2004). In the upper reaches of the River Drava numerous dams and reservoirs, mostly for hydroelectric production, had been constructed during 20th century. On the River Drava watercourse up to the River Mura mouth there are 22 barrages and reservoirs with hydroelectric power plants: 11 in Austria, 8 in Slovenia and 3 in Croatia (Table 8.2). In Austria and upper Slovenia, channel slope is high and the valley is narrow. Therefore, relatively small reservoirs were established. In lower Slovenia and throughout

Table 8.2 Hydropower plants on the Drava River

HPP	Country	Nameplate capacity (MW)	Annual generation (million kWh)	Year of completion
Strassen-Amlach	Austria	60	219	1988
Paternion	Austria	24	95	1988
Kellerberg	Austria	25	96	1985
Villach	Austria	25	100	1984
Rosegg-St. Jakob	Austria	80	338	1973
Feistritz-Ludmannsdorf	Austria	88	354	1968
Ferlach-Maria Rain	Austria	75	318	1975
Annabrücke	Austria	90	390	1981
Edling	Austria	87	407	1962
Schwabeck	Austria	79	378	1942
Lavamünd	Austria	28	156	1944
Dravograd	Slovenia	26.2	142	1943
Vuzenica	Slovenia	55.6	247	1954
Vuhred	Slovenia	72.3	297	1956
Ožbalt	Slovenia	73.2	305	1960
Fala	Slovenia	58	260	1918
Mariborski Otok	Slovenia	60	270	1948
Zlatoličje	Slovenia	126	577	1968
Formin	Slovenia	116	548	1978
Varaždin	Croatia	86	476	1975
Čakovec	Croatia	75.9	400	1982
Dubrava	Croatia	84	385	1989

Croatia the landscape is relatively flat, which enable creation of long, relatively large reservoirs with moderate depth (Bonacci and Oskoruš 2008).

Massive construction in the Drava River basin and on the river itself during the last centuries as well as recent climate change and/or variability has caused many different and possibly dangerous changes to its hydrological and ecological regime (Bonacci and Oskoruš 2008 and Chap. 9).

8.2.8 Current Situation

In the General River Regulation Plan of the Drava River, Hungary and Yugoslavia established common guidelines for regulation: to convey water, sediment and ice, as well as to comply with the regulations related to the navigation route by the United Nations Economic Commission (category II downstream of the bridge of Répás) (Table 8.3). A detailed regulation plan was elaborated between 1983 and

Table 8.3 Bankfull channel width of the Hungarian section of the Drava River

Reach	Width (m)
Zákány—Barcs	160
Barcs—Drávaszabolcs	170
Drávaszabolcs—Osijek	180
Osijek—Danube confluence	220

1985 for the reach from 70.2 to 150.0 rkm which is still in force in 2017 (György 2017). It prescribes the parameters of regulating structures (Table 8.4).

By the beginning of the 20th century the essential components of the flood-protection system downstream of Barcs had been completed (Takács and Kern 2015). There have been several development projects on the Dráva dykes in recent years. Though the lower, 25 km long section the height of the flood-control dyke meets the required measure, the dyke system is not uniform and its status is not everywhere safe enough.

The Hungarian Government decree 21/2006 (31 January) ordered strict protection for floodways, riparian zones and areas endangered by springing waters. Although no exceptions are permitted, illegal constructions in the floodway are spreading. On the other hand, the regulation is too strict at some points (Ministry of Interior 2013). The new regulation of floodway management planning (Government decree 83/2014 (14 March) prescribes the process of delimitation of floodways (for the Lower Drava see Czigány et al. 2016), preparation and implementation of their

Table 8.4 General parameters of regulating structures

Type of structure	Parameters
Bank protection	Base: 30 cm hurdles, 20 cm riprap
	Slope: 1:1.5
Guide bunds	Base: 30 cm hurdles, 20 cm riprap
	Body: riprap
	Crown width: 1.00 m
	Slope on waterfront: 1:1.5
	Slope on protected front: 1:1
Spoil banks	Body: riprap
	Width: 4.00–6.00 m
	Height: 1,00 m
	Slope: 1:1–1:1.5
Crossdams and groynes (spur dykes)	Base: 30 cm hurdles, 20 cm riprap
	Body: riprap
	Crown width: 1.00 m
	Slope on waterfront: 1:1
	Slope on protected front: 1:2

management plan. The defence during recent floods on the Danube provided justification for the new decree.

The flood protection on the left bank of the Drava embraces the Drávaszabolcs section (earth embankments of 31,709 km length, continuation of the Croatian system) and the Drávasztára section (33,720 km), where embankments are 0.6 m lower on the average compared to required security level (1.2 m above the highest water stage of the river measured to date). Further upstream high bluffs make flood protection superfluous. Secondary defence lines (localization dykes) in the Drava floodplain only amount to a couple of kilometres.

8.3 Conclusions

As a consequence of river regulation and land drainage measures on the one hand and the impact of global climate change on the other, the alternation of floods and drought periods became typical of the environment of the Hungarian Drava floodplain. Extremities in water availability strongly effect the quality of life of the local population, also suffering from the adverse influences of socioeconomic changes in the last decades (Lóczy et al. 2014). On the Drava, as along many other regulated rivers, these changes have significantly altered the physical habitat and ecological functioning of the natural systems. Given the importance of diverse freshwater habitats in the provision of ecosystem services, there is a need for restoration that can maintain sustainable functioning. To counter the influence of river regulation, restoration efforts should focus on reestablishing dynamic connectivity between the channel and floodplain water bodies (Ward and Stanford 1995).

Increasingly, river managers are turning from hard engineering solutions to ecologically based restoration activities. There is a growing interest in applying river restoration techniques to solve environmental problems, yet little agreement exists on what constitutes a successful river restoration effort (Palmer et al. 2005).

References

Archival Sources

CNA—Croatian National Archive, Zagreb (Croatia) Generalkomanda Križevci County

SAV—State Archives Varaždin (Croatia), Archives of the City of Koprivnica, Archives of the City of Varaždin

The RC (Roman Catholic) Parish Office Drnje (Croatia), Liber memorabilium parochiae Drnje

The RC Parish Office Legrad (Croatia), Liber memorabilium parochiae Legrad

The RC Parish Office Ferdinandovec (Croatia), Liber memorabilium parochiae Ferdinandovec

Franciscan monastery Virovitica (Croatia) HDCV – Historia Domus Conventus Viroviticensis
The RC Parish Office Vizvár (Hungary)
Book of Baptisms (1738–1783); Book of Weddings (1735-1806)

Edited Sources

Androić M (1968) Varaždinski trgovac Vinko Perger (The Varaždin merchant Vinko Perger). Kaj, br. 10, Zagreb (in Croatian)

Bedekovich J (1752) Natale solum magni ecclesiae dostoris sancti Hieronymi. Neostadii Kanonske vizitacije Zagrebačke (nad)biskupije. Hrg M, Kolanović J (eds). Zagreb, 1989 PG (1958–1980) Prothocolla generalium congregationum statum Regnorum Dalmatiae, Croatiae et Slavoniae, vol 1–12. Zagreb

Bonacci O, Oskoruš D (2008) The influence of three Croatian hydro-electric power plants operation on the river Drava hydrological and sediment regime. 24th Conference of the Danubian Countries on Hydrological forecasting and hydrological bases of Water management, Belgrade 2006

Borzavári B (no date) A Dráva árvízvédelmi rendszerének fejlődése (Development of the flood protection system of the Drava River). South Transdanubian Water Management Directorate, Pécs. 50–57 (in Hungarian)

Buchberger P (1975) A Dráva-völgy árvédelmének története (History of flood control in the Drava Valley). Vízügyi közlemények 57(1):103–113 (in Hungarian)

Cvekan P (1974) Župa i selo Ferdinandovac (Ferdinandovac county and village). Ferdinandovac (in Croatian)

Cvekan P (1996) Ferdinandovac. Od Broda na Dravi do Ferdinandovca u močvari (Ferdinandovac: from Brod on the Drava to Ferdinandovac in the swamp). Ferdinandovac (in Croatian)

Czigány Sz, Pirkhoffer E, Halmai Á, Lóczy D (2016) Floodway zoning and flood modelling using remote sensing and DEM along the lower Hungarian Drava section. In: Šulc-Michalková M, Miřijovský, Lóczy D, Zgłobicki W (eds) Interdisciplinary studies of river channels and UAV mapping in the V4 region. Comenius University, Bratislava. 177–200

Donauinseln (2012) Record discharge measured on the Drava! Blogspot 8 November 2012. http://donauinseln.blogspot.hu/2012/11/record-discharge-measured-on-drava.html

Dynesius M, Nilsson C (1994) Fragmentation and flow regulation of river systems in the northern third of the world. Science 266:753–762

van Duivendijk J, te Slaa J (2004) River training structures. UNESCO-IHE Lecture notes

ES (1995) Poplave (Floods). In: Enciklopedija Slovenije (Encyclopaedia of Slovenia) 9. Plo–Ps, Mladinska knjiga, Ljubljana. 416 pp (in Slovenian)

Gavrilović S (1977) Slavko, Sremska, virovitička i požeška županija u godinama gladi 1813–1817 (Slavonia, Srem and the Virovitica and Požega counties in the famine years of 1813–1817). Acta historico-oeconomica Iugoslaviae 4:59–88 (in Croatian)

Giller PS (2005) River restoration: seeking ecological standards. Editor's introduction. J Appl Ecol 42:201–207. https://doi.org/10.1111/j.1365-2664.2005.01020.x

Grabowski RC, Surian N, Gurnell AM (2014) Characterizing geomorphological change to support sustainable river restoration and management. WIREs Water 1:483–512. https://doi.org/10.1002/wat2.1037

György Zs (2017) Vízpótlás- és víz visszatartás biztosítása a barcsi téli kikötőben (Water supply and retention in the winter port of Barcs). Manuscript thesis. Institute for Hydraulic Engineering and Water management, Faculty of Water Sciences, National University of Public Service, Baja, Hungary (in Hungarian)

Hartl T, Hofbauer V (eds) (2013) Manual on Danube Navigation. ViaDonau Grasl Druck & Neue Medien GmbH Vienna, ISBN 3-00-009626-4

Horvat R (1933) Hrvatska Podravina (The Drava region in Croatia). P. Kvaternik i dr, Zagreb (in Croatian)

Horvat R (1941) Župe u Hrvatskoj Podravini (Counties in the Croatian Drava region). Hrvatska prošlost, knj. 2, Zagreb 1941 (in Croatian)
Horvat R (1993) Povijest grada Varaždina (History of the town of Varaždin). Varaždin (in Croatian)
Horvat-Levaj K (2006) Prilog istraživanju barokne sakralne arhitekture Podravine – župna crkva Sv. Martina u Martijancu (Contributions to Croatian sacral architecture: St. Martin Church in Martijanc). Podravina 5(10):80–99 (in Croatian)
Ihrig D (1973) A magyar vízszabályozás története (The history of river regulation in Hungary). OVH Budapest (in Hungarian)
Al-Khafif SM (1970) River training and sediment transportation. Manuscript, VITUKI, Budapest, Hungary
Kiss A (2018) Floods and long-term water-level changes in medieval Hungary. Springer International Publishing, Cham, Switzerland, p 499
Kiss G (1996) Ekologija Podravine – Podaci iz Ormánsága (Ecology of the Drava region: Data on the Ormánság). Etnografija Hrvata u Mađarskoj 3:85–104 (in Croatian)
Krivošić S (1983) Stanovništvo Podravine 1659-1859. godine (Population of the Drava region, 1659–1859). Podravski zbornik 9:147–164 (in Croatian)
Lóczy D, Dezső J, Czigány Sz, Gyenizse P, Pirkhoffer E, Halász A (2014) Rehabilitation potential of the Drava River floodplain in Hungary. In: Gâştescu P, Marszelewski W, Breţcan P (eds) Water resources and wetlands. Proceedings of the 2nd International Conference, 11–13 September 2014, Tulcea, Romania. Transversal Publishing House, Târgoviste, 21–29
Lóczy D, Dezső J, Sz Czigány, Pirkhoffer E (2016) Hydromorphological assessment of the Lower Hungarian Drava section and its floodplain. Landscape & Environment 10(3–4):109–116. https://doi.org/10.21120/LE/10/3-4/2
Mikoš M, Brilly M, Ribičič M (2004) Floods and landslides in Slovenia. Acta hydrotechnica (Ljubljana) 22–37
Ministry of the Interior (2013) High-water channel management. High Authority for Water Management, Budapest. https://www.vizugy.hu/index.php?module=content&programelemid=37
Moser J (2014) Hochwasserchronik Kärnten. Amt der Kärtner Landesregierung, Klagenfurt. 77 pp. http://www.ktn.gv.at/KAGIS_DOKUMENTE/UWN/Hydro/HWChronik__11_2014.pdf
Obadić I (2007) Međuodnos ljudi i rijeke Drave na području varaždinske Podravine u novom vijeku (People and the Drava in the Varaždin region in modern times). Radovi Zavoda za znanstveni rad, Varaždin 18:301–325 (in Croatian)
Palmer MA, Bernhardt ES, Allan JD, Lake PS, Alexander G, Brooks S, Carr J, Clayton S, Dahm CN, Follstad Shah J, Galat DL, Loss SG, Goodwin P, Hart DD, Hassett B, Jenkinson R, Kondolf GM, Lave R, Meyer JL, O'Donnell TK, Pagano L, Sudduth E (2005) Standards for ecologically successful river restoration. J Appl Ecol 42:208–217. https://doi.org/10.1111/j.1365-2664.2005.01004.x
Petrić H (2000) Općina i župa Drnje (The municipality and parish of Drnje). Meridijani, Koprivnica. 212 pp (in Croatian)
Petrić H (2005) Utjecaj rijeke na pogranična naselja. Primjer rijeke Drave u 18. i 19. stoljeću (Influence of the river on riverine settlements: an example of the Drava in the 18th and 19th centuries). Ekonomska i ekohistorija 1:37–62 (in Croatian)
Petrić H (2014) Some aspects of the interrelationship between humans and the River Drava in the pre-industrial times with an emphasis on the late 18th and early 19th century. In: Štih P, Zwitter Ž (eds) Man, Nature and Environment Between the Northern Adriatic and the Eastern Alps in Premodern Times. Zbirka Zgodovinskega časopisa 48:260–289
Petrić H, Obadić I (2007) Drava river flooding in Varaždin and Koprivnica parts of Podravine (Drava river region) in the period of 17th–19th centuries. Podravina 12:136–147
Peturnig CF (ed) (1969) Hochwasser in Kärnten. Eine Dokumentation. Amt der Kärtner Landesregierung, Klagenfurt. 81 pp. http://www.interpraevent.at/palm-cms/upload_files/Publikationen/Hochwasser_in_Kaernten/Hochwasser_in_Kaernten.pdf

Remenyik B (2004) A Dráva szabályozása a XVIII. századtól a XIX. század végéig (Regulation of Drava from the 18th century till the end of the 19th century) In: Proceedings of the 2nd Hungarian Conference on Geography, Szeged. Available online: http://geography.hu/mfk2004/mfk2004/cikkek/remenyik_bulcsu.pdf, last accessed on 31.08.2017 (in Hungarian)
Remenyik B (2005) Adatok a Dráva-szabályozás történetéből (Data on the history of the regulation of the Drava River). Hidrológiai Közlöny 85(3):27–30 (in Hungarian)
Rohner K, Berger I, Steiger H, Moser J, Pichler F, Sereinig N (2004) Hochwasserchronik Kärnten 792–2003. Amt der Kärntner Landesregierung, Klagenfurt
Rosenberg DM, McCully P, Pringle CM (2000) Global-scale environmental effects of hydrological alterations: introduction. Bioscience 50:746–751
Sever Z (2000) Hidroelektrane u Hrvatskoj (Hydropower plants in Croatia). HEP, Zagreb (in Croatian)
Slukan Altić M (2001) Hidrografske karte rijeka kao izvori za proučavanje dinamike hidrografskih odnosa (Hydrographic maps of rivers as sources for studying the dynamics of hydrographic relations). Hrvatske vode 34(9):15–29 (in Croatian)
Slukan Altić M (2002) Hidroregulacije Drave i njezini utjecaji na transformaciju prirodnog i kulturnog pejsaža Podravine (The regulation of the Drava and its impact on the transformation of the natural and cultural landscape of Podravina). Podravina (Koprivnica) 1(2):128–152 (in Croatian)
Tadić L, Bonacci O, Dadić T (2016) Analysis of the Drava and Danube rivers floods in Osijek (Croatia) and possibility of their coincidence. Environmental Earth Sciences 75(18). https://doi.org/10.1007/s12665-016-6052-0
Takács K, Kern Z (2015) Multidecadal changes in the river ice regime of the lower course of the River Drava since AD 1875. J Hydrol 529:1890–1900. https://doi.org/10.1016/j.hydrol.2015.01.040
Težak S (2004) Varaždinci i rijeka Drava (Varaždin people and the Drava River). In: Život uz Dravu nekad i danas (Life on the Drava now and then). City Museum, Varaždin. 35–46 (in Croatian)
Veliki atlas Hrvatske (Great Atlas of Croatia) (2002) (ed. Borovac I) Mozaik knjiga, Zagreb (in Croatian)
VITUKI-VIZITERV (1977) A víziutak műszaki paramétereinek meghatározása 2. Dráva (Determination of the technical parameters of waterways, vol 2. Drava). Budapest (in Hungarian)
Vujasinović B (2000) Radovi na uređenju voda u Hrvatskoj u 17. stoljeću (Water regulation works in Croatia in the 17th century). Hrvatska vodoprivreda, br. 96, Zagreb (in Croatian)
Ward JV, Stanford JA (1995) Ecological connectivity in alluvial river ecosystems and its disruption by flow regulation. Regulated Rivers: Research and Management 11:105–119. https://doi.org/10.1002/rrr.3450110109
Zorkóczy Z, Károlyi Z (1993) Folyó- és tószabályozás (River and lake regulation). Department of Water Management, Pollack Mihály Technical College, Baja, Hungary (in Hungarian)
Zorković A (1995) Povijesnica općine Petrijanec (History of Petrijanec Municipality). Općina Petrijanec, Varaždin (in Croatian)

Chapter 9
Human Impacts on Water Regime

Ognjen Bonacci and Dijana Oskoruš

Abstract River water regime is continuously in motion and in interaction with many natural and human-induced factors. Human interventions, especially over the last two centuries, caused the natural water regime in most rivers to become substantially and irreparably disturbed. River training, floodplain reduction by construction of levees, building of numerous hydrotechnical structures, especially dams and hydroelectric power plants, and other large structures along the Drava River, as well as in its catchment in Italy, Austria, Slovenia, Croatia, and Hungary, have greatly influenced its water regime. This section plays an important role because 167 km of the Drava River course represents the boundary between Croatia and Hungary. At the end of 19th century, major regulation works were executed along the whole Drava River course. The last massive intervention on the Drava River in Croatia and Hungary was carried out at the end of the 19th and the beginning of the 20th centuries. The changes in water regime along the lowland part of the Drava River from the boundary between Slovenia and Croatia to its confluence with the Danube River are examined in this chapter. The findings of hydromorphological and sediment transport regime analyses are also presented.

Keywords Human impact · Floods · River regulation · Floodplain
Suspended sediment · River hydromorphology

O. Bonacci (✉)
Faculty of Civil Engineering, Architecture and Geodesy,
University of Split, Matice hrvatske 15, 21000 Split, Croatia
e-mail: obonacci@gradst.hr

D. Oskoruš
Hydrometeorological Service of Republic Croatia, Grič 3, 10000 Zagreb, Croatia
e-mail: oskorus@cirus.dhz.hr

D. Lóczy (ed.), *The Drava River*, Springer Geography,
https://doi.org/10.1007/978-3-319-92816-6_9

9.1 Introduction

Many substantially different issues may be considered under the notion of 'river water regime', for instance: (1) hydrology, (2) hydrogeomorphology, (3) climate, (4) relief, (5) soil, (6) vegetation cover, (7) transport of sediment, (8) water temperature regime, (9) ice regime, (10) ecological characteristics, etc. All these issues are strongly interconnected. The fluctuations of a river water regime are caused by natural forces as well as human activities. Over the past two centuries, because of intensive and generally uncontrolled human interventions, rivers have suffered the single most intensive impact of all the world's ecosystems. Competition for water between users is a serious problem in all parts of the Drava River basin, too.

Human activities which result in alteration of natural river water regimes and their floodplains and wetlands are (1) damming of rivers, (2) extracting of surface water from the river and groundwater connected with the river surface water, (3) floodplain reduction by construction of levees, (4) change of the catchment drainage patterns, (5) river regulation and channelization, (6) extraction of gravel and alluvial sands from the channel, and dredging (Erskine et al. 1985; Kondolf 1997). Alteration in natural flow regimes can be caused by directly modifying flows, altering seasonality and the frequency, duration, magnitude, timing, predictability and variability of flow events, altering surface and subsurface water levels and modifying the rate of rise or fall of water levels.

'*River training*' refers to the structural measures which are taken to improve a river and its banks. River training is an important component in the prevention and mitigation of floods.

Frequently the changes in river water regime are simultaneously stimulated by natural and human-induced causes. Due to this fact, it is challenging (in many cases impossible) to strictly define which of them plays a more important role in the changes of the actual river water regime. In relevant scientific analyses of human impact on river water regime the crucial problem is the fact that numerous global, regional and local human interventions instantaneously act in combination with natural forces in large rivers. Some consequences of human activities occur after a short time, whereas others present themselves much later, after several decades.

Considering human impacts on the Drava River water regime, it should be noted that the river has two markedly different sections. The natural water regime of the upstream river section, from the spring to the outlet of Dubrava hydroelectric power plant (HEPP) in Croatia, is considerably and completely altered by human interventions, especially by construction and operation of HEPPs, regulations of river course and substantial interventions on the adjacent catchment (urbanization, industrialization, intensive agricultural production, excessive pumping of groundwater, deforestation, etc.). In the upstream part of the Drava River, up to the confluence with the Mura River, 22 hydroelectric power plants have been built; 11 in Austria, 8 in Slovenia and 3 in Croatia.

The situation in the lower Drava River is completely different, citing the opinion of Schwarz (2007): "*The lower Drava and Mura Rivers along the border between*

Croatia, Slovenia and Hungary represent one of the last remaining continuous riverine landscapes in Central Europe, with all typical natural river elements, such as large natural islands, gravel and sand banks, side channels, meanders, loam cliffs, oxbows and soft woods. Together with its main tributary, the Mura River, the Drava represents a unique 'river corridor' of about 380 km without dams." In the following, it will be shown that this statement is only partially true.

In the Drava River stretch in Croatia, the following three HEPPs were constructed: (1) Varaždin HEPP, (2) Čakovec HEPP, (3) Dubrava HEPP (Table 9.1; Fig. 9.1). All three mentioned barrages and their reservoirs are located upstream of the Drava-Mura confluence. Old meanders still exist only in the Drava River stretch

Table 9.1 Basic data for three Croatian HEPPs

HEPP	Varaždin	Čakovec	Dubrava
Start of operation	1975	1982	1989
Installed discharge ($m^3 s^{-1}$)	450	500	500
Reservoir volume ($10^6 m^3$)	8.0	51.0	93.5
Reservoir water surface (km^2)	3.0	10.5	16.6

Fig. 9.1 The Dubrava dam with the canal which provides water to the Dubrava HEPP power house and the old Drava channel

Table 9.2 Main characteristics of five Croatian hydrological stations along the Drava River watercourse

Station name	Datum plane H (m a.s.l.)	Distance from the mouth L (km)	Basin area A (km^2)	Mean discharge Q ($m^3\ s^{-1}$)
Varaždin	166.06	288.5	15,616	347
Botovo	121.55	227	31,038	510
Terezino Polje	100.67	152	33,916	516
Donji Miholjac	88.57	80.5	37,142	537
Osijek	81.48	19	39,982	562

between the settlements of Botovo and Terezino Polje. The analyses have shown that they become natural retention areas during high water regime.

The total length of the three Croatian reservoirs is 23.4 km, of three canals is 29.9 km and of old Drava channels is 54.1 km.

In the old Drava reaches, downstream of the three reservoirs, during 90% of the year discharge varies between 10 and 12 $m^3\ s^{-1}$. The discharge called '*biological minimum*' is agreed between Croatian Waters Company and Croatian Electricity Company and amounts to 8 $m^3\ s^{-1}$. Under natural conditions, mean annual discharge was about 335 $m^3\ s^{-1}$. During high waters when discharges of the upstream Drava River exceed 500 $m^3\ s^{-1}$ (installed discharge of HEPPs) the discharges through the old Drava channel reach are higher. Two or three times per year, maximum discharges can reach 1,500 $m^3\ s^{-1}$ and higher. The discharge of 'biological minimum' is too low; hence, it does not meet the criteria to be considered *ecological flow* (EU 2015).

The lower part of the Drava River is influenced by human interventions, but they are significantly less noticeable than upstream, from the spring in Italy to Dubrava HEPP. Downstream of Dubrava HEPP until Donji Miholjac, some large meanders, oxbows, and river islands are preserved. Downstream of Donji Miholjac to the Drava mouth, most of the large meanders were cut-off, but braided river sections are preserved.

Due to the limited space of this chapter, only human impacts on hydromorphological and sediment transport of the lower part of the Drava River water regime will be discussed.

Table 9.2 presents the main characteristics of five Croatian hydrological stations along the lover Drava River, whose different data will be analysed in this chapter.

9.2 Hydromorphological Changes

Hydromorphology is a science dealing with physical characteristic of the riverine environment, which serve as habitats based on hydrologic-hydraulic and morphologic-sedimentological parameters, including the channels, banks and floodplain. River morphology is an important control on flow, which can substantially affect flood intensification or mitigation. When rivers flow on an alluvial plain, they often meander or braid, which is typical of the lower section of the Drava River.

Schwarz's (2007) preliminary evaluation of the whole Drava River channel (749 km) indicates that about 35% of all river stretches fall into class two or better (mostly along the lower stretches in Hungary and Croatia), whereas the remaining 65% belong to classes 3–5 (over 26% are completely modified). It should be noted that human activity affecting channel morphology and fluvial processes in the analysed river section is quite varied. Large, yet uncontrolled and therefore unknown, amounts of groundwater are being pumped from the analyzed river catchment section. Indirect influences, including land use and management, urbanisation, massive deforestation, elimination of marshes and wetlands, alter the river hydrological regime, sediment yield, environmental characteristics, and ecological equilibrium. A wide range of direct impacts, as, for example, embankments which disrupt channel-floodplain connectivity (see Chap. 14 in this volume), grade-control structures, channelization, meander cut-offs and stream rectification, installation of groynes, artificial bank stabilization, etc., influenced the river channel characteristics and its stability.

In the natural state, approximately 250 years ago, the lowland part of the Drava River was full of meanders, which resulted from the natural river erosion and deposition processes (Mantuáno 1973, 1976; Bognar 1995, 2008; Biondić 1999). The predominant intervention in this part of the Drava River regulation and channelization was cutting off meanders. Due to the cutoffs, both the length and sinuosity of the lowland part of the Drava River strongly decreased, and the lengths of straight sections and the river's slope strongly increased. Figure 9.2 shows the changes in the longitude of the Drava River section between the mouth of the Mura River and the confluence with the Danube River over the 1784–1990 period. After 1990, there have not been any significant interventions. The length of the analyzed river section was decreased by 120.8 km or 34.3%. These interventions in morphological system substantially, severely, and unpredictably changed the natural dynamics of the lower Drava River with strong influences on its ecological properties. The precise consequences of these human interventions have not been studied and explained until now, because adequate complex and continuous monitoring was missing.

The best documented human influence on the Drava River hydromorphological regime is in the river section from the city of Osijek to the confluence with the Danube River. Figure 9.3 presents the changes in the Drava channel from 1796 to 2000. The shortening of the Drava River and the reduction of wetland area on the left bank is evident. In 1796, Osijek was 33 km away from the Drava mouth, whereas in 1898 the distance was only 20 km. During the period between 1880 and 1898, major river regulation works were implemented to improve navigation conditions. Two of the largest meanders were cut off; the river course was significantly shortened and straightened by 13 km. The natural Drava River delta disappeared, as well as the part of the floodplain (Biondić 1999). During the 20th

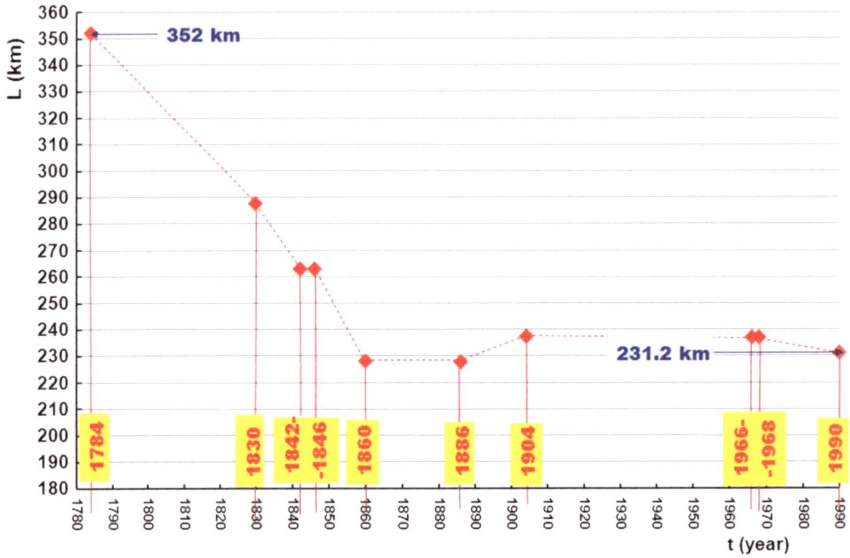

Fig. 9.2 Changes in the length of the Drava River section between the mouth of the Mura River and the confluence the Danube River, 1784–1990 (modified after Bognar 1995, 2008; Biondić 1999)

century, some interventions were also performed. On the left bank, existing dykes were heightened and connected to a dyke along the right Danube River bank. In that way, the area primarily used for flood retention was significantly reduced and replaced by agricultural fields with a land drainage system.

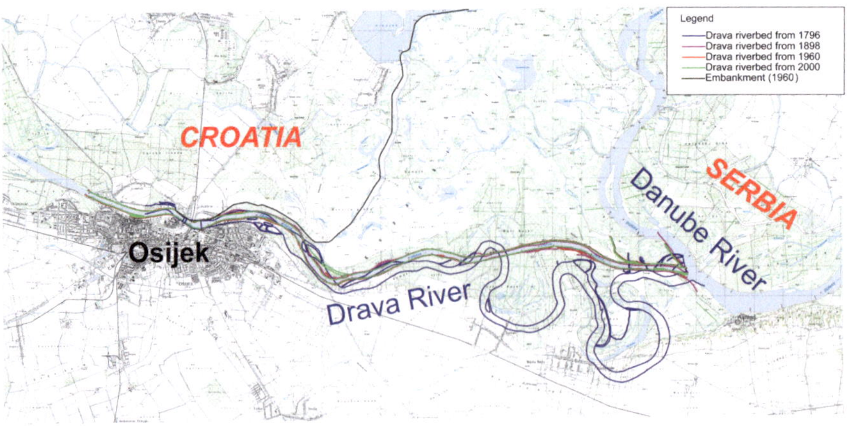

Fig. 9.3 Map of the Drava River channels around the city of Osijek between 1796 and 2000 (modified after Dadić et al. 2015)

From the engineering and ecological point of view, the following three zones of river environments play the most important role: (1) channel, (2) river banks and riparian zone, (3) floodplain. The floodplain represents the aquatic/terrestrial transition zone, i.e., an ecotone. During high water, the floodplain is incorporated into the surface flow system. The extremely important function of the floodplain is to retain high water during floods, protecting thus downstream areas along the river from flooding (Bonacci 2016). During the period of low water, significant amount of groundwater, which feeds baseflow, is stored underground. Schwarz (2007) indicates that the overall floodplain loss for the entire area of Drava and Mura Rivers can be assessed to about 75% with large regional differences.

Fast rising and falling of discharges and water levels, caused by the development of Dubrava HEPP, resulted in the erosion of the banks and channel in the downstream section of the Drava River. Daily water-level fluctuations amount to 150 cm, but do not pose a risk to the stability of river morphology. However, they represent a great risk to environmental and ecological processes.

Figure 9.4 presents eight Drava riverbed cross-sections measured at the Terezino Polje gauging station from 1st July 1977 to 13th March 2014. It is discernible that the morphology of this profile has changed rather rapidly over a short time. It should be emphasized that a similar situation is typical for all other hydrological profiles controlled by the Croatian Meteorological and Hydrological Service. At the Terezino Polje profile, and virtually along the entire lower Drava River section downstream of the last Croatian HEPP, Dubrava, there is a simultaneous process of riverbed deepening and general bottom erosion (Bonacci et al. 1992; Bonacci and

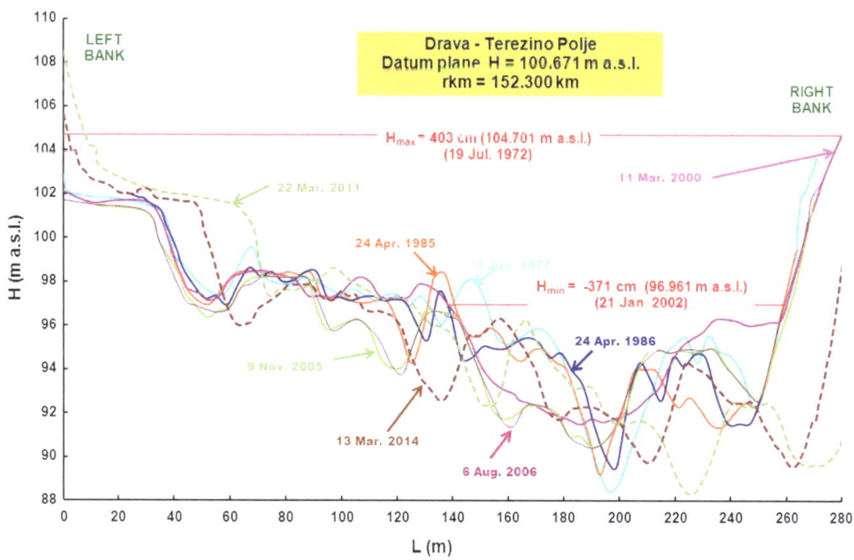

Fig. 9.4 Eight Drava riverbed cross-sections measured at the Terezino Polje hydrological station from 1 July 1977 to 13 March 2014

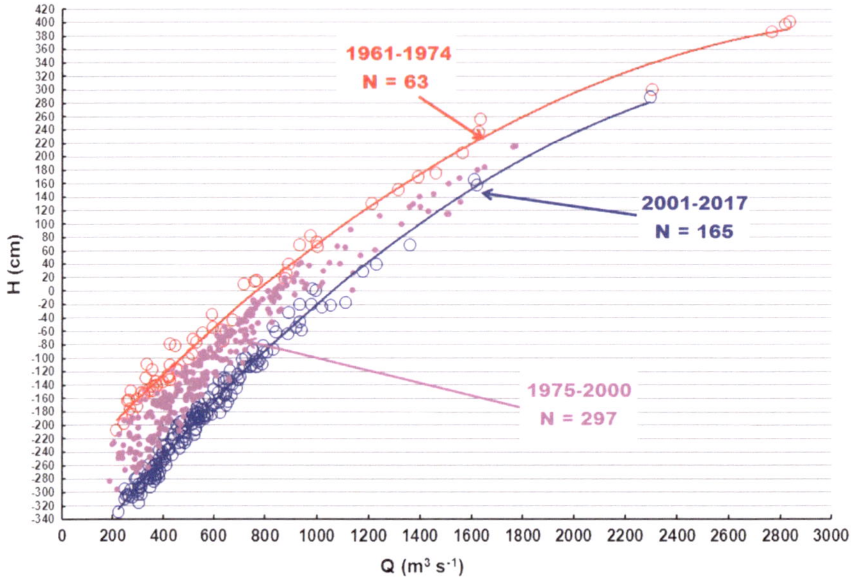

Fig. 9.5 All (565) measured discharges at Terezino Polje in the period from 25 September 1961 to 8 March 2017 with two rating curves

Oskoruš 2010). Due to this fact, the rating curves of the Drava River along the analysed stretch are not stable—particularly in the low water period. Figure 9.5 shows 565 measured discharges at Terezino Polje in the period from 25th September 1961 to 8th March 2017. The minimum measured discharge was 187 m^3 s^{-1} (on 15 February 1989) and the maximum discharge was 2,833 m^3 s^{-1} (on 19 July 1972). The rating curve for the 1961–1974 period is highlighted in red and rating curve for 2001–2017 in dark blue. During the 1975–2000 period, changes in rating curves were very fast, especially for low flows.

From a hydrogeomorphological point of view, river mouths are particularly dynamic and unstable environments. The Drava–Mura confluence changes its place every year. About 40 years ago it was located 900 m higher upstream the Mura River than today. In the mouth area, there are two enormous meanders at the Drava River. In the case of the superior meander, the Drava River almost reaches the Mura River. The two rivers are only 50 m away from each other, but this distance has reduced rather rapidly during the recent years. In September 2013, a new confluence was formed about 200 m upstream of the old site (Fig. 9.6).

Kopački Rit is located at the confluence of the Drava with the Danube. It represents one of the most important and well-preserved natural wetlands in Europe, formed during Pleistocene and Holocene epochs. The Kopački Rit Nature Park was organized 45 years ago and it was protected by law in 1967 as a nationally valuable area. In 1993, Kopački Rit was declared a Ramsar Convention site and included into the list of internationally important wetlands (Tadić et al. 2014, 2016).

Fig. 9.6 New Drava-Mura confluence formed in September 2013

Fluctuating water levels in the area create a wide variety of habitats and generate high biological diversity. Compared to the wetland area of about 37,000 ha in the 18th century, there is a substantial reduction of flood retention capacity along the left bank downstream to Osijek. Tadić et al. (2014) warned that regulation works have shortened the river channel and have increased the hydraulic gradient, which resulted in more intensive erosion. Water balance components change according to water quantity and flood intensity, which critically affect the valuable and vulnerable ecological system.

9.3 Sediment Regime

Sediment transport is critical to understanding river functioning. Analyses of suspended sediment load along the river course and its changes over time are of crucial importance to river water management and its environmental protection. Water and sediment inputs are fundamental drivers of river ecosystems, but river management tends to emphasize flow regime at the expense of sediment regime (Wohl et al. 2015). Dammed rivers are subject to changes in their flow, water-quality, and sediment regimes.

Changes of *suspended sediment* regime measured at three Croatian stations (Varaždin, Botovo and Donji Miholjac) located along the lower Drava River downstream of Croatian HEPP Dubrava will be discussed in this section.

Figure 9.7 shows the time series of annual suspended sediment load (G 10^3t y^{-1}) measured at the Varaždin station for the period 1960–1981 (Bonacci and Oskoruš 2010). Owing to the construction of Čakovec HEPP in 1982, the measurement of suspended sediment in this profile was terminated. This station was located a few kilometres downstream of the Varaždin Reservoir dam. In the first subperiod (1960–1967), before the commencement of operation of the upstream Zlatoličje HEPP in Slovenia, the sediment load was 2.3 times higher than after its construction (1968–1974 subperiod). The Varaždin Reservoir caused a drop in suspended sediment in the third subperiod (1975–1981) by a further 2.4–5.5 times less than in the first (1960–1967).

The Botovo profile is located 68 km downstream of the Varaždin Reservoir dam, 48 km from the Čakovec dam and 28 km from the Dubrava dam. The construction of the Varaždin Reservoir caused a 17% decrease in the suspended sediment transport during the 1975–1981 subperiod (Fig. 9.8). The construction of the Čakovec Reservoir decreased the suspended sediment transport by 2.7 times compared to the first subperiod (1967–1974) and by 2.2 times compared to the second (1975–1981). The construction of the Dubrava Reservoir decreased the suspended sediment transport by 3.4 times compared to the first subperiod (1967–1974), by 2.8 times compared to the second (1975–1981) and by 1.3 times compared to the third (1982–1988).

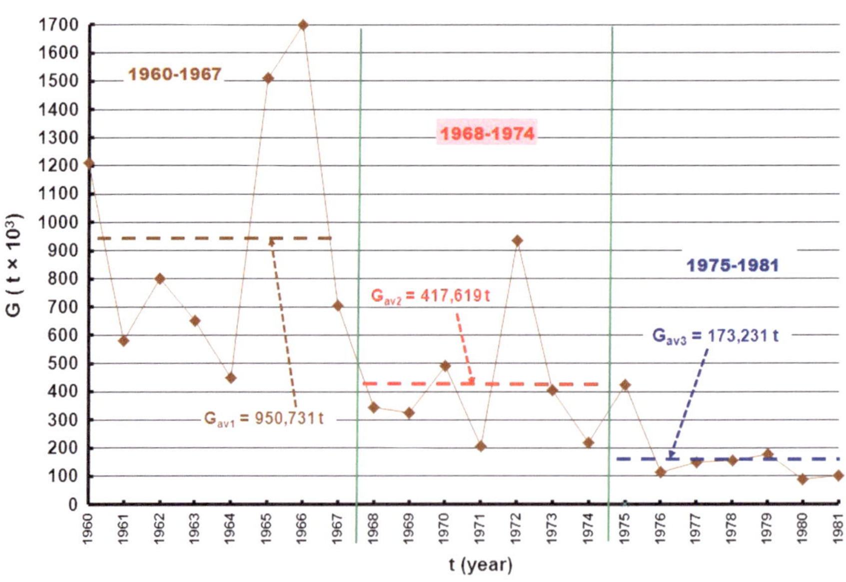

Fig. 9.7 Annual suspended sediment load at Varaždin, 1960–1981

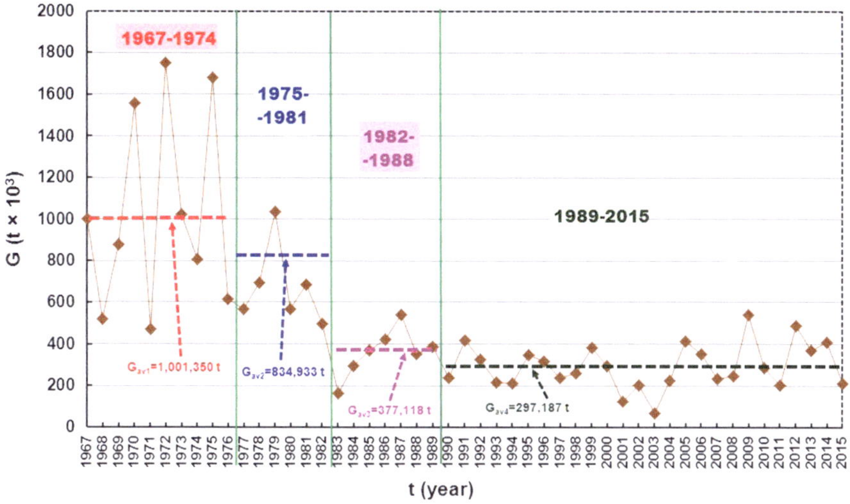

Fig. 9.8 Annual suspended sediment load at Botovo, 1967–2015

The Donji Miholjac profile is located 218 km downstream of the Varaždin dam, 198 km from the Čakovec dam and 178 km from the Dubrava dam. The construction of the Varaždin Reservoir did not cause any change in suspended sediment transport in the 1968–1981 subperiod (Fig. 9.9). The establishment of the

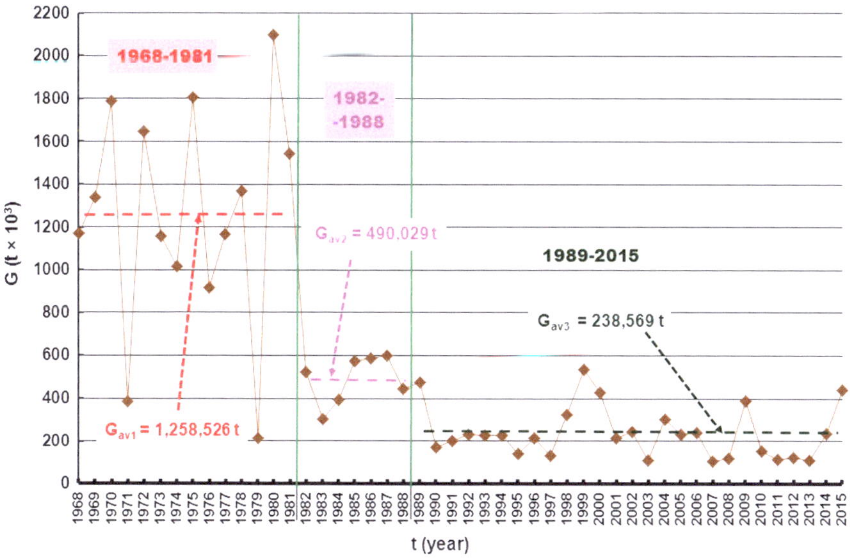

Fig. 9.9 Annual suspended sediment load at Donji Miholjac, 1968–2015

Čakovec Reservoir decreased suspended sediment transport by 2.6 times compared to the previous subperiod (1968–1981). The construction of the Dubrava HEPP decreased it by 5.3 times compared to the first subperiod (1967–1981) and by 2.1 times compared to the second (1982–1988). Today, only about 19% of suspended sediment of the 1968–1981 interval flows through the Donji Miholjac profile.

It should be noted that only a small quantity of suspended sediment is stored in the three Croatian reservoirs. The cause of suspended sediment reduction downstream of the Dubrava Reservoir could be explained mostly by the exclusion of about 80 km of the Drava bed and the adjacent catchment from sediment production.

Large amounts of *gravel and sand* for civil engineering works were and still are *excavated* from the studied section of the Drava riverbed an in its neighboring area. These activities are not under adequate control, and despite major efforts, it was not possible to obtain official and reliable data on them. A rough estimate of the volume of official and more or less controlled excavated materials in Croatia and Hungary during the past decades is about 300.000 m^3 y^{-1}, but it could be much higher in some years. The assessment of illegal excavations is not possible.

9.4 Conclusions

The management of the Drava River water regime has to consider local, national, and international boundary conditions. Sustainable development is possible only through an integral planning and interdisciplinary management approach for the considered area. Regarding future human activities special attention should be attributed to the confluence of the Mura with the Drava, and the Drava with the Danube. These two locations are very important from the hydrogeomorphological point of view and their role in global environmental processes is inestimable.

Natural flow regime today presents a paradigm for river conservation and restoration. The ecological integrity of river ecosystems depends on their natural dynamic characteristics. Human activities substantially, frequently, and severely change the naturally balanced water and ecological regime. The alteration of the natural flow regimes of rivers and streams and their floodplains and wetlands is recognized as a major factor contributing to the loss of biological diversity and ecological function in aquatic ecosystems, including floodplains. The ichtyofauna composition of the Drava River in the area and downstream of three Croatian HEPPs is considerably and adversely affected (see Chap. 16). Hydroelectric development leads to an unnatural variability of the ichthyofauna.

A definite conclusion is that the water regime of the Drava River is substantially influenced by human activities even on the lower section. If we really wish to preserve this invaluable and extremely vulnerable landscape and environment, more efficient international cooperation and interdisciplinary efforts should be invested.

References

Biondić D (1999) Erozija korita rijeke Drave (Erosion of the Drava River channel). Građevinar 51 (5):321–328 (in Croatian)

Bognar A (1995) Regulacije i njihov utjecaj na geomorfološko oblikovanje Drave i Dunava u Hrvatskoj (Regulations and their influence on geomorphological regime of Drava and Danube in Croatia). Zbornik radova 1. Hrvatske konferencije o vodama. R 8–03:449–461 (in Croatian)

Bognar A (2008) Geomorfološka obilježja korita rijeke Drave i njenog poloja u širem području naselja Križnica (Geomorphologic characteristics of the Drava River bed and its floodplain in wider area of the settlement Križnica). Hrvatski Geografski Glasnik 70(2):49–71 (in Croatian)

Bonacci O (2016) River—the bloodstream of landscape and catchment. Acta Hydrotechnica 29 (50):1–12

Bonacci O, Oskoruš D (2010) The changes in the Drava River water level, discharges and suspended sediment regime. Environ Earth Sci 59(8):1661–1670

Bonacci O, Tadić Z, Trninić D (1992) Effects of dams and reservoirs on the hydrological characteristics of the Lower Drava River. Regulated Rivers: Res Manag 7(4):349–357

Dadić T, Tadić L, Bonacci O (2015) Utjecaj Drave i Dunava kroz povijest na poplave u Osijeku (The Drava and Danube impacts on the floods in Osijek throughout history). Hrvatske Vode 23 (94):287–294 (in Croatian)

Erskine WD, Geary PM, Outhet DN (1985) Potential impacts of sand and gravel extraction on the Hunter River, NSW. Aust Geogr Stud 23:71–86

EU (2015) Ecological flows in the implementation of the water framework directive. Office for Official Publications of the European Communities, Luxembourg. 108 pp (Guidance document no 31). https://doi.org/10.2779/775712

Kondolf GM (1997) Hungry water: effects of dams and gravel mining on river channels. Environ Manag 21:533–551

Mantuáno J (1973) Információ a Dráva és Mura folyókat érintő vízgazdálkodási elgondolásokról (Information on the water management concepts related to the Drava and Mura Rivers). Manuscript. VIZITERV, Budapest (in Hungarian)

Mantuáno J (1976) Đurđevac-Barcsi vízlépcső-rendszer (The Djurdjevac-Paks barrage system). VIZITERV D-1, 15.13/1, Budapest (in Hungarian)

Schwarz U (2007) Pilot study: hydromorphological survey and map of the Drava and Mura Rivers IADS-Report prepared by FLUVIUS, Floodplain ecology and river basin management, Vienna. 141 pp. http://iad.gs/docs/reports/HydromorphIAD_Mura_Drava2007.pdf

Tadić L, Bonacci O, Dadić T (2014) Dynamics of the Kopački Rit (Croatia) wetland floodplain water regime. Environ Earth Sci 71(8):3559–3570

Tadić L, Bonacci O, Dadić T (2016) Analysis of the Drava and Danube rivers floods in Osijek (Croatia) and possibility of their coincidence. Environ Earth Sci 75(18):1238

Wohl EE, Bledsoe BP, Jacobson RB, LeRoy Poff N, Rathburn SL, Walters DM, Wilcox AC (2015) The natural sediment regime in rivers: broadening the foundation for ecosystem management. Bioscience 65(4):358–371

Chapter 10
Channel Incision Along the Lower Drava

Alajos Burián, Gábor Horváth and László Márk

Abstract The Drava channel has not yet reached a hydromorphological equilibrium since the major human interventions two centuries ago. According to present knowledge, the fundamental reasons for prolonged channel entrenchment can be summarized in the following:

1. Among river regulation activities, the cutoff of bends should be emphasized. River length was reduced by about 40% and, as a consequence, current velocity increased. In addition, regulation structures narrowed down the channel and further accelerated flow.
2. Hydroelectric plants were built along the Upper and Middle Drava. In the reservoirs, the arriving masses of bedload and suspended load settle and this causes a sediment deficit on the lower section.
3. Channel dredging along the Hungarian Drava section (between Őrtilos and Drávaszabolcs) have also contributed to deepening of the channel. The rate of incision is not uniform all along the length of the river: upstream Barcs it is 4–6 cm y^{-1}, while downstream, at the Drávaszabolcs gauge it is only 1–2 cm y^{-1}. The average rate for the entire Hungarian section is 3 cm y^{-1}.

There are several approaches to establish the vertical displacement of the riverbed. It is estimated from water level changes and also from the shift of discharge curves in time. It is somewhat more difficult to reconstruct entrenchment from measuring the morphological alterations of channel cross-section. Channel changes have major environmental impact. The subsidence of the position of the river and the reduced frequency of high-water levels seriously influences the depth of the groundwater table in the riverine environment, including vegetation and wildlife. The water-bound plant communities and animal species are replaced by those

A. Burián (✉) · G. Horváth · L. Márk
South Transdanubian Water Management Directorate, Köztársaság tér 7, Pécs 7623, Hungary
e-mail: burian.alajos@ddvizig.hu

G. Horváth
e-mail: horvath.gabor@ddvizig.hu

L. Márk
e-mail: mark.laszlo@ddvizig.hu

D. Lóczy (ed.), *The Drava River*, Springer Geography,
https://doi.org/10.1007/978-3-319-92816-6_10

favoring drier conditions. It is rather difficult to find a solution for the problem. There are possible but expensive engineering options to reduce the gradient of the riverbed and to recharge deficient bedload, but on the long term, the consequences of human interventions should be eliminated.

Keywords River regulation · Hydroelectric plants · Gradient · Bedload
Suspended load · Cross-section · Sediment extraction · River discharge
Water level · Groundwater table

10.1 Introduction

The Drava catchment comprises a wide range of geological formations from the oldest rocks in the central Alpine ranges to the alluvia of the Lower Drava plains (see Chap. 2 in this volume). An event-based regional investigation (Merz and Blöschl 2009), however, revealed that land use, soil types and geology are not so major control on runoff as climatic elements such as mean annual precipitation and the long-term ratio of actual evaporation to precipitation. The latter influences soil saturation and flood generation (see Chap. 4).

Compared to other European rivers, the Drava shows a relatively equable water regime. The river regime is primarily controlled by precipitation fallen on the high-mountain section of the drainage basin. The low water/mean water ratio is 1–2.6, while the mean water/high water ratio is between 1 to 3.7 and 1 to 5 (Lovász 1972). Average specific runoff is 20 L s^{-1} km^{-2}. From the long-term average of monthly mean water stages, it is clear that the largest discharges are usually observed in June, when snow cover in the high mountains of the upper catchment melts.

Due to Atlantic and Mediterranean influences, annual precipitation in the Drava catchment amounts to 660–1,530 mm with an average of 990 mm. Precipitation shows two basic patterns (Auer et al. 2001):

a. Atlantic air masses create a maximum in summer, from June to August, primarily in the Lavant and Gurk basins.
b. In addition, to the effect of Mediterranean cyclones an autumn maximum (in October and November) is typical, first of all, in the Gail Valley.

Moving downstream the Drava, the secondary autumn maximum in river discharge is becoming increasingly common. The reason is a strong Mediterranean influence. As good examples, the flood waves of November 2012 and September 2014. The latter only fell a mere 10 cm below the maximum observed to date at the Drávaszabolcs river gauge.

Climatic conditions in the future (see Chap. 4) may fundamentally change the river regime of the Drava, which involves hydromorphological and far-reaching environmental consequences (Majer 1994).

10.2 Riverbed Entrenchment

The entrenchment of the riverbed or channel incision along the Hungarian Drava section, between Őrtilos (236.000 river km) and Drávaszabolcs (70.200 river km) was first described by the experts of the Scientific Institute for Water Management during the preparation of the second Hydrographic Atlas of the Drava in the first years of the 21st century. At that time, the rate of riverbed entrenchment was estimated at 3 cm y^{-1}. Experts of the South-Transdanubian Water Management Directorate and the trustee responsible for the Drava River have been investigating the entrenchment process and made suspended and bedload measurements.

Few studies have been concerned directly with the explanation of riverbed entrenchment. Therefore, only alternative theories without appropriate scientific justification can be outlined. The relative contribution of the individual processes (counted in centimeters) responsible for the entrenchment are difficult to estimate. It is an important task of the future to underpin theories and actions of prevention by detailed measurements and analyses.

10.2.1 Tectonic Processes

One of the long-term processes, which could explain the subsidence of the Drava riverbed, is regional tectonics, the deformation of the Pannonian lithosphere (Dombrádi et al. 2010). Along with other sub-basins of the Pannonian (Carpathian) Basin, the Drava trough (depression) has been continuously subsiding ever since the Early Miocene, when the formation of the Pannonian Basin began. The total thickness of Neogene and Quaternary alluvial sequences may reach 7,000 m in the centre of the depression (Velić 2007). The southward shifting of the axis of the Drava graben subsidence zone (accommodating the meander belt of the river) has been known to scientists for a long time (Lovász 1972; Bognár and Schweitzer 2003), supported by geodetic measurements (Joó 1979) and explained by the compaction of the thick sedimentary fill (see Chap. 2). The rate of ground subsidence, however, was estimated at merely 1–2 mm y^{-1} (Joó 1992), i.e. one order of magnitude lower than the rate of the Drava incision. Therefore, tectonic processes obviously do not provide a satisfactory explanation for the rate of incision of the riverbed.

10.2.2 Flow Regulation

Although it is not easy to reconstruct the paleogeographic conditions of the Drava valley centuries ago, documents from the 107 villages of the Drava Plain in the Middle Ages supply us with some (descriptive) information on the environment.

Much later, the map sheets of the Military Survey (1784) depict the Drava Plain as a swamp area with interwoven with small watercourses and spotted with forests. The river channel itself used to be highly different from that what we see today. Multithread meandering was the typical river mechanism at that time. This pattern was highly dynamic with the continuous emergence and cutoff of meanders.

Flow regulation with the purpose of flood control began in the end of the 18th century (see Chap. 8). Between 1784 and 1848 62 artificial cutoffs were completed downstream the Mura confluence. These interventions reduced river length by 75 km, which meant ca 40% reduction and resulted in increased stream power and channel incision. The section first affected was between Drávagárdony and Felsőszentmárton and in the 19th century huge meanders were cut off between Drávasztára and Szaporca, too. This approach to regulation was practiced until 1885.

Regulation using stone and brushwood faggots–meant to be final solution to flood and navigation problems–started in the lowermost section between the confluence and Eszék (now: Osijek, Croatia) in 1887. Until World War I the regulation of the sections around Barcs, between Tótújfalu and Budakovac as well as between Zaláta and Kisszentmárton had been completed. In the interwar period, only maintenance operations were carried out. Following a treaty with Yugoslavia regulation works could continue, using stone and also involving some cutoffs.

Regulation has largely modified channel gradient, which is now 0.002 on the mountain section, 0.0005–0.00045 from the Mura confluence to Vízvár, 0.00025–0.0002 to Barcs, while only 0.00012–0.00007 at Drávaszabolcs. In parallel, the grain size of the bed material changes from gravels (16–52 mm diameter) at Őrtilos, to sandy gravel (20–0.5 mm) at Barcs and fine to medium sands in 20–30 m thickness at Drávaszabolcs.

A marked knickpoint in the longitudinal profile is observed at Vízvár (190 rkm). This could indicate the upper limit of regressive bed deepening due to the increased current velocity of the regulated river.

10.2.3 *Flow Impoundment on Upper Sections*

Although water levels show a clear decreasing trend over the recent decades, hardly any trend is observed in river discharges. Consequently, large-scale changes of the hydrological system in the Drava drainage basin cannot be evidenced. Since much of the bedload and a large part of the suspended load are retained by the 22 hydroelectric plants on the Austrian, Slovenian and Croatian river sections (see Chaps. 7 and 9), the majority of sediment load in the lower Drava channel derives from the Mura River. Out of the hydroelectric plants, the lowermost plant in Croatia, at Dubrava (254.000 rkm) has major impact on the hydromorphology of the Lower Drava. This is a peaking plant with considerable water (and energy) storage in its reservoir (Lake Dubrava, area: 17.1 km^2, the largest in Croatia; storage capacity: ca 100 million m^3). Peak-time operation results in diurnal cycles of water level fluctuation, amounting to maxima of 120 cm and average ranges of 60–80 cm at Őrtilos and 30–40 cm at Paks.

The reservoirs store large amounts of suspended and bedload and, thus, create significant sediment deficit on the downstream reaches and induce 'clearwater erosion' there (Rollet et al. 2013).

10.2.4 Gravel Mining from Riverbed

The gravel and sand accumulations in the riverbed are common sources of building materials. Part of the local sediment surplus in the Drava channel is mined and this also contributes to sediment deficit downstream. At the end of the 20th century dredging affected an estimated amount of 500,000–700,000 t y^{-1} in Croatia and 160,000–180,000 t y^{-1} in Hungary (Horváth 2002). Calculated for the period 1982–2011, Croatian actors were responsible for the removal of 6,292,000 m^3 extracted material and Croatians for 2,658,000 m^3 on the joint Croatian-Hungarian section. This means the sediment deficit was concentrated on Croatian state territory.

10.3 Sediment Budget of the Hungarian Section

Regular suspended sediment measurements on the Drava began in 1961 and bedload measurements in 1969 at Barcs and Drávaszabolcs on 4–6 occasions per year. In 1998 measurements started at Botovo and Bélavár, too (Fig. 10.1). At present, the frequency of sediment is conducted parallel with water discharge gauging on 8–10 occasions annually (see Chap. 9).

The recharge of coarse bedload is a crucial factor of channel incision (Simon and Rinaldi 2006). In addition to the water regime, the bedload transport of the Drava River is largely influenced by the operation of upstream hydroelectric plants, first of all, the plant built at Dubrava in 1989. The transported bedload declines between Botovo and Barcs and remarkably grows from Barcs to Drávaszabolcs (Table 10.1).

Suspended load transport represents a larger volume and less dependent on the operation of hydroelectric plants. It derives from the erosion of the riverbed and banks. Some suspended material is not trapped in the reservoirs but can pass through the turbines. The trends of suspended load are similar to those of bedload (Table 10.2).

Fig. 10.1 River gauging stations on the Lower Drava River. Red dots: Hungarian (DDVÍZIG) stations; yellow dots: Croatian (Hrvatske vode) stations

Table 10.1 Bedload transport along the Lower Drava River (after Bonacci and Oskoruš 2010)

Measured section		Annual average bedload transport (t y^{-1})		Water discharge/bedload transport relationship (g s^{-1})
Site	River, km	1986–2003	2004–2011	
Botovo	227.200	109,038	24,452	$G_g = 4.5279 \times 10^{-10} \times Q^{4,552}$
Bélavár	198.700	37,261	n.a.	$G_g = 3.8113 \times 10^{-11} \times Q^{4,755}$
Barcs	152.700	77,909	22,692	$G_g = 8.7286 \times 10^{-10} \times Q^{4,407}$
Drávaszabolcs	77.700	194,024	71,845	$G_g = 3.1623 \times 10^{-8} \times Q^4$

G_g bedload yield; Q water discharge

Table 10.2 Suspended sediment transport along the Lower Drava River (after Bonacci and Oskoruš 2010)

Measured section		Annual average suspended sediment transport (t y^{-1})		Water discharge/ suspended sediment transport relationship (kg s^{-1})	Water discharge/ suspended sediment concentration relationship (g s^{-1})
Site	River, km	1986–2003	2004–2011		
Botovo	227.200	578,173	313,270	n.a.	$C_k = 136.01 \times Q^{0.3679}$
Bélavár	198.700	533,128	n.a.	n.a.	$C_k = 226.43 \times Q^{0.2818}$
Barcs	152.700	457,161	649,840	$G_L = 256,42 \times G_L^{0.2906}$	$C_k = 196.39 \times Q^{0.3191}$
Drávaszabolcs	77.700	583,573	599,755	$G_L = 220,78 \times G_L^{0.3321}$	$C_k = 169.81 \times Q^{0.3316}$

G_L suspended sediment yield; C_k suspended sediment concentration; Q water discharge

Table 10.3 Gauging stations for the determination of the rate of riverbed entrenchment

Gauging station					
Name	River, km	Base level (m above Baltic level)	Operating organization	Year of installment	Observations
Őrtilos	235.900	125.94	DDVÍZIG	1957	H, Q, T
Heresznye-Vízvár	187.590	101.195	DDVÍZIG	2012	H, Q, T
Barcs	153.800	98.14	DDVÍZIG	1976	H, Q, T, Ck
Szentborbás	133.100	94.74	DDVÍZIG	1934	H, Q, T
Drávaszabolcs	77.700	86.75	DDVÍZIG	1935	H, Q, T, Ck

H water level, *Q* water discharge, *T* temperature, *Ck* suspended load concentration

10.4 Trends in the Markers of Riverbed Entrenchment

For determining the vertical displacement of the Drava bed, data pointing to riverbed entrenchment from the gauging stations in Hungary were analyzed (Table 10.3).

10.4.1 Changes in Water Level

The analyzed time series (47 year, 1970–2016) reflects the impact of the three hydroelectric plants on the Croatian section (Varaždin at 302 river km, built in 1975; Čakovec at 278 river km, built in 1982 and Dubrava at 254 river km, built in 1989) (Figs. 10.2, 10.3, 10.4, and 10.5).

The changes in low and medium water stages are most marked at Őrtilos, where a 1.5–2 m drop (3–4 cm y^{-1}) was observed over the studied period (Fig. 10.2), while in high water levels 0.4 m decrease was found. The difference is explained by the water spread out over the floodplain. The linear trend of low and medium water levels at Barcs shows 1.3–1.5 m (2–3 cm y^{-1}), while that of high water levels 1 m lowering (Fig. 10.3). At the Szentborbás station the corresponding values are 0.65–0.75 m (low and medium water levels, 1.5 cm y^{-1}) and 0.6 m (high water levels) (Fig. 10.4) and at Drávaszabolcs 0.4 m and 0.2 m, respectively (Fig. 10.5) with only 1 cm y^{-1} annual lowering.

In summary, it is claimed that the reduction of low and medium water levels is more marked at all stations than that of the more fluctuating high water levels.

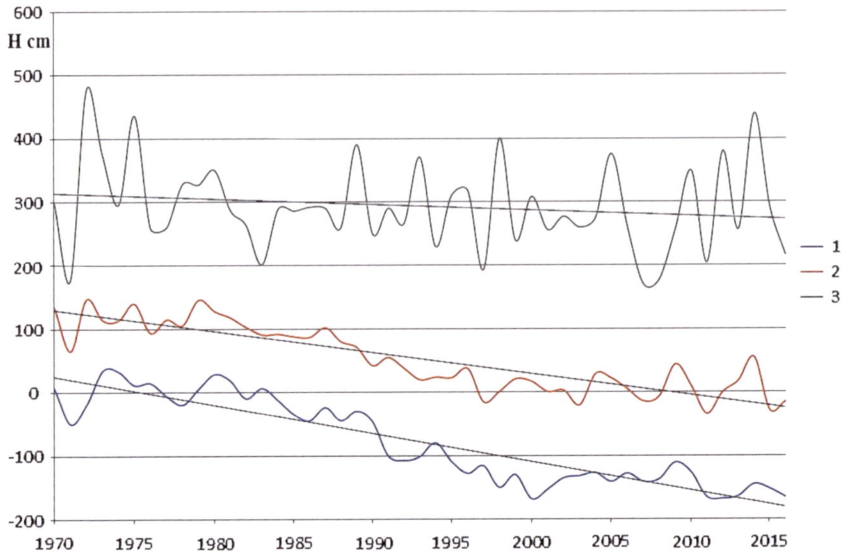

Fig. 10.2 Annual maximum, mean and minimum water levels at the Őrtilos gauging station, 1970–2016. 1, low water stages (cm); 2, medium water stages (cm); 3, high water stages (cm)

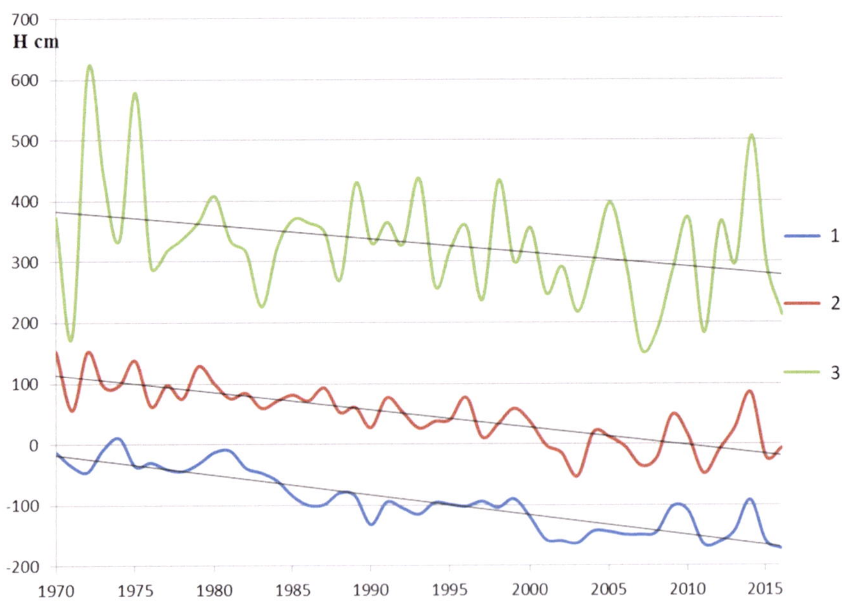

Fig. 10.3 Annual maximum, mean and minimum water levels at the Barcs gauging station, 1970–2016. For legend see Fig. 10.2

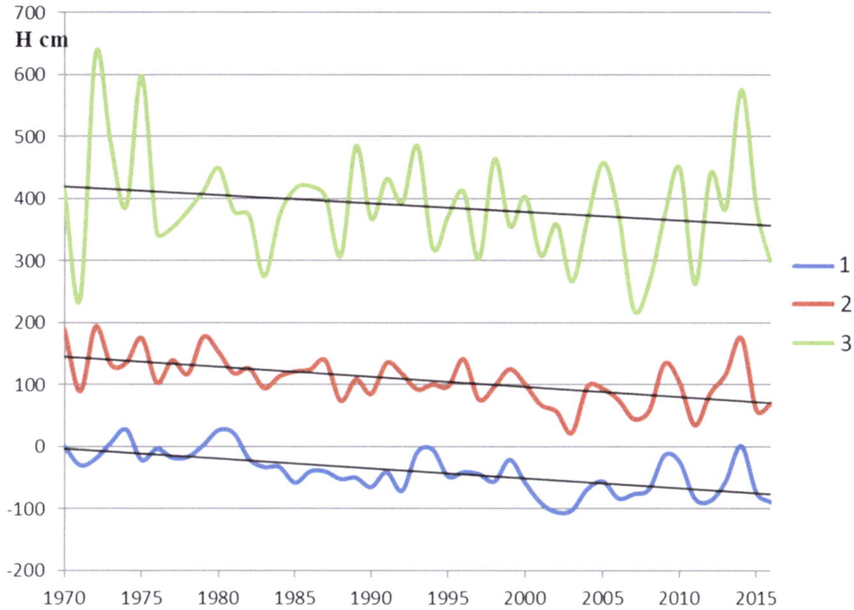

Fig. 10.4 Annual maximum, mean and minimum water levels at the Szentborbás gauging station, 1970–2016. For legend see Fig. 10.2

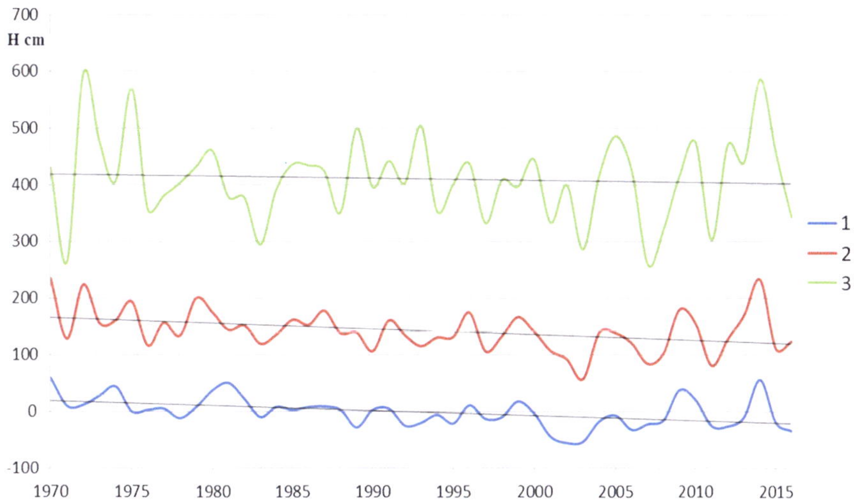

Fig. 10.5 Annual maximum, mean and minimum water levels at the Drávaszabolcs gauging station, 1970–2016. For legend see Fig. 10.2

10.4.2 Changes in Water Discharge

The length of the time series only allows the analysis of discharge trends in the case of two gauging stations, Barcs (observations since 1970) and Drávaszabolcs (since 1970) At Őrtilos and Szentborbás measurements only started in 2004. Thus, we cannot rely on their data.

The Barcs data are not unequivocal: minimum and medium water discharges show minimal growth and maximum water discharges a major decrease by ca 400 $m^3 s^{-1}$. It is to be noted that the large floods of the 1970s (in 1972 and 1975) have a decisive influence on the trend, which is compensated by the recent flood in 2014 (Fig. 10.6). At Drávaszabolcs the minor rise in low and medium discharges is negligible, but high water discharges were reduced by 100 $m^3 s^{-1}$, also influenced by the above mentioned flood events (Fig. 10.7).

In summary, water discharges on the Őrtilos-Drávaszabolcs section have not been modified over the 47 years of study, but high water discharges have dropped by 400 to 100 $m^3 s^{-1}$, more remarkably on the upper reaches.

10.4.3 Changes in Stage/Discharge Rating Curves

In recent decades, the stage/discharge curves tended to shift to the right at both Barcs and Drávaszabolcs (Figs. 10.8 and 10.9). It means that ever lower water levels corresponded to the same water discharge. In the cross-section at Barcs, the extent of the right shift was 250 $m^3 s^{-1}$ (100 cm drop in water level) for low stages, 500 $m^3 s^{-1}$ (120 cm drop) for medium water stages and 2,000 $m^3 s^{-1}$ (50 cm drop)

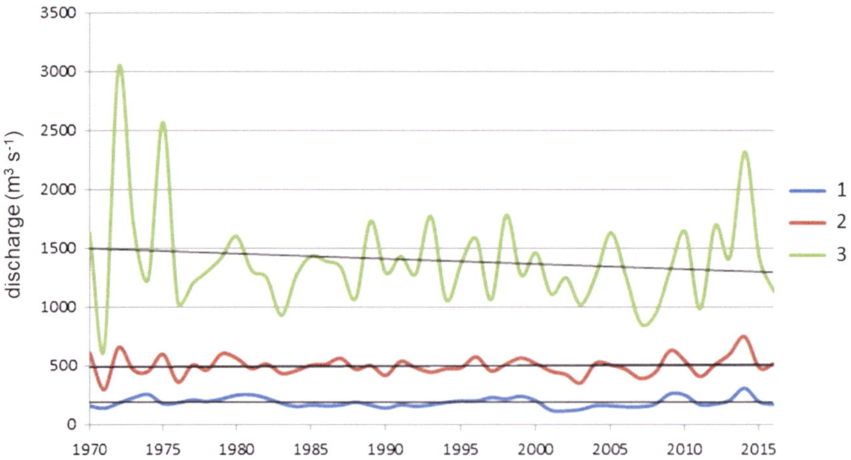

Fig. 10.6 Annual maximum, mean and minimum discharges at Barcs, 1970–2016. 1, low water discharges ($m^3 s^{-1}$); 2, medium water discharges ($m^3 s^{-1}$); 3, high water discharges ($m^3 s^{-1}$)

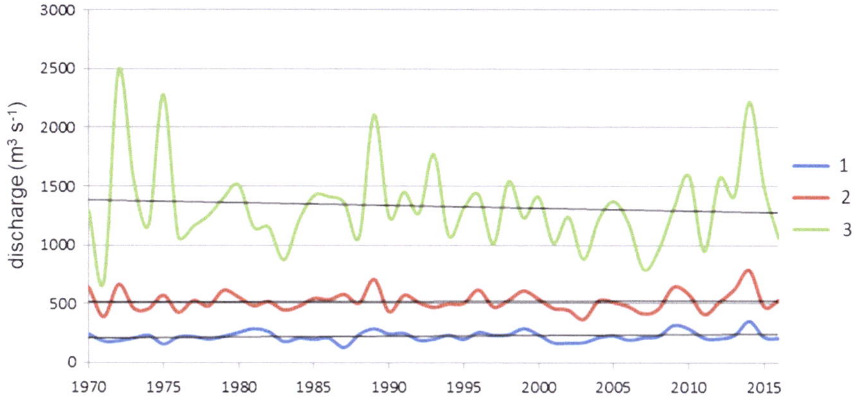

Fig. 10.7 Annual maximum, mean and minimum discharges at Drávaszabolcs, 1970–2016. For legend see Fig. 10.6

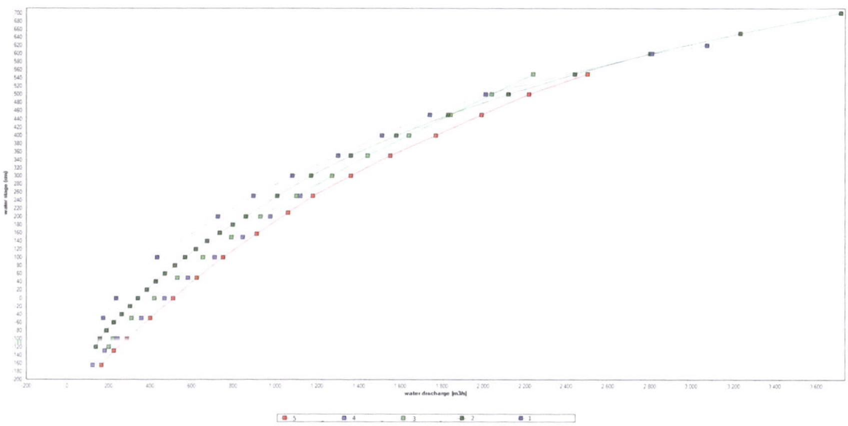

Fig. 10.8 Water discharge/water level relationships at Barcs. 1, curve #3: 1972–1974; 2, curve #13: 1988–1992; 3, curve #18: 1998–2000; 4, curve #21: 2003–2007; 5, curve#24: 2011–2017

for high water stages (Fig. 10.8). The respective ranges of values for the Drávaszabolcs station were 50, 50–60 and 0–10 cm (Fig. 10.9).

10.4.4 Changes in Cross-Section

The morphological changes of a river channel over time are best demonstrated by the repeated surveying of marker points along a fixed profile. Here the Barcs

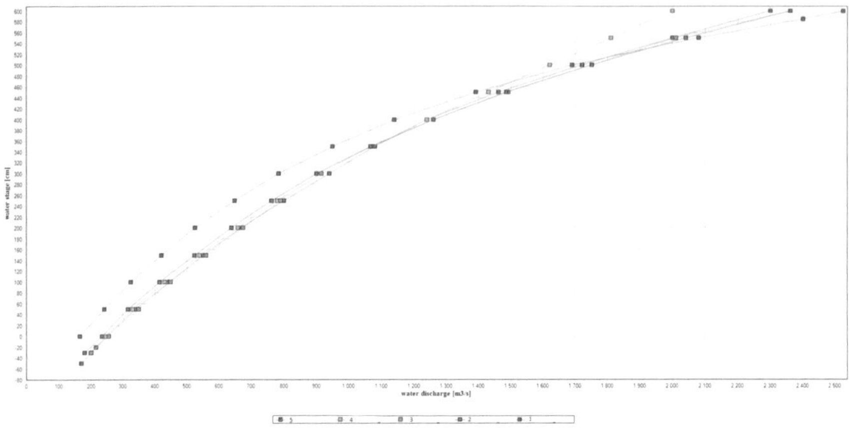

Fig. 10.9 Water discharge/water level relationships at Drávaszabolcs. 1, curve #2: 1971–1974; 2, curve #11: 1988–1993; 3, curve #15: 1997; 4, curve #18: 2000; 5, curve #23: 2009

cross-section is selected for analysis. Clear riverbed subsidence of varying rate is found for about two-thirds of the profile since 1970. The maximum value is 1.5 m. In correspondence with the observations of changes in water level (Sect. 10.4.1), this results in 3.5 cm incision in the most rapidly subsiding part of the cross-section (Fig. 10.10).

10.5 Discussion

The above presented analyses of data provided clear evidence for riverbed entrenchment on the Őrtilos–Drávaszabolcs section of the Drava River. The findings from various parameters and for the gauging stations are summarized in Table 10.4.

10.6 Impact of Damming on River Ice

Through modifying current velocity, the winter freezing conditions of rivers may also influence the rate of channel incision (Shen and Yapa 1986). Ice conditions on the Drava are presented on the basis of observations conducted–with shorter interruptions–at Barcs and Drávaszabolcs since the 1930s. The ice conditions at Drávaszabolcs altered significantly in 1975, when the first Croatian hydroelectric plant on the Drava was installed. Between 1936 and 1975 frozen river was observed on 18 occasions, drifting ice on three occasions and ice on banks on four occasions.

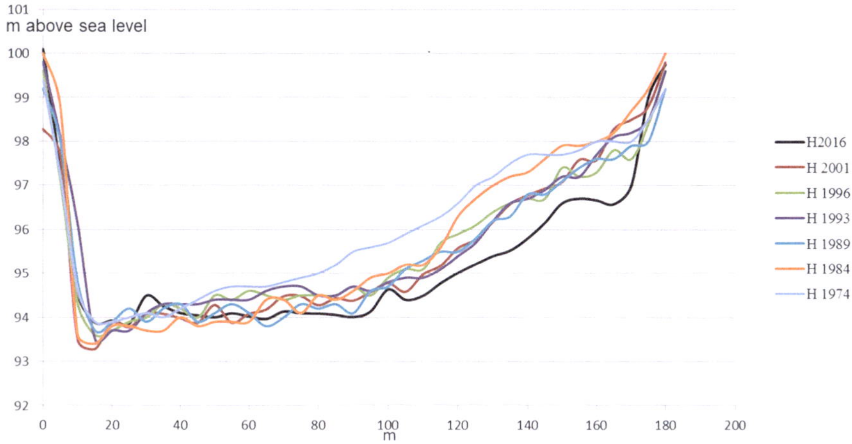

Fig. 10.10 Changes in the riverbed cross-section at Barcs, 1974–2016. 1, 2016; 2, 2001; 3, 1996; 4, 1993; 5, 1989; 6, 1984; 7, 1974

Table 10.4 Rates of channel incision reconstructed from various parameters for the period 1970–2016

Gauging station		Channel incision (cm y^{-1}) based on				
		Changes in water level		Changes in water discharge		Changes in cross-section
Name	River, km	Low and medium waters	High waters	Low and medium discharges	High discharges	
Őrtilos	235.900	3–4	1			
Barcs	153.800	2–3	2	2–2.5	1	3.5 cm
Szentborbás	133.100	1.5	1.5			
Drávaszabolcs	77.700	1	<1	1	<0.5	

After 1975 ice occurrence was drastically reduced (Lovász 2016). In the Barcs cross-section, the river was frozen over on 26 occasions between 1936 and 1975, in 15 years drifting ice was observed and the number of ice-free years was only four. After 1975, the river froze over in only 2 years and ice drifted in 9 years. Probably to the impact of global warming and increased current flow during medium water stages (caused by river regulation), ice formation was often inhibited and for 30 years no ice cover was detected on this river section (Fig. 10.11).

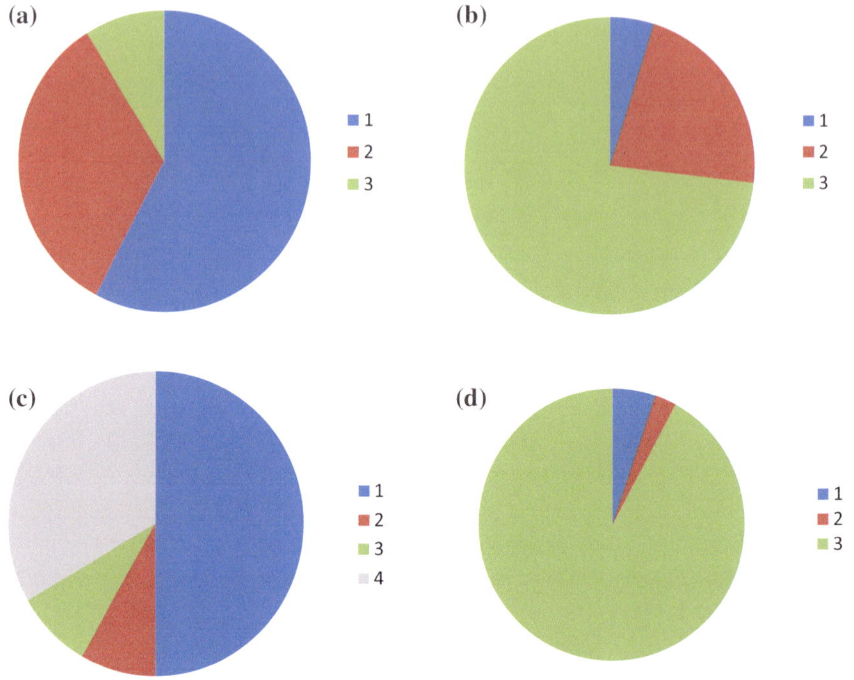

Fig. 10.11 Ice conditions (percentage of years in the given period) on the Drava at Barcs (A and B) and at Drávaszabolcs (C and D). A. 1936–1975; B. 1976–2016; C. 1936–1975; D. 1976–2016. 1, frozen river; 2, drifting ice; 3, ice on banks; 4, no ice

10.7 Environmental Impact of Channel Incision

Human impact is detectable in all directions of change in channel morphology: vertical (channel incision), horizontal (bank erosion, meander formation) and longitudinal (shifting knickpoints, channel regression) alterations. The environmental impact covers both channel and floodplain modification (Timár and Telbisz 2005). In this chapter, the direct influences on geometrical parameters and groundwater conditions are presented, but, naturally, there are far-reaching effects on the physical environment, also involving changes in vegetation and animal life.

Riverbed entrenchment reduces the duration of water stages related to the absolute level on the river gauge. The process is demonstrated through the chart of water level exceedance curves at the Barcs and Drávaszabolcs gauging stations for the period 1950–2016 (Figs. 10.12 and 10.13).

The charts show time series split up into two: one for 1950–1983 and the other for 1984–2016. The chart of 50% duration shows about 60 cm decrease of water stage, i.e. of riverbed level, for Drávaszabolcs and about 100 cm for Barcs over the

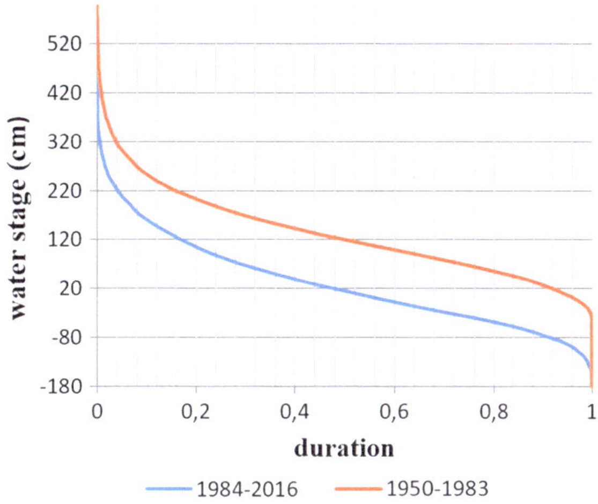

Fig. 10.12 Water level exceedance curves at Barcs, 1950–1984 and 1984–2016

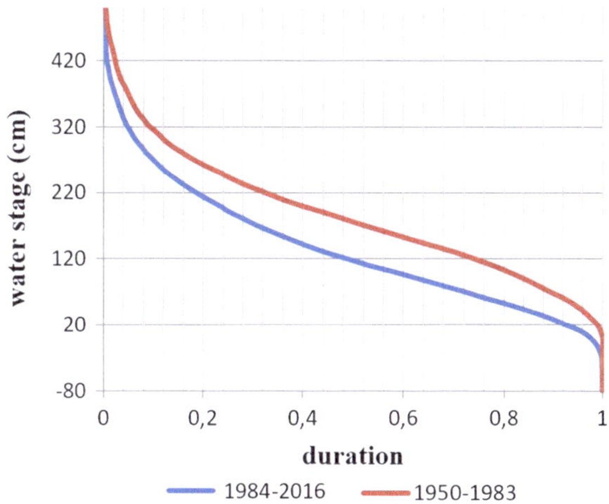

Fig. 10.13 Water level exceedance curves at Drávaszabolcs, 1950–1984 and 1984–2016

67-year period. This is, however, only a rough indication of river incision since other processes not treated here may also be influential.

An observation well which shows typical groundwater conditions is located at Lakócsa village, at some kilometers distance from the Drava River (Fig. 10.14). The analysis presents a slightly declining trend for the 1980–2016 period. With regard to the close communication between the Drava channel and groundwater, this trend allows the conclusion that the riverbed is entrenching. Since observations took place twice a week before 2010 and hourly reading after 2010, the sections of the chart in Fig. 10.14 before and after 2010 are different in detail.

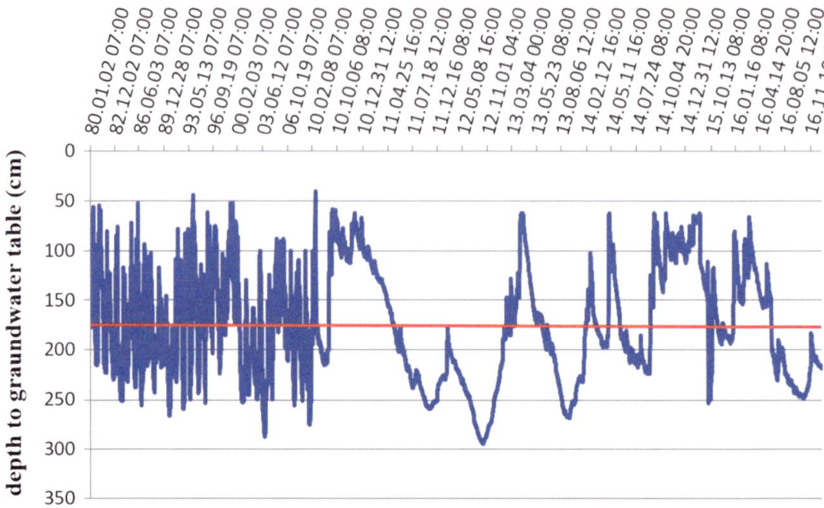

Fig. 10.14 Groundwater levels at the observation well of Lakócsa, 1980–2016

10.8 Conclusion: Possible Solutions

At first, river regulation increased the channel gradient of the Drava River, increased current velocity and induced channel deepening. Later, the construction of hydroelectric plants on the upper and middle sections and sediment mining from the channel generated a sediment deficit, which further intensified riverbed entrenchment. The interplay of the three human interventions results in channel incision of about 3 cm y^{-1} rate on the Őrtilos–Drávaszabolcs river section. This is a detrimental process to the physical environment.

To plan preventive engineering action against riverbed entrenchment, detailed numerical data on the process should be available. Natural and anthropogenic components of incision should be separated and characterized quantitatively. It is also questionable whether the consequences of human interventions could be eliminated by another anthropogenic effect, which may also start yet unknown detrimental processes. This may sound a philosophical question, but before venturing on solving the problem, it has to be answered. Although the engineering response is not too complicated (reducing the curve through establishing an equilibrium in sediment budget), the involvement of all stakeholders is necessary if we want to arrive at a solution to the satisfaction of (almost) everybody affected.

References

Auer I, Böhm R, Schöner W (2001) Austrian long-term climate 1767–2000. Österreichische Beiträge zu Meteorologie und Geophysik 25. Zentralanstalt für Meteorologie und Geodynamik (ZAMG), Wien

Bognár A, Schweitzer F (2003) Geomorphological and tectonical problems of the Drava River Plain in Croatia (right bank). In: Halasi-Kun GJ (ed) Pollution and water resources. Columbia University Seminar Proceedings (1999–2003), vol 33–34. Studies of Environmental Protection in the Carpathian Basin. Pécs, Hungarian Academy of Sciences, Department of Earth Sciences, pp 339–351

Bonacci O, Oskoruš D (2010) The changes in the lower Drava River water level, discharge and suspended sediment regime. Environ Earth Sci 59:1661–1670. https://doi.org/10.1007/s12665-009-0148-8

Dombrádi E, Sokoutis D, Bada G, Cloetingh S, Horváth F (2010) Modelling recent deformation of the Pannonian lithosphere: lithospheric folding and tectonic topography. Tectonophysics 484:103–118

Horváth G (2002) Hydrological analysis of the Hungarian-Croatian section of the Dráva. In: "Az EU VKI bevezetése a Dráva vízgyűjtőjén" conference abstract, WWF, Pécs (in Hungarian)

Joó I (Editor-in-Chief) (1979) Map of recent vertical crustal movements in the Carpatho-Balkan Region. Scale 1:1 000 000. Cartographia, Budapest

Joó I (1992) Recent vertical surface movements in the Carpathian Basin. Tectonophysics 202:129–134

Kontur I, Koris K, Winter J (1993) Hidrológiai számítások (Hydrological calculations). Akadémiai Kiadó, Budapest (in Hungarian)

Lovász Gy (1972) A Dráva-Mura vízrendszer vízjárási és lefolyási viszonyai (River regime and runoff in the Drava-Mura water system). Akadémiai Kiadó, Budapest. 158 pp (in Hungarian)

Lovász Gy (2016) The impact of climate change on the ice conditions of rivers in Hungary, 1901–2000. Unpublished manuscript, Pécs

Majer J (1994) Az ökológia alapjai (Fundamentals of ecology). Szaktudás Kiadó, Budapest (in Hungarian)

Merz R, Blöschl G (2009) A regional analysis of event runoff coefficients with respect to climate and catchment characteristics in Austria. Water Resour Res 45:W01405. https://doi.org/10.1029/2008WR007163

Remenyik B (2005) Adatok a Dráva-szabályozás történetéből (Data on the history of Drava regulation). Hidrológiai Közlöny 85(3):27–30

Rollet AJ, Piégay Dufour S, Bornette G, Persat H (2014) Assessment of consequences of sediment deficit on a gravel river bed downstream of dams in restoration perspectives: application of multicriteria, hierarchical and spatially explicit diagnosis. River Res Appl 30(8):939–953. https://doi.org/10.1002/rra.2689

Shen HT, Yapa PD (1986) Flow resistance of river ice cover. J Hydraul Eng 112(2):142–156. https://doi.org/10.1061/(asce)0733 9429(1986)112:2(142)

Simon A, Rinaldi M (2006) Disturbance, stream incision, and channel evolution: the roles of excess transport capacity and boundary materials in controlling channel response. Geomorphology 79:361–383

Stelczer K (2000) A vízkészlet-gazdálkodás hidrológiai alapjai (Hydrological foundations of water resources management). ELTE Eötvös Kiadó, Budapest (in Hungarian)

Szlávik L, Sziebert J, Tamás EA (2007–2013) A Dráva morfológiai monitoringja. Hordalékvizsgálat (Morphological monitoring of the Drava: sediment load analyses). Hungary-Croatia IPA Cross-border Co-operation Programme, Institute of Hydrological Construction and Water Management, Faculty of Engineering and Economics, Eötvös József College, Baja (in Hungarian)

Timár G, Telbisz TA (2005) A meanderező folyók medrerváltozása és az alakváltozás sebessége (Channel change rate for meandering rivers). Hidrológiai Közlöny 85(5):48–54 (in Hungarian)

Velić (2007) Geologija ležišta nafte i plina (Geology of oil and gas reserves). Udžbenik, University of Zagreb. 342 pp (in Croatian)

VITUKI (2003) A Dráva hordalékjárásának vizsgálata a legfrissebb adatok figyelembevételével (Most recent data on the sediment regime of the Drava). Summary report of Project no 714/1/6128-01. Project leader: János Szekeres (in Hungarian)

Zsuffa I (1996) Műszaki hidrológia (Engineering hydrology). Műegyetemi Kiadó, Budapest (in Hungarian)

Chapter 11
Evolution of the Drava Floodplain in Hungary in the Last 100 Years

Tímea Kiss and Gábor Andrási

Abstract The channel and floodplain development of the Drava is highly influenced by the operation of reservoirs and hydroelectric power stations established on upstream sections. As a result, the typical overbank floodplain processes terminated and almost totally replaced by horizontal accretion. During the last approximately 150 years the floodplain area increased by several thousand hectares simultaneously with the loss of channel area. The side channels gradually lose their water conductivity. Therefore, the islands are merging into the floodplain. Meander development has also accelerated as the stability of the eroding banks decreases due to the drop of stages. Therefore, the rate of bank erosion increases and, simultaneously, point-bar development intensifies, leading to accelerated lateral aggradation.

Keywords Meander development · Bank erosion · Point-bar development
Missing floods · Lateral floodplain accretion

11.1 Introduction

The development of floodplains is determined by two basic processes: lateral and vertical accretion. Lateral aggradation is mostly connected to the continuous horizontal displacement of meanders (Nanson and Croke 1992; Brierley and Fryirs 2005), when point-bars develop simultaneously with the erosion of the concave banks. Lateral aggradation also appears in case of channel narrowing, as a consequence of hydrological changes (Kiss and Blanka 2012). The vertical aggradation of floodplains takes place during overbank flow, when rivers transport greater

T. Kiss (✉) · G. Andrási
Department of Physical Geography and Geoinformatics, University of Szeged,
Egyetem u. 2-6, Szeged 6722, Hungary
e-mail: kisstimi@gmail.com

G. Andrási
e-mail: andgab86@gmail.com

D. Lóczy (ed.), *The Drava River*, Springer Geography,
https://doi.org/10.1007/978-3-319-92816-6_11

volume of sediment load. Overbank aggradation is triggered by the high friction of floodplains (due to decreasing height of water column, higher roughness, dense vegetation), therefore the velocity of the floods drops and the sediment is deposited (Wolfert et al. 2002). These aggradational processes can coexist, their ratio, however, depends on the energy conditions of the river, and the discharge and grain-size of the transported material (Nanson 1986; Nanson and Croke 1992; Brierley and Hickin 1992; Zwoliński 1992). Chorley et al. (1985) identified a third, intermediate way of floodplain aggradation, in connection with oxbow lake accumulation and island development (merging into the banks).

These processes result in characteristic floodplain landform assemblages (Nanson and Croke 1992). Lateral accretion is indicated by point-bars, swales, oxbows and their sediment plugs. Vertical accretion builds natural levees and crevasses near the banks, but the farther forms (e.g. oxbows) and the entire floodplain will be buried by sediments, therefore they slowly disappear (Lóczy 2013). On those floodplains where island and oxbow development dominates, the landform assemblage is characterized by side channels and oxbow lakes filled up to a different extent, and old island surfaces merged into the banks.

Although lateral accretion can take place at any stage, even during low-stage or flood periods, vertical aggradation is exclusively associated with floods. Thus, the development of the above mentioned floodplain features is highly dependent on the regime of the river (Fryirs and Brierley 2013; Lóczy 2013). Such processes usually dominate on the wide lowland floodplains, though on narrow floodplains in valleys landforms due to slope processes also appear, but they can be destroyed by the river or slope wash (Dietrich et al. 1986).

The rate of vertical and lateral aggradation and the ratio between them can vary spatially and temporally. Several factors influence them, most importantly, the quality and the quantity of water and sediment transport. In general, on a catchment scale it is determined by all those processes which influence runoff (e.g. climate change, land-use change), the erosion of slopes and the transport of the eroded material. However, reach-scale processes determine the amount and location of floodplain sedimentation, by influencing the hydraulics of overbank flow (Nicholas and Walling 1997). Such local factors are microrelief and the landform assemblage of the floodplain (Oroszi 2009), distance from the channel (Walling and He 1998), sediment load and flow conditions of the flood (Sándor 2011), vegetation cover of the floodplain (Steiger et al. 2001; Kiss and Sándor 2009), and floodplain width (Gábris et al. 2002).

Floodplain development can be highly influenced by human impact. In general, the floodplain development rate may increase by at least one order of magnitude due to human impact (e.g. Knox 1987; Florsheim and Mount 2003; Benedetti 2003). It could be explained by human activities on the catchment which increase the amount of transported material, for example, mining (Knox 2006), land-use change (Florsheim and Mount 2003), intensive agriculture (Mücher et al. 1990; Owens and Walling 2002; Lecce and Pavlowsky 2004; Knox 2006), forest clearance (Constantine et al. 2005) and human impact on the channel itself (Hohensinner et al. 2004; Owens et al. 2005). Floodplain aggradation can also be accelerated by creating narrow artificial floodplains (Kiss et al. 2011), impeding the lateral movement of the channel (Károlyi 1960; Brown 1983), engineering works and river regulations (Ten Brinke et al. 1998). However, the rate of floodplain aggradation can decrease too by sustainable land use and protection of soils (Knox 1987; Benedetti 2003), afforestation (Keesstra 2007) and channel incision (Wyżga 2001).

The hydrology and morphology of the Drava River have altered significantly during the last ca one and a half century. Therefore, the aim of this paper is to present the short-term (ca 150-year) floodplain evolution along the lowland section of the Drava River (downstream of the Mura confluence) and to determine its main influencing factors by applying geoinformatical and dendro-geomorphological methods.

11.2 Main Characteristics of the Drava Floodplain

The 10–45-km-wide floodplain if the Dráva stretches from NW to SE direction along the river, and it is getting wider toward the confluence with the Danube (Fig. 11.1). At some locations, the floodplain is bordered by a 10–20-m-high escarpment (locally, e.g. between Berzence and Bélavár, almost 30 m high), which is actively eroded by the river at several sites. The broader floodplain areas are characterized by large number of abandoned channel fragments and oxbow lakes (Lóczy 2001). Today the floodplain is artificially confined by flood-protection levees. Thus, active fluvial processes could only operate on the 1–10-km-wide artificially confined floodplain.

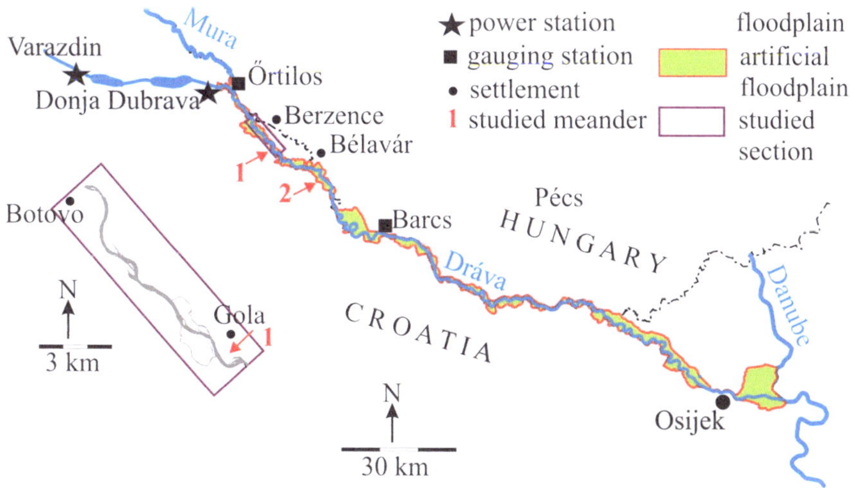

Fig. 11.1 The natural and artificially confined floodplains of the Drava and the location of the studied section and meanders (1, Gola; 2, Heresznye)

Structurally, the floodplain is a series of subsiding basins, which were filled up by the Danube and its tributaries until the early Quaternary (Lóczy 2001). However, due to tectonic movements in Transdanubia (Hungary), only small streams reach the river from north ever since (Somogyi 1967). Later, terraces were formed as the result of Late Pleistocene incision of the Drava. However, they have been mostly destroyed by the intensive lateral erosion of the river, or subsided and covered by younger fluvial deposits (Marosi 1970). The crenulated edge of the escarpments and the paleo-channels at their foot point to intensive Quaternary floodplain widening by the Drava. The lowermost layers of the escarpments are usually coarse-grained sand and gravel, covered by loess and aeolian sand (Marosi 1970). Where the river is actively eroding the slopes of the escarpment, they are very steep. Mass movements also contribute to their development. As soon as the river abandons the area, however, the slope angle of escarpments is getting ever gentler due to slope wash.

11.3 Study Areas

The horizontal floodplain processes were studied in detail along the lower, 236-km-long section of the Drava (Andrási 2015) and in five meanders (Kiss and Andrási 2015; Kiss and Balogh 2015). In the present paper only shorter, 13 km-long section between Botovo and Gola (215.4–202.2 river km) will be analyzed in detail (Fig. 11.1), and the meander development of two bends at Gola (215–216 rkm) and Heresznye (187–188 rkm). The evolution of the channel and the floodplain is highly influenced by the Donja Dubrava Hydroelectric Power Plant, which generates ‘mini flood waves’ daily. The height of these flood waves is 1.0–1.2 m at Gola and reducing downstream to only 0.6–0.8 m at Heresznye. At Gola the 1.48-km-long eroding bank is only 2–3 m high, whilst at Heresznye the upstream part of the bank (along 520 m length) is 20–22 m high and downstream (along 850 m) only 3.0–3.5 m high.

11.4 Methods

Vertical floodplain aggradation is basically determined by floods. Therefore, the water levels (1901–2014) at the Barcs gauging station (154.1 rkm) were analyzed in detail. Here bankfull level is at 420 cm stage and floods develop above this level. The annual number of overbank flood days, the yearly number of flood waves and the return period of floods were calculated.

To analyze the long-term horizontal floodplain development by meander migration, channel narrowing and island development, a series of maps (Third Military Survey, 1882; Hydrological Atlas of the Drava, 1968; Croatian topographical maps from 1979, 1982, 2003 and 2006), Google Earth images (2007) and a Croatian aerial photo (2011) were geo-referenced. On all sources the banklines, the islands and bars we digitized under ArcGIS 10.1.

Bank erosion and point-bar development in the selected meanders were evaluated applying dendro-geomorphology and geodetic surveys. The eroding banklines were annually resurveyed between 2011 and 2015 by Topcon HiPer Pro RTK GPS. The amount of eroded material (in m^3 y^{-1}) was calculated by multiplying the eroded area with mean bank height. On the studied point bars geomorphological maps were drawn, and along cross-sectional profiles their height conditions were determined by RTK GPS. The spatial and temporal evolution of the point-bar series were analyzed applying dendro-geomorphology, determining the age of trees colonizing the surfaces. Through combining the results of dendrology, geomorphological mapping and GPS measurements, isochrone-maps were created, thus the aggradational periods could be determined. Applying this method, the minimum age of the surface was determined, and the spatial and temporal characteristics of meander development (see Everitt 1968), thus lateral floodplain evolution could be analyzed.

11.5 Hydrological Background to Floodplain Development

11.5.1 Slope and Flow Velocity

The development of floodplains is highly dependent on river slope and stream power. The Drava has a relatively high slope (0.00021–0.00005) upstream of the Mura confluence, while on the lowland section it drops gradually (0.00004–0.000005) towards the confluence with the Danube (Mantuáno 1974). Accordingly, at Őrtilos flow velocity is 1.0–1.8 m s^{-1}, which decreases to 0.5 m s^{-1} on the way to the Danube (Schwarz and Bloesch 2004). Considering the Hjulström diagram, it means that during overbank flows the chance for overbank deposition from suspended sediment (clay, silt) is quite limited, it is only possible in the neighbourhood of the Drava confluence with the Danube or in standing water bodies.

11.5.2 Sediment Transport

At the Őrtilos gauging station the volume of the bedload transport is 116,000 t y^{-1} (Horváth 2002). In the upstream part of the study area bedload is dominated by gravel (≤ 6 cm diameter), towards Barcs it is getting finer (sandy gravel), and downstream of Barcs sand fraction dominates and gravels disappear (Somogyi 1967; Mantuáno 1974; Horváth 2002). From the point of view of overbank accumulation, the amount of suspended sediment load is important. Its discharge at

Őrtilos is only 0.48–0.65 million t y^{-1}. (In comparison, the Tisza River transports 30 times more [18.7 million t y^{-1}] suspended sediment and the annual rate of overbank aggradation is ca 1.0 cm—Kiss 2015).

The actual sediment transport of the Drava is highly influenced by the 22 reservoirs established on the upstream section. For example, in 1968 the Zlatoličje Hydroelectric Power Plant (Slovenia) started to operate and reduced the pre-construction suspended sediment yield from 0.95 to 0.41 million t y^{-1} (Bonacci and Oskoruš 2008). Later, the construction of Varaždin reservoir (Croatia, 1975) decreased this value further to 0.17 t y^{-1}, and this trend continued with the reservoirs of the last two power stations at Čakovec (1982) and Donja Dubrava (1989) (Bonacci and Oskoruš 2010; see also Chap. 9 in this volume).

Bank erosion can slightly recharge suspended sediment to the channel. The amount of local sediment input into the river by bank erosion was calculated for four meanders. On average, 10.8–76.5 thousand m^3 y^{-1} material reached the river channel (Fig. 11.2). The rate of bank erosion and the amount of input sediment load depends on

- the location of the meander: the upstream bends are characterized by higher bank erosion rate, just because of greater stream power;
- the height of eroding banks: high banks retreat at lower rates (e.g. at Heresznye), just because of the different mode of erosion (landslides instead of particle drag), although the amount of sediment input is almost the same;
- the hydrological properties of the given year: in 2014 a flood wave intensified bank erosion.

However, the material eroded from the banks, fine-grained sand and silt (loess), is different from the sediment trapped in reservoirs, which is mostly bedload. This points to major changes in the sedimentary environment of the Drava since the construction of reservoirs. Considering flow velocity, the recharged material is hardly an important factor in floodplain processes.

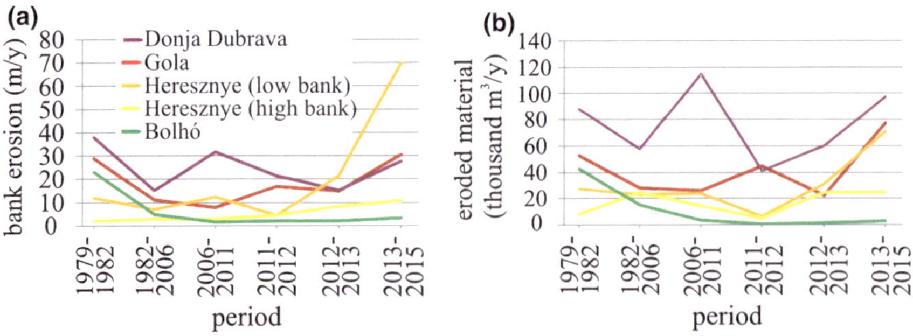

Fig. 11.2 Maximum rate (m y^{-1}) of bank erosion (**a**) and the amount of eroded material (m^3 y^{-1}) in four studied meanders of the Drava (**b**) (based on the data of Andrási 2015)

11.5.3 Water Regime

From the point of view of overbank aggradation, flood duration and characteristics are the most decisive factors, although the sedimentation of side channels and the resulted merging of islands are also influenced by lower-than-bankfull stages. Water level changes and floods were evaluated applying the dataset obtained at Barcs (Fig. 11.3).

At the beginning of the 20th century (1901–1917) no reservoir existed on the Drava yet. Therefore, floods appeared in each year, covering the floodplain for 2–3 weeks in average. The longest flood lasted for 54 days in 1904. Usually several flood waves occurred within a year. The return interval of floods was 4–6 months. Later (1918–1967), on the upstream (Austrian and Slovenian) section of the Drava, reservoirs and hydroelectric power stations were constructed. As a consequence, water levels gradually dropped by 90–100 cm, which is reflected by their shifting frequency distribution curves too. Although the height of floods (1% frequency stage) did not change considerably, but their duration decreased, and in some years they even disappeared. Therefore, at the beginning of the period within a year (1926) even 42 days with flood stage were observed, but by the end decreased to 26 days (1965) and their return interval lengthened to 1.4–2 years. In recent decades (since 1968), as a result of the construction of the last Slovenian and three Croatian reservoirs, these processes accelerated. Low and medium stages decreased further by 130 cm and high stages by 174 cm. The frequency distribution curves shifted further down. Today the level of the 50% frequency stages is below the 90%

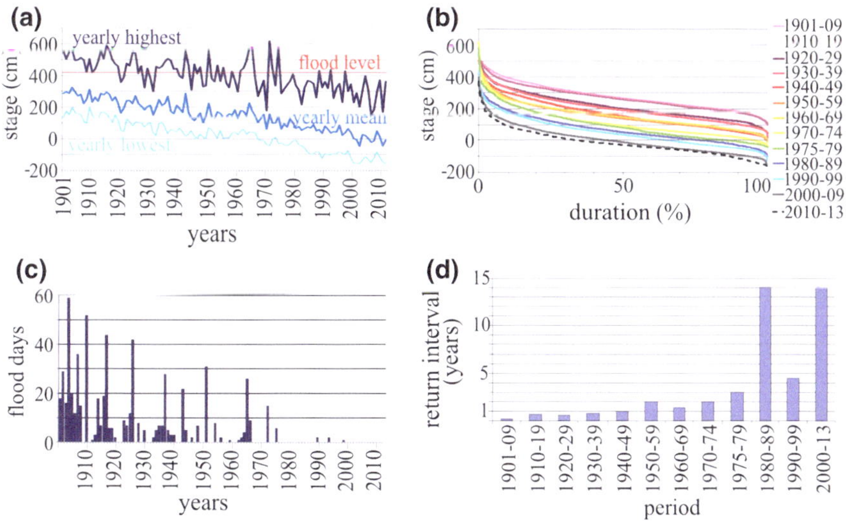

Fig. 11.3 Hydrological changes of the Drava based on the data of the Barcs gauging station. **a** Annual minimum, medium and maximum stages; **b** Decadal stage curves (frequency distribution). **c** Annual duration of overbank flow (above 420 cm); **d** Return interval of floods

frequency stage measured at the beginning of the 20th century. During the last four decades floods almost totally disappeared, as floods entered onto the floodplain just in 5 years, and their duration was only 1–2 days. The return interval increased to 5–15 years. The role of reservoirs in the disappearance of floods was well reflected in 2013, when extremely large amounts of snow covered the catchment, but, despite rapid snowmelt, only bankfull stage developed on the downstream section.

Thus, it can be concluded, that floods gradually disappear from the Drava, resulting in the termination of overbank fluvial processes (vertical accretion) on the floodplains. The low and medium stages continuously dropped in the last decades without decreasing discharge values, caused by the incision (2.8 cm y^{-1}) of the thalweg (VITUKI 2003). This process will intensify the lateral accretion of the floodplain.

11.6 Processes of Floodplain Development

In historical times the floodplain of the Dráva was dominated by marshlands (Buchberger 1975; Mérey 2002). In the late 18th century the society required flood safe arable lands, therefore the first local flood-protection levees were built (Ihrig 1973), and some meanders were artificially cut off. However, very often the artificial levees were destroyed by floods, especially if they were located close to the actively meandering channel (Buchberger 1975; Remenyik 2005). In the 19–20th centuries an uniform artificial levee system was built. The levees are still not continuous as at high escarpments there is no need for flood protection. As a result, the originally 10–45-km-wide natural floodplain was artificially restricted to a 1–10-km active floodplain. Therefore, fluvial processes can only prevail within this narrow zone (Fig. 11.1).

Now overbank floodplain processes (vertical aggradation) are limited due to the missing floods. Floodplain aggradation must have been very limited even under natural circumstances because of high channel slope, great flow velocity and low amount of suspended sediment load. Thus, especially in the last 40 years, no overbank accretion happened and lateral floodplain processes dominate within a narrow strip along the channel and the floodplain. The process will be introduced within a short, 13-km-long section and at two meanders.

11.6.1 Channel Narrowing

The channel of the Drava has continuously incised since the 19th century and this provided space for lateral floodplain development. The water surface is shrinking, the channel is narrowing, and islands rather merge into the bank than newly establish. The process was analyzed in detail along a short section between Botovo and Gola (Fig. 11.4). The effect of the low number of groynes built into the channel

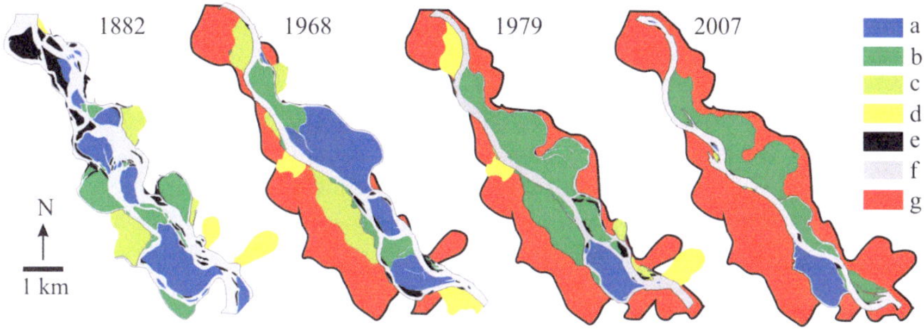

Fig. 11.4 Lateral floodplain accretion by channel narrowing and islands merging into the bank. **a** Mid-channel island; **b** island located close to the bankline or another island; **c** island partially merged with the bank; **d** island totally merged with the bank; **e** bare gravel-bar; **f** water surface; **g** floodplain evolved by lateral accretion since 1882

is surpassed by the effects of the nearby (at ca 40 km distance) reservoir and hydroelectric power station at Donja Dubrava, as they trap the sediment and create daily 'mini flood waves'.

During the First Military Survey (1882) the Drava was a large, multi-channelled river (area of water surface: 829 ha; channel width: 1,241 m). In the wide channel several (43) islands with a large area (786 ha) existed. Most of them (39%) split the thalweg, some (45%) drifted close to the banklines or to another island and only relatively few (16%) merged into the banks. In 1882 a large (61 ha) floodplain island also existed, but it was already connected to the banks and added to the floodplain area. At that time, frequent floods could re-arrange channel morphology creating new gravel-bar and island surfaces and actively eroding and aggrading banks.

Until the next survey (1968), the channel of the Drava changed considerably. The area of the water surface declined (by 64.4%), while the area of the floodplain increased by 6.2 ha y^{-1} through lateral accretion. The mean width of the channel halved (to 699 m, at 6.3 m y^{-1} rate). The number of mid-channel islands decreased by 40% (to 26 islands) and their total area also dropped (by 312 ha). Two large (340 ha) floodplain islands rose above the channel, probably developed from smaller merging islands, as the side channels between them became dry due to the drop of water stages. Several islands were connected partially or totally to the banks. These processes were characteristic not only on the short introduced section, but all along the Drava between the Mura and Danube. Therefore, between 1882 and 1968 the floodplain area increased by over 3,000 ha (12.5 ha rkm^{-1}), the mean channel width decreased from 513 to 361 m, thus it became narrower by 1.8 m y^{-1} with a rate of lateral floodplain accretion of 1.8 m y^{-1}. The process was especially rapid (23.7 ha rkm^{-1}) on the upstream reach (241–160 rkm), while on the downstream reach (160–0 rkm) it was slower (6.9 ha rkm^{-1}). These data suggest that the wide upstream reach of the Drava had a more sensitive and dynamic response to hydrological changes (drop of stages, incision, lack of sediment) caused by power plants than of the lower reach. As a consequence, along the upper reach of the Drava floodplain evolution was also more rapid.

Until the next survey (1979), along the studied short section, the area of the water surface continued to decrease (by 6.4%, to 276 ha) and floodplain area increased by 1.7 ha y^{-1}. The channel became narrower by 33 m (4.7%, narrowing rate: 3 m y^{-1}). Compared to the previous period (1882–1968), less islands (18) merged totally into the banks, increasing the floodplain area by 50 ha. The two large floodplain islands—which existed in 1968—merged, and their total territory increased (by 19 ha). The former mid-channel islands shifted towards the banks and they partially merged. The simplifying channel morphology and the continuously expanding floodplain area could be connected to the already operating Varaždin power station (1975) which influenced the regime and sediment yield more effectively than the farther Austrian or Slovenian hydroelectric plants. Considering the whole section of the Drava between the Mura and the Danube, similar processes can be detected. Between 1968 and 1979 the floodplain area increased by 3.0 ha rkm^{-1} on average. This was the period of the most intensive channel narrowing (3.6 m y^{-1}), as channel width dropped from 361 m to 321 m on average. On the upstream part of the reach the rate of lateral accretion decreased (1.0 ha rkm^{-1}), downstream it remained relatively high (4.0 ha rkm^{-1}), indicating downstream propagation. Along the whole section, lateral accretion was the result of channel narrowing and island merging into the banks. The latter process was supported by the narrowing of side channels, which filled up quickly or sediment plugs impeded water inflow into them.

Until the last survey (2007) the channel became even narrower in the studied short reach, as the water surface decreased to 227 ha (by 17.8%), which equals 1.8 ha y^{-1} horizontal floodplain accretion. Mean channel width was reduced to 355 m (by 46.7%), rate of narrowing: 11.1 m y^{-1}. It can be explained by the existence and operation of the Čakovec (1982) and Donja Dubrava (1989) power stations, which further decreased sediment yield and water levels. Compared to the 1979 survey, more mid-channel islands (22) developed in the channel, but their area decreased. The side channel between the large floodplain island and the banks gradually lost its function and the island almost merged to the bank. The incision of the channel and the lack of bedload intensified lateral erosion within the channel. As a consequence, the area (309 ha) of the floodplain island decreased. Almost 70% of the islands already merged or partially merged to the banks, their side channels lost their water supply. In this period (1979–2007), channel width further decreased (from 321 to 256 m) along the whole Drava reach and the floodplain area increased by 24 ha y^{-1}. As the whole river incised, most of the side channels lost their water and sediment supply and desiccated and the neighbouring islands merged with the floodplain. The process was faster again on the upstream reach, whilst downstream it slowed down in the wake of river engineering works (cutoffs, construction of groynes and revetments).

Thus, it can be claimed that between 1882 and 2007 the floodplain area extended remarkably through lateral floodplain accretion, i.e. channel narrowing and island-merging to the banks. The originally island-braided channel turned into meandering. Today lateral meander migration prevails.

11.6.2 Lateral Meander Migration

Meander migration is an important process of lateral accretion and floodplain evolution. The spatial and temporal changes of a meander and its forms reflect the fluvial environmental changes too. To present the process, two meanders were selected for the present chapter, which are located on different morphological units of the Drava. The meander at Gola is located at the downstream end of the section presented above, whilst the bend at Heresznye is located ca 30 km downstream. The meander at Gola is a good example of meander development in a formerly braided channel, under the heavy influence of the Donja Dubrava Hydroelectric Power Station and its reservoir, which is located at ca 40 km distance. These effects are weaker at Heresznye, where meander development is determined by a high bank.

11.6.2.1 Meander Development at Gola

The point-bar system of the meander evolved from mid-channel islands, just like at other meanders nearby (Andrási 2015; Kiss and Andrási 2015; Kiss and Balogh 2015). The map made in 1979 represents a multi-thread channel with large islands (Fig. 11.5). According to our dendro-geomorphological study, the gravel-bars were deposited by the 1972 1975 floods. Later these bar surfaces were colonized by trees, creating islands. Afterwards elongated point bars connected to the sides and downstream ends of the islands (Fig. 11.6), but they were separated by secondary channels and swales. As the water levels dropped due to the operation of power stations, even these side channels and swales became dry and the islands and the point bars connected to the floodplain. Simultaneously with the drop of stages and the incision of the channel, point-bar surfaces developed at lower altitudes (Fig. 11.6). Thus, today the relief between the original bankline and the active gravel bar is 1.3 m. The trees colonized the point-bar surfaces at different elevations almost at the same time, indicating very rapid lateral floodplain development.

As the braid transformed into a bend after 1982 and the thalweg gradually shifted towards the south, bank erosion became very intensive and balanced the rapid point-bar development. As a consequence, the length of the eroding bank increased (Fig. 11.5): in 1979 the bank erosion affected ca 0.5-km-long bankline, but now it is 1.5 km. The rate of bank erosion was the greatest (max. 38 m y^{-1}) between 1979 and 1982. This could be explained by the higher stages and the 1979–1980 floods. Between 1982 and 2006 bank erosion decreased by two thirds (Fig. 11.2) and this temporal change was typical for the other meanders too.

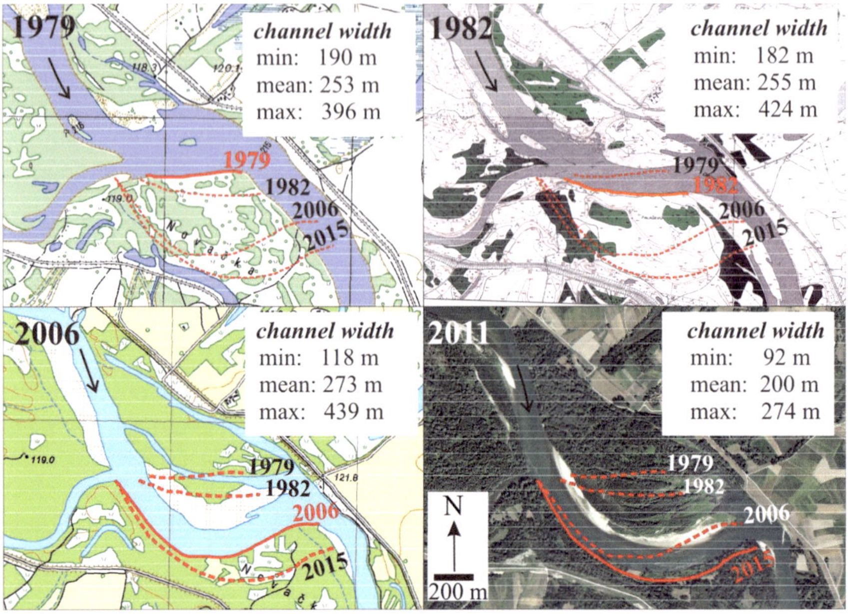

Fig. 11.5 Meander development and channel width changes at Gola. The location of the eroding bank is indicated by red lines

It was in connection with the operation of two new power stations started in 1982 and 1989, resulting in the further drop of stages and the lack of floods. The mean rate of bank erosion (5–12 m y^{-1}) was high at Gola, explained by the great slope and stream power and also by the loose bank material (unconsolidated deposit of the 19th century braided channel). In 2014 a large flood developed on the Dráva, and it increased bank erosion, reworking large amount of transported material.

At Gola, the eroding banks rapidly migrate towards the artificial flood protection levee (in 2011 their distance was 105 m; in 2015 it was only 75 m) and levees will be endangered in the near future. At other places, valuable arable fields along the river are lost by erosion (at 0.1–0.2 ha y^{-1}).

11.6.2.2 Meander Development at Heresznye

Downstream of the island-braided reach of the Dráva sinuous then meandering reaches developed. The studied bend at Heresznye evolved on the sinuous channel (Fig. 11.7). The core of meander development was at Gola island, where point bars were connected to the bank. However, at Heresznye meander development is partially confined by the high escarpment.

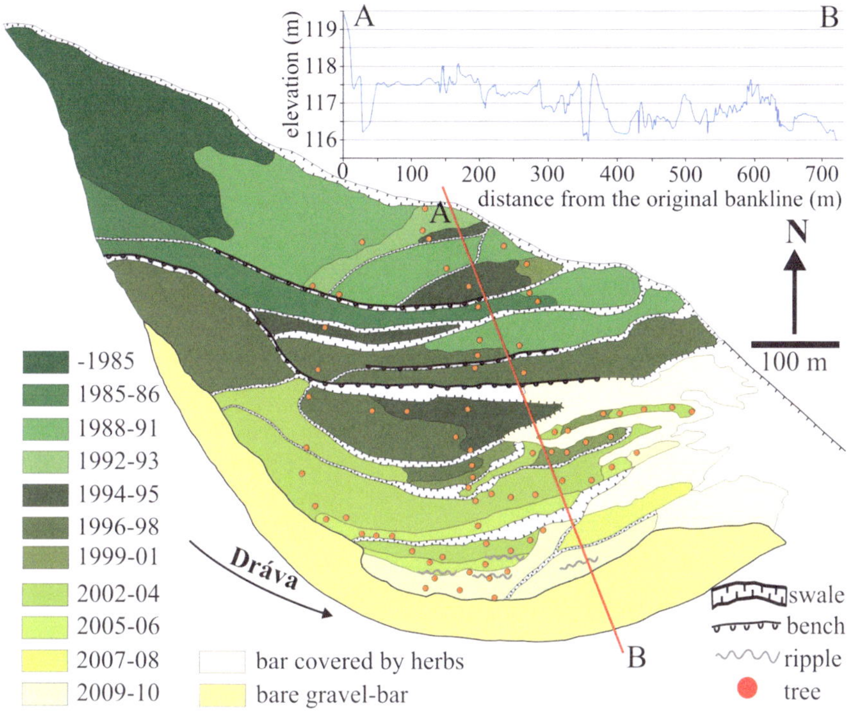

Fig. 11.6 Evolutionary periods of the point-bar system at Gola based on our dendro-geomorphological survey and the height conditions of the point bars along an A-B profile

During the 1979 survey the sinuous channel was wide, though as the meander started to develop its width decreased, thus the area of the floodplain increased by lateral accretion. The process is triggered partly by dropping water levels, and partly by a nearby cutoff upstream (at Vízvár, 197–193 rkm). Probably the cutoff increased local sediment load by eroding the artificial channel, and the provided large amounts of sediment and the decreasing stages accelerated the development of bars, which diverted the thalweg and plugged the side channels and swales. The oldest point bar was connected to the bankline in 1960 (Fig. 11.7). This ca 5 ha area was colonized by poplars and willows in the 1960s. Later trees stabilized the bars formed in 1978–1979 upstream of the axis of the bend, and finally the outmost point bars were colonized. The swales in between them were occupied by trees as late as 2005–2006. The isochrone map suggests that point-bar development was continuous, and similarly to the meander at Gola, the higher heads of the bars were colonized first and later the body of the bars also became dry and suitable for tree growth. The largest area was occupied by trees in 2002–2005. This extensive point bar diverted the thalweg and the direction of the meander development changed: previously it migrated to east, but after 2002 towards the southeast.

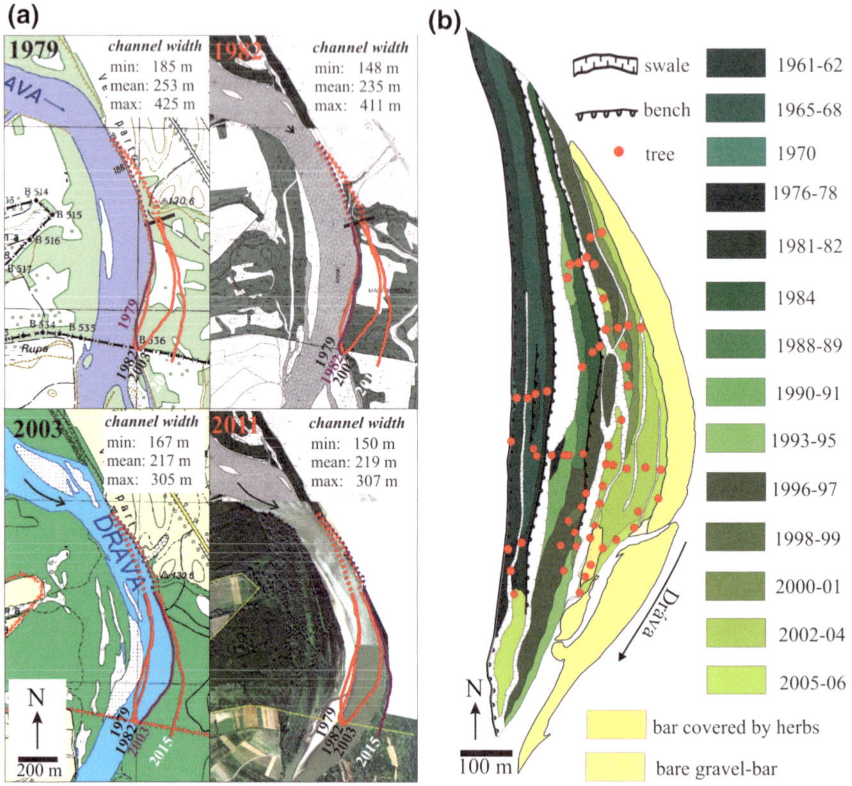

Fig. 11.7 Meander development and channel width changes at Heresznye (**a**) and the evolutionary periods of the point-bar system based on the dendro-geomorphological survey (**b**)

Meander development at Heresznye is similar to the evolution of those Drava sections where the river is actively eroding a high escarpment or terrace rim. Here the upper part of the meander is bordered by a 20–22 m high escarpment, which is eroding along 520 m, whilst along the lower part the bank is only 3.0–3.5 m high and it is eroding along 850 m. The bank erosion rate of the lower bank is 2–8 times higher than that of the high bank (Fig. 11.2). Similar to the Gola meander, bank erosion rate (ca 70 m y^{-1}) was very high in 2014. (It was the largest flood since the Varaždin Hydroelectric Power Plant was built in 1975). Greater volume of sediment was derived from the low bank than from the high escarpment, which can be explained by (1) the longer low bank, (2) more effective erosion of the low bank even during lower stages, (3) more consolidated material of the high bank, thus slope failure occurs only if very steep walls are eroded by the river, and (4) the colluvium from the escarpment temporally protects the high bank from further erosion.

11.7 Discussion

The power plants and their reservoirs altered channel and floodplain development of the Drava. Major hydrological consequences are

(1) missing floods;
(2) decreasing river stages;
(3) development of small (max. 1.5 m high) daily flood-waves created by peak operation and
(4) trapping bedload in the reservoirs.

The evolution of the studied meanders has some similarities. On the upper section of the Drava (at Donja Dubrava: Kiss and Andrási 2015; Sigetec: Kiss and Balogh 2015; and at Gola presented here), the meanders developed in the originally island-braided channel: gravel-bars connected to the islands, and as the water levels dropped they gradually became stabilized by woody vegetation and became part of the floodplain. In contrast, on the lower section (at Bolhó: Andrási 2015, and at Heresznye presented here) the point-bar series developed in an originally slightly sinuous channel, where the thalweg shifted towards the outer banks. The farther is a meander from the series of reservoirs and power stations, the lower is the rate of channel incision. Thus, the elevation differences within the same point-bar series decrease.

The process of meander migration is driven by point-bar stabilization by vegetation. As a first step, islands develop at the head of point bars, and later the lower point-bar surfaces gradually connect to them. The age distribution of the colonizing trees suggests that on the lower section of the Drava meander development started earlier (1940–1960s) and slowly propagated upstream, where it did not start before the 1980s.

Meander development is a typical way of lateral floodplain evolution. Its spatial evolution is reflected by the emergence of point bars and swales and its temporal evolution by the age of trees colonizing the point-bar surfaces. The largest surfaces were stabilized by trees and became part of the floodplain in the 1980s (e.g., at Gola 7 ha between 1979–1984; and 3.1 ha between 1985–1986). It could be explained by the decreasing water stages after the hydroelectric plants at Varaždin (1975) and Čakovec (1982) started to operate (Fig. 11.8). The process continued after the Donja Dubrava Plant was installed. At Gola another 8 ha became part of the floodplain between 1988 and 1991.

The dendro-geomorphological analysis and the morphology of side-channels and swales reflect that the point-bar development is more dynamic on the upper section with greater slope, and the aggradational process is compensated by very intensive bank erosion. The rate of bank erosion is fundamentally influenced by the height of the bank, its material and stream power. On the upper Drava the very intensive (mean: 4.9–27.9 m y^{-1}) bank retreat is connected to high channel slope, unconsolidated bank material and clear-water erosion downstream of the reservoirs. On the lower reach, the rate of bank erosion is lower (mean: 0.8–12.9 m y^{-1}),

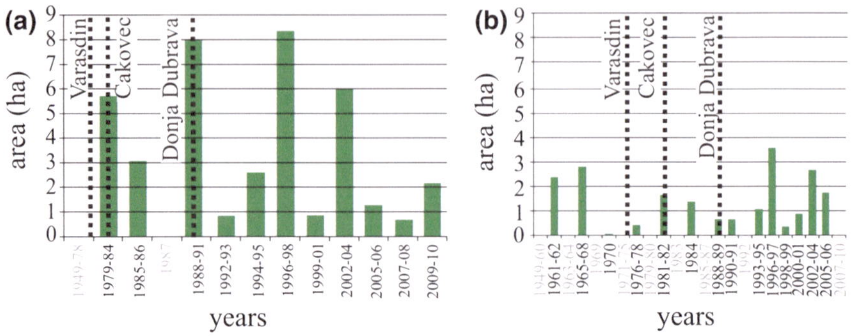

Fig. 11.8 Area colonized by vegetation at the meander of Gola (**a**) and Heresznye (**b**) The grey years show no increase in floodplain area

as here the banks are higher, consist of more compact Pleistocene sediments, and stream power is lower. The retreat rate of the high bank (0.4–1.9 m y^{-1}) was much lower than the bank erosion (1.3–12.9 m y^{-1}) of a lower bank at the same meander. It suggests that longer time is needed for the effective undercutting of the high banks and for the transport of the colluvium. In addition, due to their more stable sedimentary structure, the walls of the escarpments are less erodible.

11.8 Conclusions

To the combined effect of the hydrological changes induced by reservoirs and dams, the typical floodplain processes terminated. Thus, today there is almost no chance for overbank floodplain aggradation, for the development of natural levees and crevasses and the inorganic material transport into the distal marshes. The latter are only filled up by autogenous organic material. Besides, the continuous drop of stages and the incision of the channel (at 2.8 cm y^{-1} rate) are also unfavourable for the water balance of the floodplain (dropping groundwater table) with its oxbow lakes and marshes (Lóczy et al. 2016 and Chap. 12 in this volume).

At present, overbank (vertical) aggradation is almost totally replaced by horizontal accretion. Over the last almost 150 years, floodplain area has increased by several thousand hectares simultaneously with the loss of channel area. The side-channels gradually lose their water conductivity, therefore the islands are merging to the banks, adding to the floodplain area. The process of island merging accelerates the narrowing of the channel and the expansion of the floodplain. As the stability of the eroding banks decreases due to the drop of river stages, meander development is also speeded up. Therefore, the rate of bank erosion increases and simultaneously point-bar development also becomes more intensive and, consequently, lateral accretion is accelerating. However, rapid lateral meander development induces erosion risks locally.

The eroded material of the banks can contribute to local sediment yield since large amounts of sediment are missing from the system due to sediment trapping in the reservoirs and in-channel gravel mining. However, first of all, bedload yield was altered and the recharged material mainly adds to suspended sediment supply.

References

Andrási G (2015) Hidrológiai és morfológiai változások a Dráva alföldi szakaszán (Hydrological and morphological changes of the lowland section of the Dráva). Manuscript PhD Thesis. University of Szeged, Szeged, 120 pp (in Hungarian)

Benedetti MM (2003) Controls on overbank deposition in the Upper Mississippi River. Geomorphology 56:271–290

Bonacci O, Oskoruš D (2008) The influence of three Croatian hydroelectric power plants operation on the river Drava hydrological and sediment regime. In: 24th Conference of the Danube countries on the hydrological forecasting and hydrological bases of water management. Bled, Slovenia, 2–4 June 2008

Bonacci O, Oskoruš D (2010) The changes in the lower Drava River water level, discharge and suspended sediment regime. Environ Earth Sci 59(8):1661–1670

Brierley GJ, Fryirs KA (2005) Geomorphology and river management: applications of the river styles framework. Blackwell Publishing, Oxford

Brierley GJ, Hickin EJ (1992) Floodplain development based on selective preservation of sediments, Squamish River, British Columbia. Geomorphology 4:381–391

Brown AG (1983) An analysis of overbank deposits of a flood at Blandford-Forum, Dorset, England. Revue Geomorphologie Dynamique 32:95–99

Buchberger P (1975) A Dráva-völgy árvédelmének története (History of flood protection in the Drava Valley). Vízügyi Közlemények 57(1):103–113 (in Hungarian)

Chorley RJ, Schumm SA, Sugden DE (1985) Geomorphology. Methuen, London

Constantine JA, Pasternack GB, Johnson ML (2005) Logging effects on sediment flux observed in a pollen based record of overbank deposition in a northern California catchment. Earth Surf Proc Land 30:813–821

Dietrich WE, Wilson CJ, Reneau SL (1986) Hollows, colluvium, and landslides in soil-mantled landscapes. In: Abrahams AD (ed) Hillslope processes. Allen and Unwin, Boston, pp 361–388

Everitt BL (1968) Use of the cottonwood in an investigation of the recent history of a flood plain. Am J Sci 266:417–439

Florsheim JL, Mount JF (2003) Changes in lowland floodplain sedimentation processes: pre-disturbance to post-rehabilitation, Cosumnes River, CA. Geomorphology 56:305–323

Fryirs KA, Brierley GJ (2013) Geomorphic analysis of river systems: an approach to reading the landscape. Wiley-Blackwell, Hoboken

Gábris Gy, Telbisz T, Nagy B, Bellardinelli E (2002) A tiszai hullámtér feltöltődésének kérdése és az üledékképződés geomorfológiai alapjai (Floodplain aggradation and its geomorphological background along the Tisza). Vízügyi Közlemények 84:305–322 (in Hungarian)

Hohensinner S, Habersack H, Jungwirth M, Zauner G (2004) Reconstruction of the characteristics of a natural alluvial river-floodplain system and hydromorphological changes following human modifications: the Danube River (1812–1991). River Res Appl 20:25–41

Horváth G (2002) A Dráva folyó magyar-horvát szakaszának hidrológiai, jellemzése az EU keretirányelvének figyelembevételével (Hydrological analysis of the Hungarian-Croatian section of the Drava with regard to the EU Water Framework Directive). In: Conference abstract "Introduction of the EU Water Framework Directive in the Drava catchment". 12 February 2002. WWF, Pécs (in Hungarian)

Ihrig D (1973) A magyar vízszabályozás története (History of the Hungarian water regulations). OVH, Budapest. 398 pp (in Hungarian)
Károlyi Z (1960) A Tisza mederváltozásai a folyószabályozás tükrében (Channel shifting of Tisza River as reflected in river regulation). VITUKI, Budapest. 102 pp (in Hungarian)
Keesstra SD (2007) Impact of natural reforestation on floodplain sedimentation in the Dragonja basin, SW Slovenia. Earth Surf Proc Land 32:49–65
Kiss T (2015) Fluviális folyamatok antropogén hatásra megváltozó dinamikája: egyensúly és érzékenység vizsgálata folyóvízi környezetben (Altered fluvial processes caused by human impact: equilibrium and sensitivity study of the fluvial environment). DSc dissertation, Hungarian Academy of Sciences, Budapest. 165 pp (in Hungarian)
Kiss T, Andrási G (2015) Kanyarulatfejlődés sajátosságai és antropogén hatások vizsgálata két drávai kanyarulat példáján (Study on meander development and human impacts in two meanders of the Drava). Tájökológiai Lapok 13(1):73–88 (in Hungarian)
Kiss T, Balogh M (2015) Characteristics of point-bar development under the influence of a dam: case study on the Dráva River at Sigetec, Croatia. J Environ Geograph 8(1–2):23–30
Kiss T, Blanka V (2012) River channel response to climate- and human-induced hydrological changes: case study on the meandering Hernád River, Hungary. Geomorphology 175–176: 115–125
Kiss T, Sándor A (2009) Land-use changes and their effect on floodplain aggradation along the Middle-Tisza River, Hungary. AGD Landscape Environ 3(1):1–10
Kiss T, Oroszi VGy, Sipos Gy, Fiala K, Benyhe B (2011) Accelerated overbank accumulation after nineteenth century river regulation works: a case study on the Maros River, Hungary. Geomorphology 135:191–202
Knox JC (1987) Historical valley floor sedimentation in the Upper Mississippi Valley. Ann Assoc Am Geogr 77:224–244
Knox JC (2006) Floodplain sedimentation in the Upper Mississippi Valley: natural versus human accelerated. Geomorphology 79:286–310
Lecce SA, Pavlowsky RT (2004) Spatial and temporal variations in the grain size characteristics of historical flood plain deposits, Blue River, Wisconsin, USA. Geomorphology 61:361–371
Lóczy D (2001) Geomorfológiai, tájökológiai és természetvédelmi megfigyelések a Duna-ártér Mohács alatti (bédai) szakaszán (Geomorphology, landscape ecology and nature conservation on the floodplain of the Danube downstream of Mohács, at Béda). Abstracts of the Hungarian Geographical Conference, Szeged. 1–7 (in Hungarian)
Lóczy D (2013) Az árterek geomorfológiai osztályozása a nemzetözi szakirodalomban (Geomorphological classifications of floodplains in the international literature). Földrajzi Közlemények 137(1):105–120 (in Hungarian)
Lóczy D, Dezső J, Sz Czigány, Pirkhoffer E (2016) Hydromorphological assessment of the Lower Hungarian Dráva section and its floodplain. AGD Landscape Environ 10(3–4):109–116
Mantuáno J (1974) A Dráva hidrológiája (Hydrology of the Drava). Vízügyi Közlemények 56 (3):368–401 (in Hungarian)
Marosi S (1970) Belső-Somogy kialakulása és felszínalaktana (Evolution and geomorphology of Inner Somogy). Akadémiai Kiadó, Budapest 170 pp (Földrajzi Tanulmányok 11) (in Hungarian)
Mücher HJ, Slotboom RT, ten Veen WJ (1990) Palynology and micro-morphology of a man-made soil. A reconstruction of the agricultural history since late-medieval times of the Posteles in the Netherlands. CATENA 17:55–67
Nanson GC (1986) Episodes of vertical accretion and catastrophic stripping—a model of disequilibrium floodplain development. Geol Soc Am Bull 97(12):1467–1475
Nanson GC, Croke JC (1992) A genetic classification of floodplains. Geomorphology 4:459–486
Nicholas AP, Walling DE (1997) Investigating spatial patterns of medium-term overbank sedimentation on floodplains: a combined numerical modelling and radiocaesium-based approach. Geomorphology 19:133–150
Oroszi VGy (2009) Hullámtér-fejlődés vizsgálata a Maros magyarországi szakaszán (Investigation of the evolution of active floodplain along the Hungarian section of the Maros River. PhD dissertation, University of Szeged, Szeged. 135 pp (in Hungarian)

Owens PN, Batalla RJ, Collins AJ, Gomez B, Hicks DM, Horowitz AJ, Kondolf GM, Marden M, Page MJ, Peacock DH, Petticrew EL, Salomons W, Trustrum NA (2005) Fine-grained sediment in river systems: environmental significance and management issues. River Res Appl 21:693–717

Owens PN, Walling DE (2002) Changes in sediment sources and floodplain deposition rates in the catchment of the River Tweed, Scotland, over the last 100 years: the impact of climate and land use change. Earth Surf Proc Land 27:403–423

Remenyik B (2005) A Drávaszabályozás története (History of the regulation works of the Dráva). Hidrológiai Közlöny 85(3):27–32 (in Hungarian)

Sándor A (2011) Floodplain aggradation along the Middle and lowland section of river Tisza. PhD Dissertation, SZTE, Szeged, 120 pp (in Hungarian)

Schwarz U, Bloesch J (2004) The impact of hydropower dams on the flood plains of the Dráva-Mura Rivers in Croatia/Hungary. GIS and RS in Hydrology, Water Resources and Environment. IAHS Press, Wallingford, pp 178–187

Somogyi S (1967) A Drávamenti-síkság (The Drava Plain). In: Pécsi M (ed) A dunai Alföld (The Danubian Great Plain). Akadémiai Kiadó, Budapest, pp 298–304 (in Hungarian)

Steiger J, Gurnell AM, Ergenzinger P, Snelder DD (2001) Sedimentation in the riparian zone of an incising river. Earth Surf Proc Land 26:91–108

T Mérey K (2002) A Dráva a magyar történelemben (The Dráva in Hungarian history). Agrártörténeti szemle 44(1–2):195–209 (in Hungarian)

Ten Brinke WBM, Schoor MM, Sorber AM, Berendsen HJA (1998) Overbank sand deposition in relation to transport volumes during large-magnitude floods in the Dutch sand-bed Rhine River system. Earth Surf Proc Land 23:809–824

VITUKI (2003) A Dráva hordalékjárásának vizsgálata a legfrissebb adatok figyelembevételével (Investigation of the sediment regime of the Drava based on the newest data). Research report 714/1/6128-01. Project leader: Szekeres J. VITUKI, Budapest (in Hungarian)

Walling DE, He Q (1998) The spatial variability of overbank sedimentation on river floodplains. Geomorphology 24:209–223

Wolfert HP, Hommel PWFM, Prins AH, Stam MH (2002) The formation of natural levees as a disturbance process significant to the conservation of riverine pastures. Landscape Ecol 17 (1):47–57

Wyżga B (2001) Impact of the channelization-induced incision of the Skawa and Wisłoka Rivers, Southern Poland, on the conditions of overbank deposition. Regulated Rivers: Research and Management 17:85–100

Zwoliński Z (1992) Sedimentology and geomorphology of overbank flows of meandering river floodplains. Geomorphology 4:367–379

Chapter 12
Oxbow Lakes: Hydromorphology

Dénes Lóczy, József Dezső, Péter Gyenizse, Szabolcs Czigány and Gabriella Tóth

Abstract Cut-off oxbows are the most remarkable fluvial landforms and the most valuable wetland habitats in the protected floodplain of the Lower Drava River in Hungary. Their geomorphic evolution, however, has not been studied yet. Recently, a complex hydromorphological survey of oxbows covered their geographical position, connection with the main Drava channel, water balance, hydrogeological properties, water retention capacity and groundwater flow in their environs. The purpose of the investigations was to assess the potential for oxbow lake and floodplain rehabilitation. Two zones of oxbows, possibly differing in age and geomorphological evolution (the date of cutoff), have been identified and preliminarily described. The focus of research was on the Cún-Szaporca lake system, part of the Danube-Drava National Park and a Ramsar area, where the clogging of the oxbow bed, a critical factor of transmissivity, was analyzed in detail. For planning landscape-scale rehabilitation (the Old Drava Programme) more information on the old courses of the Drava and its preserved but gradually disappearing traces (the present-day oxbow lakes) would be necessary.

D. Lóczy (✉) · J. Dezső · S. Czigány · G. Tóth
Department of Physical and Environmental Geography,
Institute of Geography and Earth Sciences, Faculty of Sciences,
University of Pécs, Ifjúság útja 6, Pécs 7624, Hungary
e-mail: loczyd@gamma.ttk.pte.hu

J. Dezső
e-mail: dejozsi@gamma.ttk.pte.hu

S. Czigány
e-mail: sczigany@gamma.ttk.pte.hu

G. Tóth
e-mail: toth.gabi90@gmail.com

P. Gyenizse
Department of Cartography and Geoinformatics,
Institute of Geography and Earth Sciences,
University of Pécs, Ifjúság útja 6, Pécs 7624, Hungary
e-mail: gyenizse@gamma.ttk.pte.hu

D. Lóczy (ed.), *The Drava River*, Springer Geography,
https://doi.org/10.1007/978-3-319-92816-6_12

Keywords Cutoff · Water balance · Evapotranspiration · Groundwater flow Floodplain connectivity · Oxbow lakes · Water replenishment · Floodplain rehabilitation

12.1 Introduction

Water availability is of primary importance for all kinds of landscape. As a consequence of human interventions and probably of global climate change, even river floodplains–rich in wetlands under natural conditions–can regularly experience water shortage (Brookes 1996; Amoros and Bornette 2002). In other parts of the world, riverine environments with alternating extreme floods and droughts are also typical (e.g. in Australia—Erskine and Warner 1988).

In this situation, an important landscape function of floodplain lakes is floodwater retention in wet years and groundwater recharge in dry years. This ecosystem service is indirectly influenced by the trophic state of floodplain water bodies (external and internal nutrient loadings and littoral macrophytic vegetation—Reckendorfer et al. 2013). Evidence have been found for the opinion that the life span of backwaters has a significant effect on ecosystem services (in the Danube basin by Hein et al. 2015; along the Rhône by Dépret et al. 2017). Since there have been dramatic losses of floodplain wetland habitat due to river regulation and land drainage, a functional degradation of these ecosystems resulted worldwide (Heiler et al. 1995; Tockner et al. 2010a, b). The reduction in area and impairment of the environment have diminished the capacity of floodplains (particularly of floodplain lakes) for water retention and flood risk was enhanced (e.g., Lóczy et al. 2009; Habersack et al. 2015), while other key services, such as groundwater replenishment, nutrient storage (Hein et al. 2004) and water purification have also suffered serious limitations. The ecosystem services of floodplain wetlands are significantly influenced by the position they occupy in the landscape (Schwarz et al. 1996), which also influences their rehabilitation potential (Dragun et al. 2014; Lóczy et al. 2017).

In the floodplain of the Hungarian section of the Drava River, traditionally *floods* represented the most serious natural hazard (see Chap. 8 in this volume). The most perilous event of the 19th century was certainly the 1827 flood. This probably involved the largest inundated area ever (28 villages on ca 25,000 ha). In the 20th century, the 1972 flood of the Drava River was a disaster of special significance (Buchberger 1975). Although his time no dyke breach happened, the event brought attention to the insufficient protection provided by defence lines. In 1975 the next flood took place, but it stayed below the 1972 water level. Nevertheless, a supervision of flood-control strategy was required. The structures previously only protected by temporary ('summer') dykes (e.g. at the Majláth-puszta manor) became encircled by main dykes by 1979. In spite of channel incision and theoretically higher floodway capacity, high waters almost reached the maximum again in 2013, and particularly in September 2014, when a Mediterranean cyclone brought heavy and prolonged rainfall (Bizjak et al. 2014). The recently built dykes, however, successfully prevented overbank flow.

Although *droughts* must have been common in earlier times, too, even less archive data are available on severe water shortage conditions (Kiss and Nikolić 2015). Following profound human interventions (river regulation, barrage constructions and enhanced water retention in the reservoirs of hydroelectric plants upstream, sand and gravel extraction from riverbed etc.), recently low water stages show an increasing duration over the year (see Chaps. 6 and 9 in this volume). Floodplain drainage related to river regulation and the drawdown effect of channel incision reduced groundwater levels by several metres (Viczián and Zatykó 2011).

Water shortage is also due to the aridification trend related to global climate change. At the same time, channel incision and increased gradient result in the more rapid conveyance of flood waves (WWF 2002), reducing opportunities for floodwater retention (Kiedrzyńska et al. 2015). The flood retention role of floodplain wetlands has been studied from several approaches in international literature (Valentová et al. 2010; Schober et al. 2015).

Although it is agricultural areas which are most severely affected by drought, the water availability of oxbow lakes under protection is also endangered. As a result of water shortage and increasing sedimentation, only traces of *vegetation* with higher naturalness value (Borhidi 1997) can be detected in wetlands. The open water surfaces of oxbow lakes are shrinking; riverweed biodiversity is declining, while riparian (reed, bulrush, tall forb) vegetation and plants characteristic of dry habitats are spreading (see Chap. 13). Invasive weeds, including the Giant Goldenrod (*Solidago gigantea*), Common Milkweed (*Asclepias syriaca*), False Indigo-bush (*Amorpha fruticosa*), Boxelder (*Acer negundo*) and Tree of Heaven (*Ailanthus altissima*), are also proliferating (WWF 2002). The water bodies are surrounded by agricultural fields and improper land use (insufficient management of grasslands, large-scale intensive arable farming in the closest proximity, separated by a very narrow buffer zone) aggravates environmental problems.

12.2 Study Area

The *Drava Plain* is a lowland at 96–110 m elevation above sea level and with 2 m km^{-2} average relative relief (Fig. 12.1). It is more undulating with sand dunes in the west and flattens out towards the east. The Hungarian Drava section is 75 km long, and the corresponding catchment extends to 1.143 km^2. On this section, there are 20 major side channels, 13 tributary streams and 19 major cut-off meanders with oxbow lakes (of ca 150 ha total area) on the floodplain (Fig. 12.1). Although the oxbow lakes along the Drava have been inventoried for general hydrological (Pálfai 2001) and ecological conditions (Ortmann-Ajkai et al. 2003), no systematization and typology have been made to date.

In Croatia, 222 objects of various size, including side-arms, cut-off meanders and largely infilled abandoned channels, were identified in an inventory and studied for biodiversity (Grlica 2008). The largest oxbow lake has an area of 101.7 ha at mean water level (the Old Drava of Križnica, an overdeveloped composite meander system on the left bank, but on Croatian territory, between 173 and 166 rkm).

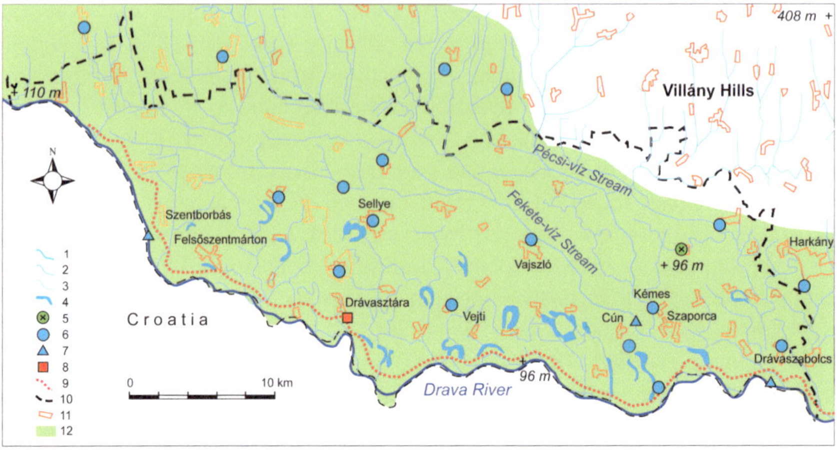

Fig. 12.1 Location of the lower Hungarian Drava section and its floodplain. 1, Drava main channel; 2, major stream; 3, small watercourse, drainage canal; 4, oxbow lake; 5, deep groundwater observation well; 6, shallow groundwater observation well; 7, river gauge; 8, hydrometeorological station; 9, main flood-defence line; 10, boundary of planning area in the Old Drava Programme; 11, built-up area; 12, morphological floodplain

The *climate* of the floodplain is moderately warm and wet in the west (under Atlantic influence) and moderately warm and moderately wet in the east (under Mediterranean influence) (WWF 2002). Mean annual temperatures are 10.2 °C in the west, 10.8 °C in the east (above 11 °C in recent decades) and in the growing season they are between 16.5 °C and 17.5 °C. Absolute minimum temperature is −17 °C and maximum temperature is 35 °C. The number of severely cold days (below −10 °C mean daily temperature) is 9–11 and that of hot days (above 30 °C) is 20–22 on the average. Annual precipitation drops from ca 800 mm in the west to below 700 mm in the east. May is the rainiest month and January and February are the driest. Absolute daily maxima rise to 104 mm. Summer water deficit has shown an increasing trend in recent decades (Blanka et al. 2013; Mezősi et al. 2016).

The genetic soil types occurring on the floodplain (in parentheses: WRB soil classes) are

- alluvial soils next to the Drava channel (Fluvisols);
- meadow soils in a zone further away from the channel and at higher elevation, locally with chernozem dynamics (Histosols);
- sandy soils near Barcs (Arenosols);
- marsh soils on the shores of oxbows (Gleysols).

Soils are more acidic in the west and calcareous in the east. Soil erosion is only observed on sandy soils (deflation) and in the vicinity of high loess bluffs (gully erosion).

The representative area selected for a more detailed study, the *Cún-Szaporca oxbow* (Fig. 12.2), is a cut-off composite meander of the Drava River of 275 ha total area (Fleit et al. 2012; DDKÖVÍZIG 2012). After artificial cutoff in the mid-19th century, it was eventually disconnected from the main channel in 1975,

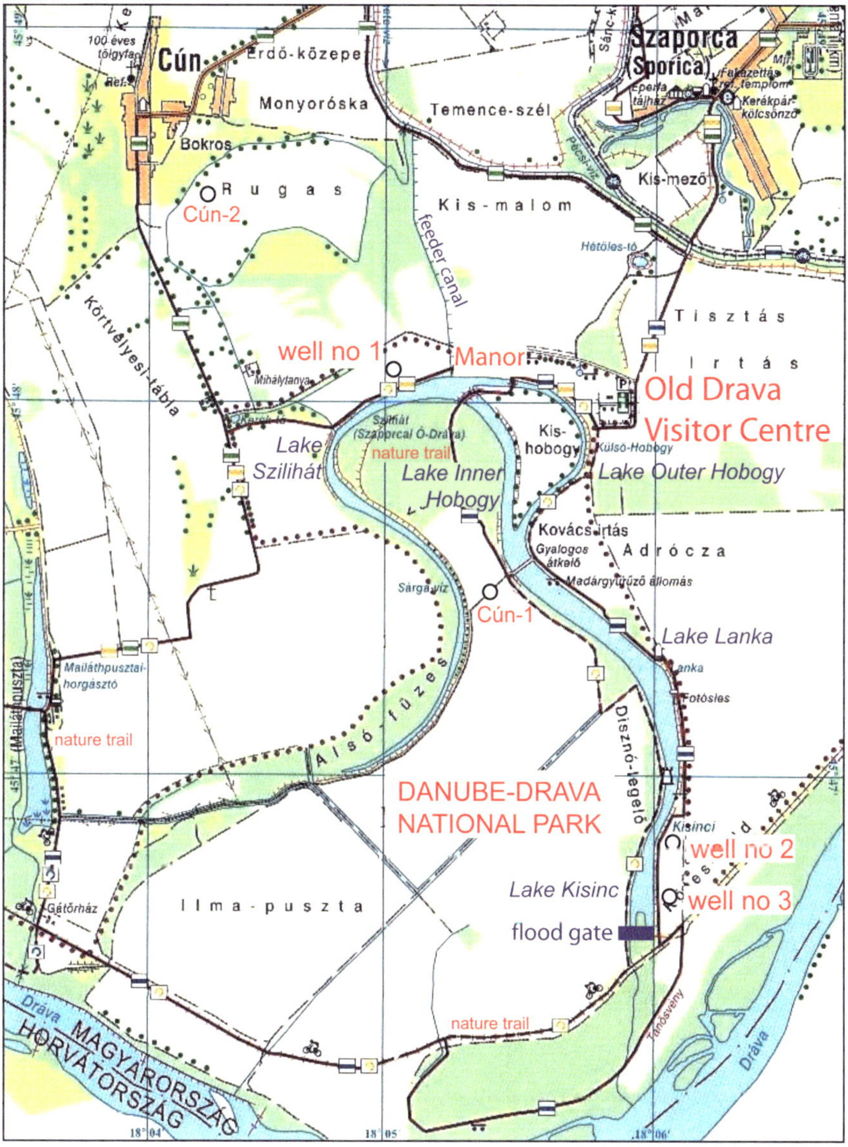

Fig. 12.2 The Cún-Szaporca oxbow with water management structures: feeder canal, groundwater observation wells, Cún-1 and Cún-2 monitoring sites and flood gate (base map: Szaporca–Kémes map, scale: 1:35,000. Cartographia, Budapest, 2016)

when the principal flood-defence line was built. Open water surface prevails on the oxbow, which certainly used to be the main channel of the Drava in 1784, over most of the year. The water surface is composed of four distinct lakes: Lake Kisinc (20 ha), being the largest, the lakes Lanka, Szilihát and Inner Hobogy (Majer 1998).

An intermittent pond (Outer Hobogy) occupies the bed of an earlier active river branch in the east. Since 1996 the oxbow area is an important waterfowl sanctuary under the Ramsar Agreement. Therefore, good water supply is not only vital for agriculture, but also for nature conservation. The maintenance of water level in the oxbow lakes, however, is difficult under increasingly arid climatic conditions, exacerbated by the long-term impacts of river regulation.

Oxbow lakes have played a significant part in the history of *human settlement* of the Drava Plain (Vajda 2001; Gyenizse and Lóczy 2010). In accordance with the wandering meandering pattern, there is much evidence that the positions of river channels shifted significantly over the floodplain in historical times. For instance, in the 16th century the village Szaporca lay on the left bank of the main Drava River branch of that time (Káldy-Nagy 1960), but permanent flood hazard forced its inhabitants to resettle on higher ground (a natural levee), where the other village of the neighbourhood, Cún, was built. (The abandoned site is called Lanka, which gave its name to one of the present-day oxbow lakes.)

12.3 Objectives

Within the frame of the four-year project "Rehabilitation potential of the Hungarian Drava floodplain" (2012–2017), the main objectives of research on the Drava oxbows were

(1) to survey the distribution of oxbow lakes in the floodplain and analyze their connectivity with the main channel;
(2) to outline geomorphic evolution;
(3) to reveal water balance with special regard to water retention capacity (cf. Dawidek and Ferencz 2014);
(4) to assess rehabilitation potential (on the example of the Cún-Szaporca oxbow).

12.4 Methods and Discussion

12.4.1 General Hydromorphological Properties

The hydromorphological character of the river and its floodplain in the Cún-Szaporca reach is presented with the help of the *indicators* elaborated for the EU project REFORM (**RE**storing Rivers **FOR** Effective Catchment **M**anagement) (González del Tánago et al. 2015) (Table 12.1). Mean river discharge (at the Barcs gauge for the period 1896–2016) is 595 $m^3\ s^{-1}$, its maximum (assumed to be of 0.1% probability, estimated for the 1827 flood) is around 3,100 $m^3\ s^{-1}$, while baseflow is around 170 $m^3\ s^{-1}$ and bankfull discharge is around 900 $m^3\ s^{-1}$ (VKKI 2010).

Table 12.1 Hydromorphological indicators of the Drava River reach in the study area (selection of indicators based on González del Tánago et al. 2015)

Key process/ features	Indicator	Literature source	Unit	Value/class at Cún-Szaporca	
				Pre-regulation	Post-regulation
Channel/ floodplain types and dimensions	Basic river type (BRT)	Rinaldi et al. (2015)		Single-thread: meandering (4)	Heavily artificial (0)
	Extended river type (ERT)	Rinaldi et al. (2015)		Unconfined, sand (+ fine gravel) bed, meandering (18)	Heavily artificial (0)
	Floodplain type	Rinaldi et al. (2015)		(Sinuous/ meandering) lateral migration (G)	
	Floodplain type	Nanson and Croke (1992)		Meandering with lateral migration (2b)	
	Planform	Richards (1982)	Dimensionless sinuosity index	3.8	1.1
	Channel bankfull width	Graf (2006)	Metres	Active channel: 350	Oxbow lakes: 200
	Channel bankfull depth	Graf (2006)	Metres	Active channel: ca 5.5	Oxbow lakes: 3.3
	Channel slope		mm^{-1}	0.00023	0.000114
Flooding extent	Morphological floodplain accessible by flood	Ward et al. (2002)	%	80	7
	Floodplain inundation frequency		Times per decade	>10	1
River energy	Specific stream power at bankfull discharge		$W\ m^{-2}$	ca 10	35 (FLUVIUS 2007)
Channel adjustment	Eroding/ aggrading channel banks		% of active channel length	ca 50/50	90/10
	Lateral bank movement	Brierley and Fryirs (2005)	$m\ year^{-1}$	>1	<0.1
	Bed incision		$cm\ year^{-1}$	n.a.	2.4 (Lovász 2013)

(continued)

Table 12.1 (continued)

Key process/features	Indicator	Literature source	Unit	Value/class at Cún-Szaporca	
				Pre-regulation	Post-regulation
Riparian vegetation	Riparian corridor		Average width (m)	>80	20
	Age structure	Corenblit et al. (2007)	% of old, mature and young forests	n.a.	Old forest: 20%; mature forest: 70%; young forest: 10%
	Dominant plant associations	Corenblit et al. (2007)	Association type	Softwood and hardwood forests	Alluvial and mixed riparian forests
Aquatic vegetation	Aquatic plant coverage	Gurnell et al. (2010, 2015)	% of channel bed	n.a.	In oxbow lake: <10
Constraints on channel adjustment	Bank revetments, embankments, artificial levees		% of channel length	<10	100
	Average width of erodible corridor for 50 years	Piégay et al. (2005)	Channel widths	n.a.	1.5 (Kiss et al. 2011)

12.4.2 *Oxbow Position Related to the Main Channel*

A first step towards the reconstruction of the geomorphic evolution of the floodplain is an inventory of hydromorphological characteristics of oxbow lakes as traces of shifting meanders. Their topographical location was determined from GoogleEarth satellite images and a Digital Elevation Model (see later). For a rough estimate of relative position (connectivity) compared to the main Drava channel, straight-line distance from and relative height difference above the channel were measured. For calculating height difference, the altitude of maximum water levels in the lakes were invariably considered.

The studied oxbows are referred into *two groups*: one is constituted by lakes whose proximal tip is within 1 km from the main Drava channel and less than 2 m above that, while the others are located at 2.8–5.7 km distances and at 2–4 m relative height (Table 12.2). The altitudinal position also indicates their association with former Drava channels which used to exist in different periods.

Table 12.2 Topographical data of oxbows in the Hungarian Drava Plain

No	Oxbow	Length (m)	Width (m)	Mean depth (m)	Mean lake level above sea level (m)	Water surface area in August 2014 (ha)	Height above the Drava channel (m)	Distance from the Drava channel (m)		Geomorphological character
								Straight line	Along drainage lines	
1	Lakócsa-Drávafok oxbow	6,310; 1,373; 2,007	150–200	n.a.	98.0	4	3	3,760	4,677	Series of composite meanders
2	Fenék swamp	2,570; 1,165; 1,642	140–185	0	98.0	0	1.5	3,934	5,953	Infilled composite meander
3	Kanszki-berek	1,210	170	0	98.0	0	2.5	2,800	3,760	Infilled simple meander
4	Mrtvica of Felsőszent-márton	1,947	110–150	0.5	96.0	16	0.2	374	408	Simple meander loop
5	Lake Sellye	3,276	245–405	0.3	96.0	2.5	3	5,677	6,424	Simple meander
6	Lake Dráva-keresztúr	2,721	130–215	0.8	97.0	0.1	1	336	336	Simple meander
7	Lake Bresztik	1,841	150–250	0.8	93.6	10	0.5	153	192	Simple meander
8	Adravica of Drávasztára	3,680	150–340	1	93.6	9	2	355	370	Simple meander
9	Lake Fekete	860	120–175	1	93.0	19	1	865	1,893	Short meander section
10	Lake Nagysziget	2,387; 1,971; 3,370	170–250	0	93.0	0	2	425	1,211	Simple meander

(continued)

Table 12.2 (continued)

No	Oxbow	Length (m)	Width (m)	Mean depth (m)	Mean lake level above sea level (m)	Water surface area in August 2014 (ha)	Height above the Drava channel (m)	Distance from the Drava channel (m)		Geomorphological character
								Straight line	Along drainage lines	
11	Lake Kápolna-puszta	5,619; 2,290	130–190	n.a.	93.0	7	3	5,421	8,015	Double meander
12	Lake Piskó (Hosszú + Kerek)	3,306; 1,817	110–240	n.a.	93.0	0.5	2	3,270	3,829	Double meander
13	Lake Verság	5,854	130–-340	1.2	92.9	12.8	4	1,438	3,160	Simple meander loop
14	Lake Kelemen-liget	1,912; 938	50–100	0.5	92.0	8.5	1	2,848	4,712	Composite meander
15	Lake Kis-szent-márton	5,890	130–190	dry	92.0	0	2	1,450	2,270	Simple meander
16	Lake Majláth-puszta	6,954	220–475	2	91.5	4	0	290	290	Simple meander
17	Cún-Szaporca oxbow	7,840; 1,400	151–290	1; 0.5; 0.5	90.15	20; 6; 6	0	280	285	Composite meander
18	Lake Matty	4,022; 963	50–200	1.2	86	10	2	2,972	3,630	Composite meander
19	Hótedra	1,846	100–170	1.5	86	4	2	1,356	2,982	Simple meander

12.4.3 Geomorphic Evolution

The geomorphic evolution of landforms in the Hungarian morphological floodplain of the Drava River has never been studied. From the above described hydromorphological properties, however, it can be at least presumed that *evolution history* has been largely different for older and younger oxbow rows (Fig. 12.3). Accepting the concept of gradual shift of the meander belt in southwestern direction over the Quaternary (Lovász 1964, 2013), this allows the conclusion that oxbows in the distal zone were cut off through natural processes during the lateral migration of the Old Drava in the Late Holocene at latest. In lack of absolute dating of oxbow deposits, no estimates can be made for the date of cutoff in the case of old oxbows. Their morphology (e.g. water depth, 'freshness' of banks and sediment plugs), thickness of organic fill and sporadic archaeological (human settlement) data (Bándi 1973) point to a loss of communication with the Drava mostly by neck cutoff some millenia ago. Most of the oxbows in the close neighbourhood of the present-day Drava channel, however, were detached from the main channel during river regulation works in the 19th century. The approximate time intervals when cutoff happened can only be estimated for the younger oxbows using map sheets of the 1st and 2nd Military Surveys as well as other archive maps (Table 12.3).

Table 12.3 shows that several meanders of the younger zone had been cut off from the present-day Drava channel shortly before river regulation started. Unfortunately, map representations before the 18th century are rare and difficult to interpret.

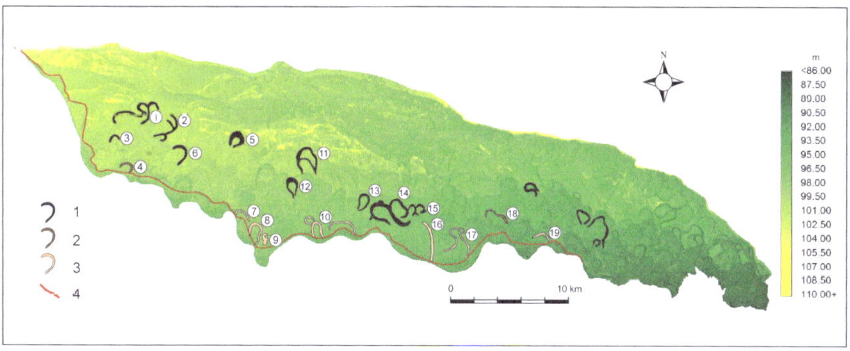

Fig. 12.3 Classification of oxbows according to the date of cutoff (by Dénes Lóczy). 1, 'old oxbows' with natural cutoff from a previous Drava channel in prehistorical times; 2, 'young oxbows' cut off by natural processes in historical times; 3, 'young oxbows' artificially cut-off in connection with river regulation; 4, main flood-control line. Numbered oxbows: 1, Lakócsa-Drávafok oxbow; 2, Fenék swamp; 3, Kanszki-berek; 4, Mrtvica of Felsőszentmárton; 5, Lake Sellye; 6, Lake Drávakeresztúr; 7, Lake Bresztik; 8, Adravica of Drávasztára; 9, Lake Fekete; 10, Lake Nagysziget; 11, Lake Kápolna-puszta; 12, Lakes of Piskó (Hosszú and Kerek); 13, Lake Verság; 14, Lake Kelemen-liget; 15, Lake Kisszentmárton; 16, Lake Majláth-puszta; 17, Cún-Szaporca oxbow; 18, Lake Matty; 19, Hótedra

Table 12.3 Estimated dates of cutoff for the younger series of oxbows

No	Oxbow	Date (register mark) of archive map	
		With meander still active	With meander already cut off
4	Mrtvica	None	1784 (1st Military Survey)
6	Lake Drávakeresztúr	1784 (1st Military Survey)	1850 (2nd Military Survey)
7	Lake Bresztik	None	1784 (1st Military Survey)
8	Adravica of Drávasztára	None	1784 (1st Military Survey)
9	Lake Fekete	None	1784 (1st Military Survey)
10	Lake Nagysziget	None	1784 (1st Military Survey)
16	Lake Majláth-puszta	1833 (06 BmT 260)	1850 (2nd Military Survey)
17	Cún-Szaporca oxbow	1833 (06 BmT 260)	1850 (2nd Military Survey)

12.4.4 Water Balance

Detailed water balance studies were performed for the representative oxbow of Cún-Szaporca (Fig. 12.2), where a monitoring system of soil and hydrometeorological conditions was established. The parameters monitored for the period between July 2013 and December 2015 covered virtually all possible factors influencing water inflow into and outflow from the lake:

(1) infiltration (measured by Drain Gauge G2 Passive Capillary Lysimeter);
(2) water potential (MPS-2);
(3) soil temperature and
(4) soil moisture content (5TM) at various depths (25 cm and 70 cm from ground surface) (sensors manufactured by Decagon Devices, Pullmann, WA, USA);
(5) depth to groundwater table (monitored by Dataqua LB 601 instruments—Dataqua Co., Balatonalmádi, Hungary).

Data collection (EM50) intervals were set for 30 min for all subsurface measurements and 10 min for the rainfall measurements. Precipitation has been recorded by an ECRN-100 tipping-bucket rain gauge (Decagon Devices Inc, Pullman, WA, USA) next to the northeastern section of the oxbow.

The obtained time series of soil moisture contents and infiltration rates, however, were too short to reveal trends. Therefore, 16 characteristic rainfall events were analyzed. In the wake of these events, considerable changes in both groundwater table elevation and shallow subsurface soil moisture content were registered at monitoring sites (Lóczy et al. 2017). The findings were spatially extended on the basis of the map of physical soil types (AGROTOPO 2013–2017). Water retention capacity in floodplain soils is estimated from the porosity of various physical soil types, generalized from the analysis of 48 soil profiles (Fig. 12.4).

To reveal the hydrogeology of alluvial deposits and the dynamics of groundwater flow, both field measurements and laboratory investigations were performed. An *aquifer test* served to establish the rate of seepage from the oxbow after refilling

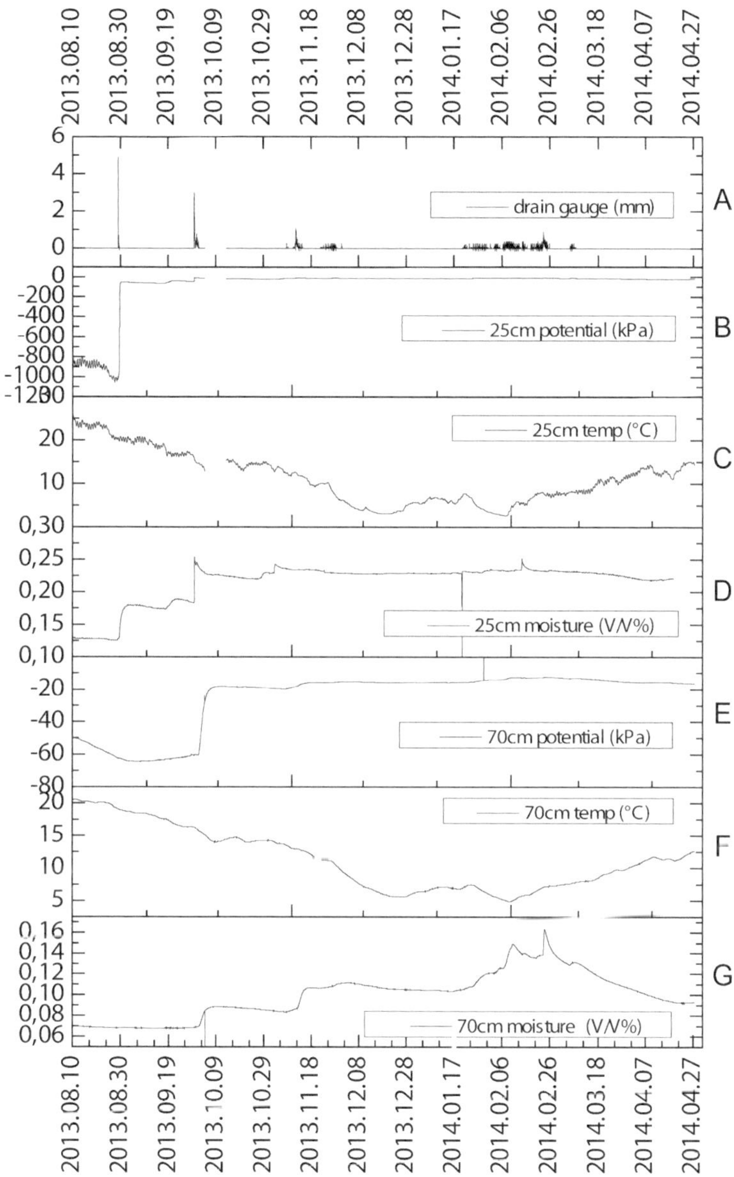

Fig. 12.4 Infiltration after various types of rainfall events (by József Dezső). A, daily rainfall (mm); B, water potential at 25 cm depth (kPa); C, soil temperature at 25 cm depth (°C); D, soil moisture content at 25 cm depth (v/v%); E, water potential at 70 cm depth (kPa); F, soil temperature at 70 cm depth (°C); G, soil moisture content at 70 cm depth (v/v%)

(cf. Halford and Kuniansky 2002; Mclin 2007). The hydraulic conductivity (k) of the deposits in the immediate environs of the oxbow was calculated. In a laboratory experiment hydraulic conductivity was established for three intact samples from the oxbow lake using the principle of water column with reducing pressure (for details see Neuman 1972 and Chap. 14).

Laboratory experiments were performed to establish water loss from the oxbow lakes by seepage. Soil moisture conditions in the alluvial deposits of complex stratification was modelled using Hydrus-1D applied for 69 profiles of various textural composition at three different groundwater table depths (modelling based on Sanford 2002—for details see Chap. 14). Oxbow beds are either clayey-silty or calcareous sandy. The clogging of the oxbow bed was found critical for groundwater flow. Water flow in impervious deposits is of the order of 10^{-8} m s^{-1}.

Both summer and winter periods are characterized by low infiltration rates. Recorded *groundwater levels* were found to drop with increasing distance from the oxbow (Fig. 12.5). This also proves that seepage from the oxbow takes place and it is the main source of groundwater recharge (cf. Winter 1999). Groundwater table dynamics in the area are rather controlled by lateral flow (hyporheic flow in the south and groundwater flow from the north) than infiltration. Through groundwater flow, the water regime of the Drava exerts a control on oxbow lake levels (Fig. 12.6).

Evaporation from open water surfaces was calculated using empirical formulas (see Lóczy et al. 2017) and was found to range between 500 and 600 mm y^{-1} adding up to 70,000–84,000 m^3 of water vapour from the average open lake surface of ca 140 ha. *Transpiration* of riparian softwood (willow and poplar) forests was estimated on the principle of analogy relying on Central European forestry literature (Čermák and Prax 2001, 2009). Cumulative evapotranspiration was calculated on the basis of the proportions of tree species in the softwood forests of the environs of the oxbow (Table 12.4).

12.4.5 Water Retention Capacity

Potentially, a major ecosystem function of oxbows is the storage of surplus water during extremely wet periods or high water stages of rivers (e.g., Thoms 2003). The volumetric estimation of maximum possible water retention capacity for all the floodplain lakes was performed (Lóczy et al. 2017). The Digital Elevation Model (DEM) used for the analysis was the Hidro-DEM, prepared for the South-Transdanubian Water Management Directorate (DDVÍZIG). It is based on LiDAR topographic survey and has 0.5-m vertical and 10-m horizontal resolution. Average volumes of water storage for 17 major Drava oxbows are calculated relying on the DEM, field checking and literature data (Pálfai 2001). Oxbow cross-sections were assumed to be trapezoid. Sedimentation in oxbow lakes since cutoff was estimated at an average of 20 cm. (For lakes used as fish and angling ponds dredging was considered.)

Fig. 12.5 Relationship between groundwater levels in three observation wells, water level of Lake Kisinc and the water regime of the Drava at the Drávaszabolcs gauging station (by József Dezső)

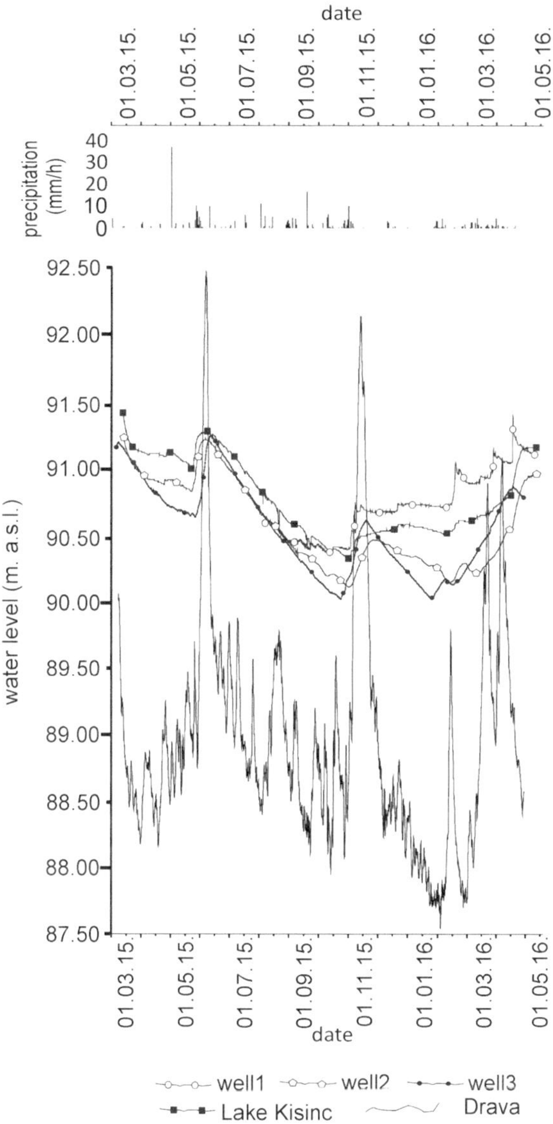

The results of volumetric estimates indicate a potential retention of almost 10 million m^3 of floodwater in the oxbows (Table 12.5). Naturally, the exploitation of this opportunity for water retention strongly depends on channel/floodplain connectivity on the surface, i.e. how easy it is to conduct floodwater and drain excess water into the oxbows.

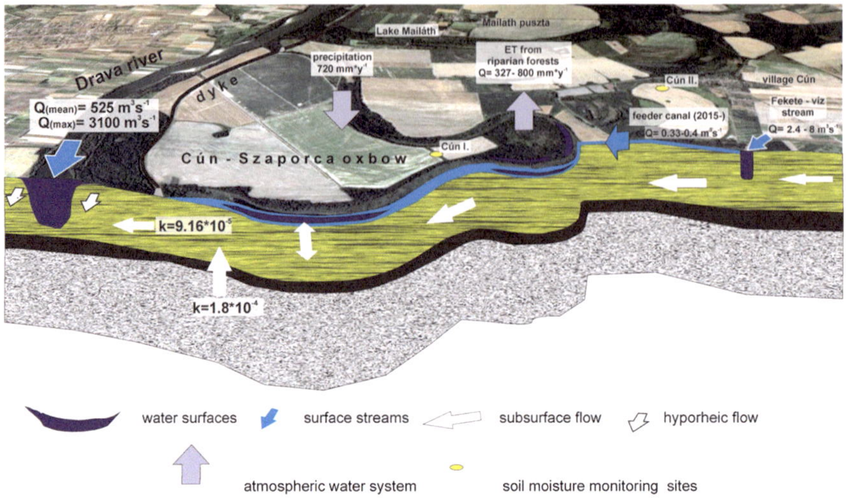

Fig. 12.6 Estimated water balance of the Cún-Szaporca oxbow (by József Dezső)

Table 12.4 Estimated water loss by transpiration from oxbow softwood forests (by Hedvig Prokos)

Tree species	Area (ha)	Share in total forest area (%)	Mean annual ETP (mm)	Total loss by ETP (m^3 y^{-1})
poplar (*Populus* sp.)	42.4	35	710	298,200
willow (*Salix* sp.)	30.4	25	750	225,000
black locust (*Robinia pseudoacacia*)	27.6	23	279	77,004
maple (*Acer* sp.)	15.6	13	ca 400	ca 60,000
sessile oak (*Quercus petraea*)	2.4	2	441	10,584
elm (Ulmus sp.)	2.4	2	ca 400	ca 10,000
Total	120.8	100		610,788 + ca 70,000
Total evapotraspiration				ca 3,660,000

12.5 Rehabilitation Potential

The assessment of restoration/rehabilitation potential covers the study of floodplain landforms (size, shape, configuration, connectivity), landscape pattern, vegetation, land use from the aspects of the maintenance of protected areas and provisioning various ecosystem services (National Research Council 1992; WWF International 2010; Waidbacher and Schultz 2005).

Table 12.5 Water retention capacities of oxbow lakes in the Hungarian Drava Plain (calculated by Péter Gyenizse)

	Oxbow, backswamp	Mean lake level above sea level (m)	Water level elevation when filled up (m)	Water surface area when filled up (total oxbow area, ha)	Planned dredging (thousand m^3)	Maximum water retention capacity (million m^3)
1	Lakócsa-Drávafok oxbow	98.0	99.0	119.7	200	0.86
2	Fenék swamp	98.0	98.5	23.3	0	0.08
3	Kanszki-berek	98.0	99.5	21.1	0	0.06
4	Mrtvica of Felsőszentmárton	96.0	97.2	22.4	30	0.26
5	Lake Sellye	96.0	98.0	70.5	100	1.10
6	Lake Drávakeresztúr	97.0	98.0	13.3	80	0.11
7	Lake Bresztik	93.6	95.0	18.2	0	0.07
8	Adravica of Drávasztára	93.6	94.0	12.6	70	0.29
9	Lake Fekete	93.0	98.0	134.7	50	1.60
10	Lake Nagysziget	93.0	96.0	45.3	100	0.26
11	Lake Kápolna-puszta	93.0	95.0	12.6	0	0.08
12	Lake Piskó (Hosszú + Kerek)	93.0	95.0	19.2	50	0.14
13	Lake Verság	92.9	93.5	27.7	100	0.25
14	Lake Kelemenliget	92.0	92.5	21.6	70	0.12
15	Lake Kisszentmárton	92.0	92.5	17.7	0	0.18
16	Lake Majláth-puszta	91.5	93.5	213.1	0	2.6
17	Cún-Szaporca oxbow	90.15	92.3	60.0	150	0.55

At present, during high water stages of the Drava River water recharge to oxbow lakes is possible through the opening of a flood gate from the direction of the Drava channel. This solution, although alleviates flood hazard downstream, is not found reliable for water replenishment. A probability analysis of replenishment options to the Cún-Szaporca oxbow shows the probability of a situation when the Drava water level reaches 360 cm (i.e., 91.50 m above sea level) is P = 2% (calculating with the summer half-year only: P = 14%). Naturally, raising water levels to 92.00 m would require even higher water levels of the Drava River: 410 cm on the Drávaszabolcs gauge – P = 1% (for the summer half-year: P = 3%) (DD-KTVF 2013).

As a consequence, water replenishment from northern direction, from the Fekete-víz Stream (mean discharge: 4.5 $m^3 s^{-1}$), was decided on (DDKÖVÍZIG 2012). The natural discharge of the stream is increased by damming the Fekete-víz Stream. This reservoir is designed to be filled up gravitationally or, in periods when it becomes necessary, by pumping with Drava water (see Chap. 21). The main water transfer canal follows an old course of the river and for the distribution of water abandoned channels and infilled oxbows and backswamps are planned to be revitalized (Lóczy et al. 2014).

The rehabilitation potential is fundamentally determined by the processes of connectivity of the hydrological system: underground water flow from the oxbow and the communication with hyporheic flow (see Chap. 14). The ongoing rehabilitation scheme strives to improve both lateral and vertical connectivity (in the sense of Roni et al. 2013) along the Drava River (Lóczy et al. 2017). A principal factor of uncertainty for the envisioned water recharge is the insufficiently known hydraulic connection between the oxbow bed and the geology of its immediate neighbourhood. It is assumed that good ecological status can only be achieved if periodical flushing of oxbow lakes is ensured.

There is also much uncertainty about the quality of the replenished water. The water will be impounded for about two weeks in order the ensure sufficient flow ($Q = 0.33–0.5\ m^3 s^{-1}$) for the required rate of replenishment. Storage might cause water quality deterioration with respect to critical parameters.

12.6 Conclusions

The hydromorphological survey of the Drava floodplain was directed to practical purposes, i.e. underpinning the assessment of rehabilitation potential with new data. Therefore, the hydrogeological properties of alluvial deposits, the water balance and retention potential of oxbows were placed in the focus of research. The grouping of oxbows according to the date of their cutoff from a Paleo-Drava or from the present-day channel is purely hypothetical. With no means for dating oxbow lake deposits, it was not possible to reconstruct the chronology of geomorphic evolution. Tracing the gradual shift of the Drava channel in space and time would require more efforts in the interpretation archive maps, Ground Penetrating Radar profiles, Digital Elevation Models as well as field checking. Such activities are partly underway and partly planned in a follow-up project to be implemented in cooperation with Croatian experts. The practical implication of such a basic research would be an improved alignment of the planned water recharge routes.

Acknowledgements Authors are grateful for financial support from the Hungarian National Scientific Fund (OTKA, contracts K 104552 and K 108755) and for help from the South-Transdanubian Water Management Directorate.

References

AGROTOPO (2013–2017) Agrotopographic Data Base. Institute for Soil Sciences and Agricultural Chemistry Centre for Agricultural Research, Hungarian Academy of Sciences, Budapest. https://maps.rissac.hu:3344/webappbuilder/apps/2/

Amoros C, Bornette G (2002) Connectivity and biocomplexity in waterbodies of riverine floodplains. Freshwater Biol 47:761–776

Bándi G (1973) Baranya megye az őskorban (Baranya County in prehistoric times). In: Bándi G, Burger A, Fülep F, Kiss A (eds) Baranya története az őskortól az Árpád korig (History of Baranya County from prehistory to the Árpáds' Age). Papers of the Janus Pannonius Museum, Pécs 15:12–17 (in Hungarian)

Bizjak A, Cunder M, Marovt L, Skrt P, Softič M, Zinke A (2014) The SEE river toolkit for facilitating cross-sectoral management of river corridors. Booklet 2: Practical applications on the Drava River. Institute for Water of the Republic of Slovenia, Ljubljana. 62 pp

Blanka V, Mezősi G, Meyer B (2013) Projected changes in the drought hazard in Hungary due to climate change. Időjárás 117(2):219–237

Borhidi A (1997) Social behaviour types, their naturalness and relative indicator values of the higher plants of the Hungarian flora. Acta Botanica Hungariae 39:97–182

Brierley GJ, Fryirs KA (2005) Geomorphology and river management. Applications of the River styles framework. Blackwell Publishing, Carlton, Victoria, p 398

Brookes A (1996) Floodplain restoration and rehabilitation. In: Anderson MG, Walling DE, Bates PD (eds) Floodplain processes, Chap 17. John Wiley & Sons, Chichester, pp 553–576

Buchberger P (1975) A Dráva-völgy árvédelmének története (History of flood control in the Drava Valley). Vízügyi Közlemények 1:103–112 (in Hungarian)

Čermák J, Prax A (2001) Water balance of a Southern Moravian floodplain forest under natural and modified soil water regimes and its ecological consequences. Ann Forest Sci 58:15–29

Čermák J, Prax A (2009) Transpiration and soil water supply in floodplain forests. Ekológia (Bratislava) 28(3):248–254

Corenblit D, Tabacchi E, Steiger J, Gurnell AM (2007) Reciprocal interactions and adjustments between fluvial landforms and vegetation dynamics in river corridors: a review of complementary approaches. Earth Sci Rev 84(1):56–86

Dawidek J, Ferencz B (2014) Water balance of selected floodplain lake basins in the Middle Bug River valley. Hydrol Earth Syst Sci 18:1457–1465. https://doi.org/10.5194/hess-18-1457-2014

DDKÖVIZIG (2012) Revitalization of the Cún-Szaporca oxbow system. Final Master Plan. South-Transdanubian Environment and Water Directorate (DDKÖVIZIG), Pécs. 100 pp. http://vpf.vizugy.hu/uploads/ddvizig/projekt/lezarult-fejlesztesek/regionalis/INTERREG_IV_vegleges_master_plan_angol.pdf

DD-KVTF (2013) Jegyzőkönyv közmeghallgatásról: a Dél-dunántúli Vízügyi Igazgatóság "A Cún –Szaporcai holtág vízpótlása az Ős-Dráva Program keretén belül" projekt környezeti hatásvizsgálati eljárása (Protocol of the Public Hearing on the EIS procedure of the project "Water replenishment of the Cún-Szaporca oxbow within the Old Drava Programme" by the South Transdanubian Water Management Directorate). South Transdanubian Inspectorate for Environmental Protection, Natural Protection and Water Management, Pécs. 3 pp. (in Hungarian)

Dépret T, Riquier J, Piégay H (2017) Evolution of abandoned channels: insights on controlling factors in a multi-pressure river system. Geomorphology. https://doi.org/10.1016/j.geomorph.2017.01.036

Dragun D, Gavran A, Car V (2014) Hydrographic survey of the River Drava branches in the process of revitalization, flood control and morphological monitoring. In: FIG Congress 2014 Engaging the Challenges - Enhancing the Relevance, Kuala Lumpur, Malaysia 16–21 June 2014. 10 pp. https://www.fig.net/resources/proceedings/fig_proceedings/fig2014/papers/ts06j/TS06J_dragun_gavran_6869.pdf

Erskine WD, Warner RF (1988) Geomorphic effects of alternating flood and drought dominated regimes on New South Wales coastal rivers. In: Warner RF (ed) Fluvial geomorphology of Australia. Academic Press, Sydney, pp 223–244

Fleit E, Márk L, Sindler Cs (2012) Restoration of Szaporca oxbow system at River Drava. South-Transdanubian Environment Protection and Water Management Directorate, Pécs. 23 pp

FLUVIUS (2007) Hydromorphological survey and mapping of the Drava and Mura Rivers. FLUVIUS, Floodplain Ecology and River Basin Management, Vienna. 140 pp

González del Tánago M, Gurnell AM, Belletti B, García de Jalón D (2015) Indicators for river system hydromorphological character and dynamics: understanding current conditions and guiding sustainable river management. Aquatic Syst 78(1):35–55. https://doi.org/10.1007/s00027-015-0429-0

Graf WL (2006) Downstream hydrologic and geomorphic effects of large dams on America rivers. Geomorphology 79(3):336–360

Grlica I (2008) Studija biološke raznolikosti rijeke Drave: Dravske mrtvice i odvojeni rukavci 2 dio (Study of biological diversity of the Drava River: Drava oxbows and side-arms). State Institute for Nature Protection, Virovitica. 79 pp. http://www.dzzp.hr/dokumenti_upload/20100423/dzzp201004231344260.pdf (in Croatian)

Gurnell AM, Corenblit D, García de Jalón D, González del Tánago M, Grabowski RC, O'Hare MT, Szewczyk M (2015) A conceptual model of vegetation-hydrogeomorphology interactions within river corridors. River Res Appl 32(2):142–163. https://doi.org/10.1002/rra.2928

Gurnell AM, O'Hare JM, O'Hare MT, Dunbar MJ, Scarlett PM (2010) An exploration of associations between assemblages of aquatic plant morphotypes and channel geomorphological properties within British rivers. Geomorphology 116(1):135–144

Gyenizse P, Lóczy D (2010) The impact of microtopography and drainage on land use and settlement development in the Hungarian Drava Plain. Hrvatski Geografski Glasnik 72(1):5–20

Habersack H, Schober B, Hauer C (2015) Floodplain evaluation matrix (FEM): an interdisciplinary method for evaluating river floodplains in the context of integrated flood risk management. Nat Hazards 75:5–32. https://doi.org/10.1007/s11069-013-0842-4

Halford KJ, Kuniansky EL (2002) Documentation of spreadsheets for the analysis of aquifer pumping and slug test data. US Geological Survey, Reston, VA (USGS Open-File Report 02-197)

Heiler G, Hein T, Schiemer F, Bornette G (1995) Hydrological connectivity and flood pulses as the central aspects for the integrity of a river-floodplain system. Regulated Rivers: Res Manag 11:351–361. https://doi.org/10.1002/rrr.3450110309

Hein T, Baranyi C, Reckendorfer W, Schiemer F (2004) The impact of surface water exchange on the nutrient and particle dynamics in side-arms along the River Danube, Austria. Sci Total Environ 328:207–218. https://doi.org/10.1016/j.scitotenv.2004.01.006

Hein T, Schwarz U, Habersack H, Nichersu I, Preiner S, Willby N, Weigelhofer G (2015) Current status and restoration options for floodplains along the Danube River. Sci Total Environ 543:778–790. https://doi.org/10.1016/j.scitotenv.2015.09.073

Káldy-Nagy Gy (1960) Baranya megye XVI. századi török adóösszeírásai (Turkish defters of Baranya County in the 16th century). Hungarian Linguistic Society, Budapest (in Hungarian)

Kiedrzyńska E, Kiedrzyński M, Zalewski M (2015) Sustainable floodplain management for flood prevention and water quality improvement. Nat Hazards 76:955–977. https://doi.org/10.1007/s11069-014-1529-1

Kiss T, Andrási G, Hernesz P (2011) Morphological alteration of the Dráva as the result of human impact. AGD Landscape Environ 5(2):58–75

Kiss A, Nikolić Z (2015) Droughts, dry spells and low water levels in medieval Hungary (and Croatia) I: the great droughts of 1362, 1474, 1479, 1494 and 1507. J Environ Geogr 8(1–2): 11–22. https://doi.org/10.1515/jengeo-2015-0002

Lóczy D, Dezső J, Czigány Sz, Gyenizse P, Pirkhoffer E, Halász A (2014) Rehabilitation potential of the Drava River floodplain in Hungary. In: Gâştescu P, Marszelewski W, Breţcan P (eds) Water resources and wetlands. In: Proceedings of the 2nd International Conference, 11–13 September 2014, Tulcea, Romania. Transversal Publishing House, Târgoviste, 21–29

Lóczy D, Dezső J, Sz Czigány, Prokos H, Tóth G (2017) An environmental assessment of water replenishment to a floodplain lake. J Environ Manag 202(2):337–347. https://doi.org/10.1016/j.jenvman.2017.01.020

Lóczy D, Kis É, Schweitzer F (2009) Local flood hazards assessed from channel morphometry along the Tisza River in Hungary. Geomorphology 113(3):200–209. https://doi.org/10.1016/j.geomorph.2009.03.013

Lovász Gy (1964) Geomorfológiai tanulmányok a Dráva-völgyben (Geomorphological studies in the Drava Valley.) Papers of the Transdanubian Scientific Institute, Hungarian Academy of Sciences, Pécs. 67–114 (in Hungarian)

Lovász Gy (2013) A jelenkori tektonika hatása a Duna, a Tisza és a Dráva hidrológiai folyamataira (The impact of Holocene tectonics on hydrological processes on the Danube, Tisza and Drava Rivers). Földrajzi Közlemények 137(3):248–256 (in Hungarian with English abstract)

Majer J (1998) A Dráva vízminőségének hosszú távú alakulása (Long-term water quality trends of the Drava River). Manuscript. Institute of Environmental Sciences, University of Pécs, Pécs. 39 pp (in Hungarian)

Mclin SG (2007) Hydrogeologic characterization of a groundwater system using sequential aquifer tests and flowmeter logs. In: Kues BS, Kelley SA, Lueth VW (eds) Geology of the Jemez Region II. New Mexico Geological Society 58th Annual Fall Field Conference Guidebook. 485–491. http://nmgs.nmt.edu/publications/guidebooks/58

Mezősi G, Blanka V, Zs Ladányi, Bata T, Urdea P, Frank A, Meyer B (2016) Expected mid- and long-term changes in drought hazard for the South-Eastern Carpathian Basin. Carpathian J Earth Environ Sci 11(2):355–366

Nanson GC, Croke JC (1992) A genetic classification of floodplains. Geomorphology 4(6): 459–486

National Research Council (1992) Restoration of aquatic ecosystems: science, technology and public policy. The National Academies Press, Washington, DC 576 pp. http://www.nap.edu/catalog/1807.html

Neuman SP (1972) Theory of flow in unconfined aquifers considering delayed gravity response of the water table. Water Resour Res 8(4):1031–1045

Ortmann-Ajkai A, Czirok A, Dénes A, Oldal I, Fehér G, Gots Zs, Kamarás-Buchberger E, Szabó E, Vörös Zs, Wágner L (2003) Dráva holtágak komplex állapotértékelése (Complex baseline survey of the Drava oxbows). In: Hanyus E (ed) Az EU Víz Keretirányelvének bevezetése a Dráva vízgyűjtőjén (Introduction of the EU Water Framework Directive). WWF Hungary, Budapest. 68–79 (in Hungarian)

Pálfai I (2001) Magyarország holtágai (Oxbows in Hungary). Hungarian Ministry for Transport and Water Management, Budapest. 231 pp (in Hungarian)

Piégay H, Darby SE, Mosselman E, Surian N (2005) A review of techniques available for delimiting erodible river corridor: a sustainable approach to managing bank erosion. River Res Appl 21:773–789. https://doi.org/10.1002/rra.881

Reckendorfer W, Funk A, Gschöpf C, Hein T, Schiemer F (2013) Aquatic ecosystem functions of an isolated floodplain and their implications for flood retention and management. J Appl Ecol 50:119–128. https://doi.org/10.1111/1365-2664.12029

Richards KS (1982) Rivers: form and process in alluvial channels. Methuen, London, p 357

Rinaldi M, Gurnell AM, González del Tánago M, Bussettini M, Hendricks D (2015) Classification and characterization of river morphology and hydrology to support management and restoration. Aquatic Stud 78(1):17–33. https://doi.org/10.1007/s00027-015-0438-z

Roni P, Pess G, Hanson K, Pearsons M (2013) Selecting appropriate stream and watershed restoration techniques. In: Roni P, Beechie T (eds) Stream and watershed restoration: a guide to restoring riverine processes and habitats. John Wiley and Sons, Chichester, pp 144–188

Sanford W (2002) Recharge and groundwater models: an overview. Hydrogeol J 10:110–120

Schober B, Hauer C, Habersack H (2015) A novel assessment of the role of Danube floodplains in flood hazard reduction (FEM method). Nat Hazards 75(Supplement 1):33–50. https://doi.org/10.1007/s11069-013-0880-y

Schwarz WL, Malanson GP, Weirich FH (1996) Effect of landscape position on the sediment chemistry of abandoned-channel wetlands. Landscape Ecology 11:27–38

Thoms MC (2003) Floodplain–river ecosystems: lateral connections and the implications of human interference. Geomorphology 56(3–4):335–349. https://doi.org/10.1016/s0169-555x(03)00160-0

Tockner K, Lorang MS, Stanford JA (2010a) River flood plains are model ecosystems to test general hydrogeomorphic and ecological concepts. River Res Appl 26:76–86

Tockner K, Pusch MT, Borchardt D, Lorang MS (2010b) Multiple stressors in coupled river-floodplain ecosystems. Freshwater Biol 55(1):135–151

Vajda T (2001) Adatok a Dráva menti középkori fokgazdálkodásról (Data on the medieval 'fok' economy on the Drava River). In: Weisz B, Balogh L, Szarka J (eds) Tanulmányok a középkorról (Studies on the Middle Ages). Szegedi Középkorász Műhely, Szeged, pp 125–137 (in Hungarian)

Valentová J, Valenta P, Weyskrabová L (2010) Assessing the retention capacity of a floodplain using a 2D numerical model. J Hydrol Hydromech 58(4):221–232. https://doi.org/10.2478/v10098-010-0021-1

Viczián I, Cs Zatykó (2011) Geomorphology and environmental history in the Drava valley, near Berzence. Hungarian Geograph Bull 60(4):357–377

VKKI (2010) Vízgyűjtő-gazdálkodási terv. Dráva részvízgyűjtő (Watershed Management Plan: Drava Partial Watershed). Vízügyi és Környezetvédelmi Központi Igazgatóság (Central Directorate for Water Management and Environmental Protection), Budapest. 162 pp (in Hungarian)

Waidbacher H, Schultz H (2005) The Floodplain Index—a new approach for assessing the ecological status of river/floodplain-systems according to the EU Water Framework Directive. Large Rivers 15(1–4):169–185. http://www.boku.ac.at/hfa/forschung/graf/publications/large_rivers.pdf

Ward JV, Tockner K, Arscott DB, Claret C (2002) Riverine landscape diversity. Freshwater Biol 47(4):517–539

Winter TC (1999) Relation of streams, lakes, and wetlands to groundwater flow systems. Hydrogeol J 7:28–45

WWF (2002) Áttekintés a Dráva alsó vízgyűjtőjének tájhasználatáról (An overview of land use in the Lower Drava River catchment). Report for the Worldwide Fund for Nature Hungary, Budapest, 142 pp (in Hungarian)

WWF International (2010) Assessment of the restoration potential along the Danube and main tributaries. Working paper for the Danube River Basin. Final Draft. World-Wide Fund for Nature,Vienna. 60 pp. http://assets.panda.org/downloads/wwf_restoration_potential_danube.pdf

Chapter 13
Oxbow Lakes: Vegetation History and Conservation

Adrienne Ortmann-Ajkai

Abstract Along the heavily regulated Drava River, oxbows were formed by meander cut-offs in the first half of the 19th century. Now they are biodiversity refugia in an environment dominated by agricultural fields. Main seminatural vegetation types of the oxbows are euhydrophyte vegetation, reed and cattail beds, tussocky and non-tussocky large sedge beds, grey willow shrubs, and riparian willow-poplar or alder forests. The majority of oxbows harbour eu- or mesotrophic habitats. Oligo- and dystrophic habitats are far less common, endangered and protected. The aquatic and riparian vegetations of the oxbows of the Drava floodplain are relatively well-preserved, due to moderate land use, state-border position for fifty years, and partly National Parks status for 20 years. Since Hungary's accession to the European Union (2004), habitats and sites of European Community importance (Natura 2000) are the priorities of nature conservation. The main threat to oxbows is hydroseral succession, accelerated by riverbed incision, drought hazard and dropping groundwater levels. Other threats are excess nutrient load, spread of invasive species and oversized large game and herbivorous fish populations. For the sustenance of floodplain wetlands intensive, multi-sided and coordinated conservation measures are needed.

Keywords Wetland habitats · Succession · Threats · Nature conservation Natura 2000 · Nutrient overload · Game damage · Invasive plants Introduced fish species

---

A. Ortmann-Ajkai (✉)
Department of Hydrobiology, Institute of Biology, Faculty of Sciences, University of Pécs, Ifjúság útja 6, Pécs 7624, Hungary
e-mail: aadrienn@gamma.ttk.pte.hu

D. Lóczy (ed.), *The Drava River*, Springer Geography,
https://doi.org/10.1007/978-3-319-92816-6_13

13.1 Introduction

Like most large rivers in Europe, the Drava is heavily regulated. It had existed in natural state as a 'water wilderness' with a broad range of highly changeable aquatic and floodplain habitats until the end of the 18th century (Fig. 13.1). This diversity was further enhanced by different ages (different developmental stages) of water bodies, due to the natural dynamics of the unregulated river-floodplain system (Ward and Stanford 1995). People lived from a '*floodplain economy*', which utilized the manifold resources of the floodplain in complex and sustainable way. Flow regime was regulated by a drainage system (notch or 'fok' system), which stored the water of spring and early summer floods, maintaining wet conditions in the floodplain. It also protected the villages, settled on natural levees, from high floods. Productive wet meadows and forests were utilized by a sylvo-pastoral cattle breeding system involving seasonal transhumance (Andrásfalvy 1974).

The majority of *water regulation* interventions took place in the 19th century (see Chap. 8 for details). At first, large meanders (the present oxbows) had been cut off and bank-protection structures were built. A total of 62 cutoffs in the Hungarian stretch shortened the river by 182 km, to 60% its original length. The training of tributaries followed in the 1830s with the drainage of the floodplain (Ihrig 1973; Remenyik 2004).

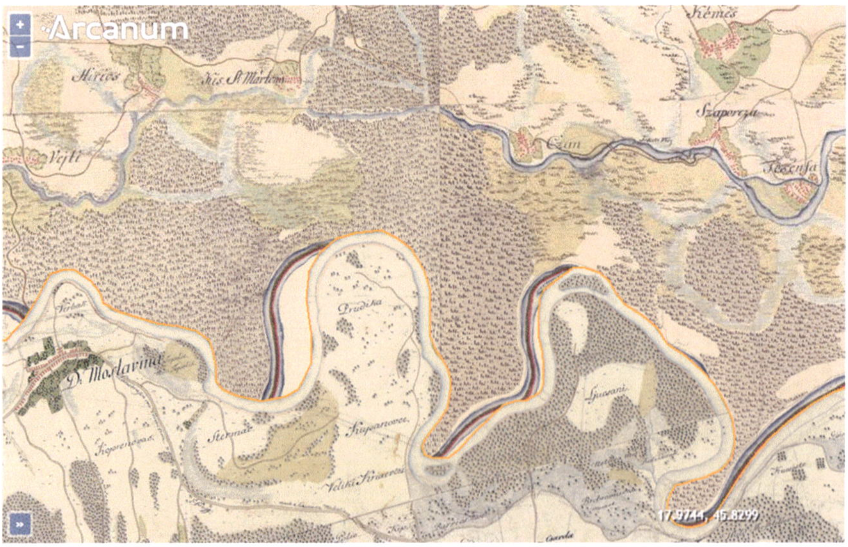

Fig. 13.1 Part of the Drava floodplain on the map of the 1st Military survey (1787), before regulations. Bottom: the large Drava meanders at Majláth-puszta and Cún-Szaporca, which became oxbows later; top: Fekete-víz Stream, yet unregulated, too. *Source* http://mapire.eu/hu/map/firstsurvey/

On the map of Fig. 13.1, the whole landscape is interwowen with eu- and plesiopotamon wetlands. Main landscape elements are forests and (wooded) pastures. Settlements and arable fields are situated on a higher sand ridge.

The present-day floodplain is divided by flood-control dykes into an active and a protected part. In the active floodplain, para- and plesiopotamon water bodies are characteristic. They are relatively undisturbed, surrounded by forests, and flooded by the Drava river in every spring. Due to sedimentation-dominated quick allogenic succession, manifested in decadal timespan, presently they are in an advanced stage of *terrestrialisation*. The active floodplain is protected as part of the Danube-Drava National Park. In recent years, several side-arm revitalisation projects started (Purger 2013) to restore connectivity (see Chap. 20 in this volume). Setting back earlier terrestrialization states, the restart of succession in some sites resulted in higher habitat diversity at floodplain level.

13.2 Vegetation of Oxbows

In the protected floodplain, oxbow lakes (paleopotamon) are the most important wetlands, formed by the meander cut-offs in the first half of the 19th century (see Chap. 12). They represent a relatively new habitat type, which had not existed before the regulations. Oxbows are biodiversity refugia in a land-use mosaic dominated by agricultural fields (Fig. 13.2). As a rule, larger oxbows of more than 15 ha area undergo slower succession dominated by autogenic processes, slowed down further by the removal of aquatic plants by local anglers. Smaller oxbows are shallower and only temporarily waterlogged at their bottom. Due to more rapid allogenic succession, they are mostly covered by woody vegetation.

The aquatic and riparian vegetations of oxbows are diverse, rich in species, and the whole hydroseral succession (Ortmann-Ajkai and Dénes 1998; Dénes and Ortmann-Ajkai 2006) and all the vegetation zones can be observed in most oxbow lakes (Fig. 13.2). The major seminatural vegetation types are euhydrophyte vegetation, reed and cattail beds, tussocky and non-tussocky large sedge beds, grey willow shrubs, riparian willow-poplar forests and, rarely, alder bog forests.

Oxbows are usually connected by two ditches to the main channel of the Drava (Fig. 13.2). Due to riverbed incision, the upper inflow canal is usually dry throughout the year (Lóczy et al. 2017). Water can flow into the oxbow through the lower ditch only at high water levels. The lower reach of the oxbow harbours diverse vegetation due to permanent waterlogging. The outer side the riparian reed-willow belt is fragmented by wooden anglers' stands. Oxbows are often bordered towards the arable fields by a treeline of native tree species (oak, ash, elm, poplar, willow), sometimes with spectacular, huge old specimen. On the inner side, this belt is wider (up to 50 m). Oxbow lakes are partly covered by floating and rooting aquatic vegetation, the open water surface in the centre is 25–100 m wide. At the lowest tip of the oxbow floor, the reed belt is wider. In the lowermost part willow-poplar forests (formerly called *Salici-Populetum*—see Chap. 17) can be

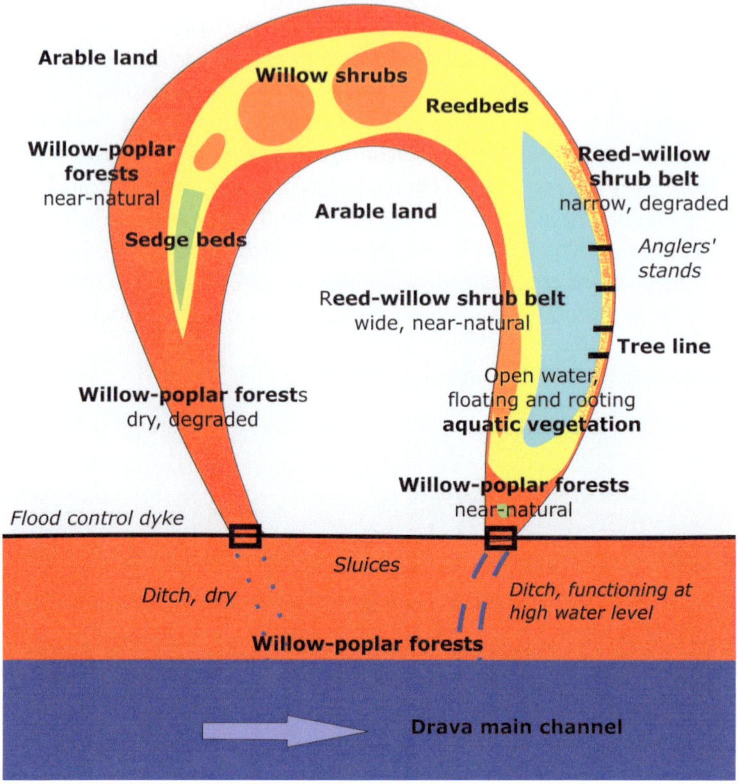

Fig. 13.2 Generalized vegetation pattern and landscape connections of oxbows (by Roland Hollós)

found, with patches of sedge beds. They are flooded in every spring for several weeks, so they are the most natural forests of the oxbows.

Towards the higher end of the oxbows, water levels and flooding intervals decrease. Reed beds and grey willow shrubs (*Leucojo aestivi-Salicetum albae*) replace aquatic communities. Sedge beds are rare, presumably because their zone is invaded by *Salix cinerea*. About half of the oxbow floor is covered by riparian softwood forests. On lower ground, where water supply is better, they can be pretty natural. With growing water deficit, they became drier and degraded.

The majority of oxbows are eu- or mesotrophic, with corresponding plant associations. Oligo- and dystrophic habitats are far less common, endangered and protected. In regulated floodplains, due to disrupted connection of water bodies with the main channel, many undrained wetlands with stagnating water have been formed after the regulation. Anaerobic conditions lead to peat accumulation, increasing acidity and lower trophic level. Under these conditions dystrophic plant communities emerged. *Caricetum elatae* were considered as a characteristic, common association in the 1960–70s (Vöröss 1964; Kovács and Kárpáti 1973), with characteristic dystrophic species such as *Carex elata, Hottonia palustris, Ranunculus lingua, Thelypteris palustris, Urtica kioviensis, Utricularia vulgaris*. Land drainage led to more aerobic conditions, peat decomposition and decreasing acidity; while landscape-scale eutrophication turned dystrophic waters to meso- or eutrophic. Dystrophic habitats get more scarce and endangered both in the European Union (Janssen et al. 2016) and in Hungary (for the Drava floodplain: Ortmann-Ajkai 2002). Presently, they are listed as habitats of Community Importance (Natura 2000 habitats: EC 2007), and all protected by law in Hungary. More dystrophic species got extinct or get extremely rare in the last decades, as e.g. *Menyanthes trifoliata* and *Pedicularis palustris* (Ortmann-Ajkai and Dénes 1999).

Botanical data from some oxbows are known since the 1960s (Vöröss 1963, 1964; Kovács and Kárpáti 1973), but detailed studies began in the 1990s, at the time of the foundation of the Danube-Drava National Park (Dénes et al. 1998; Ortmann-Ajkai and Dénes 1998).

13.2.1 Euhydrophyte Vegetation

Due to different types of oxbows and varied intensity of use, euhydrophyte vegetation is especially diverse. The majority of euhydrophyte communities belongs to the Natura 2000 habitat type "Natural eutrophic lakes with Magnopotamion or Hydrocharition vegetation" (EC 2007).

Present in all oxbows, floating aquatic communities comprise some typical aquatic species, such as *Lemna minor, Spirodela polyrrhiza, Hydrocharis morsus-rani* and *Salvinia natans*, mixed in variable quantities (*Lemnetum minoris, Lemno minoris-Spirodeletum, Lemno-Hydrocharietum morsus-ranae, Salvinio-Spirodeletem*). Under the floating plants a dense mat of *Ceratophyllum demersum* or *C. submersum* is found. *Lemnetum trisulcae* occurs in more shaded waters, often in the active floodplain.

More spectacular emergent vegetation consists of different assemblages of large, rooting species (*Nymphaea alba, Nymphoides peltata, Trapa natans* and *Nuphar lutea*). The first three species are protected in Hungary. The *Nymphaetum albo-luteae* association is subdivided into monodominant subassociations of *Nymphaea alba* and of *Nuphar lutea,* both accompanied by floating plant species and a submersed layer of *Ceratophyllum demersum, C. submersum* or *Myriophyllum spicatum* (Borhidi et al. 2012). *Trapa natans*, whose large, starch-containing fruits were consumed in the past (called "water chestnut"), forms almost monospecific *Trapaetum natantis* community in some oxbows with population size fluctuating (Balázs and Erdős 2011). Possibly due to selective plant removal by anglers (they spare, even introduce these showy plants), to resistance to grazing of introduced herbivorous fish species and to their large stature, they are not endangered, stable over the last 20 years.

Natural dystrophic lakes and ponds (NATURA 2000 code: 3160) are rare. Their aquatic vegetation is characterized by *Lemno-Utricularietum vulgaris* and *Hottonietum palustris* communities, besides other common communities of *Lemna* spp., *Spirodela polyrrhiza, Hydrocharis morsus-ranae and Salvinia natans*. *Lemnetum trisulcae* occurs more often in slightly dystrophic waters, often with the water moss *Riccia fluitans*. They are especially endangered by nutrient overload, their frequency is decreased in the last decades (Ortmann-Ajkai et al. 2017b).

13.2.2 Marsh Vegetation

13.2.2.1 Reed and Cattail Beds

Reed and cattails stands represent the next stage of succession. Dense stands of *Phragmites australis*, *Typha latifolia* and *T. angustifolia* are common everywhere in the riparian zone of oxbows. They form large beds (up to 5–10 ha) in terrestrialized (sections of) oxbows. The only habitat type not of Community importance, it is most common in the Drava floodplain, not only in oxbows, but also in artificial lakes, canals, ditches and all kinds of secondary wetlands. High closed stands côntain individuals or patches of *Schoenoplectus lacustris, Carex riparia, C. acutiformis, Glyceria maxima* or *Acorus calamus,* whose frequency decreased detectably in the last two decades. Common accessory species are *Calystegia sepium, Lysimachia vulgaris, Lythrum salicaria, Solanum dulcamara, Solidago gigantea, Stachys palustris*. A lower layer is formed by the common aquatic plants, *Lemna* spp., *Spirodela polyrrhiza, Hydrocharis morsus-ranae, Salvinia natans*.

Oligotrophic reed and Typha beds *Thelypteridi-Typhaetum angustifoliae* (oligotrophic floating fen community, with *Carex pseudocyperus, Thelypteris palustris, Urtica kioviensis*) occur only in few dystrophic oxbows, usually together with *Lemno-Utricularietum vulgaris* (Ortmann-Ajkai and Dénes 1998).

13.2.2.2 Large Sedge Beds

Stands of high, non-tussocky sedge species (*Carex riparia, C. acutiformis*) are also common. *Caricetum ripariae* and *C. gracilis* usually do not form a zone in steep banks outside of the reed belt, but occurs in the more terrestrialized upper reaches of oxbows and in small gaps of willow-poplar forests. *Carex* species are usually present in the *Phragmitetum australis* stands. Furthermore, often embedded in meadows, sedge beds can cover several hectares in fully terrestrialized plesiopotamon wetlands. *Carex gracilis* occurs in smaller stands (*Caricetum gracilis*), usually in temporarily waterlogged sites in forests or in forest edges. Accessory species are similar to those of reedbeds: *Calystegia sepium, Lysimachia vulgaris, Lythrum salicaria, Solanum dulcamara, Stachys palustris).* In optimal conditions, these habitats are inundated in spring and dry out only in summer. In years with no inundation in spring, the alien invasive *Solidago gigantea,* and the native-invasive grey willow (*Salix cinerea)* overgrow and supress sedge stands.

Tussocky sedge stands (*Caricetum elatae*), belonging to Natura 2000 category "Alkaline fens (7230)", are pretty rare (known from three sites only) and deteriorating. Their loss of area is due to the spreading of *Salix cinerea,* facilitated by terrestrialization.

13.2.3 Shrubs and Forests: A Short Summary

Salix cinerea shrubs show massive increase in the last two decades. They are present in all oxbows, forming a belt (often between angling stands), and apparently spread over herbaceous marsh habitats. Riparian forests: willow (*Salix alba*)—poplar (*Populus alba*) and alder (*Alnus glutinosa*) groves; 91E0) also showed a slight increase (Ortmann-Ajkai et al. under review).

Bog shrubs and forests (*Calamagrostio-Salicetum cinereae* with *Thelypteris palustris, Urtica kioviensis, Hottonia palustris and Carex elata* as indicator dystrophic species) and *Fraxino pannonicae-Alnetum* (alder mires, part of 91E0) occur at few sites. For details on floodplain shrubs and forests (see Chap. 17).

13.3 Conservation and Sustainable Use

13.3.1 Conservation Status

Due to land use of moderate intensity, position on the national border for fifty years and partly to the National Park status for 20 years, both aquatic and riparian oxbow vegetations of the Drava are relatively well-preserved. Around 2000, eight of ten large oxbows were qualified in the 2nd, two in the 3rd naturalness class on a five-class scale (Ortmann-Ajkai et al. 2003). Since Hungary's accession to the European Union in 2004, habitats and sites of European Community importance (Natura 2000—EC 2007) are the priorities of nature conservation. Three special areas of conservation (SACs; below: Natura 2000 sites) are designated since 2010 in the Drava floodplain, which include oxbows: East Drava, Forest of Ormánság, and Wetlands and meadows of Ormánság (http://natura2000.cea.europa.eu). Oxbows are especially rich in Natura 2000 habitats. Dystrophic communities are protected by law in Hungary, but without buffer zones legal protection did not exclude most of the threats, e.g. nutrient overload, desiccation and game damage. The best-known and highlighted site is the Cún-Szaporca oxbow, which is protected since 1969, designated as Ramsar area in 1979, and part of the National Park since 1996 (see Chap. 12). Nevertheless, no special conservational measures were taken till the 2010s. In 2016, a water replenishment project was implemented and a visitors' centre was also built (Lóczy et al. 2017 and Chap. 20).

13.3.2 Land Use and Impacts Since 1950

In the socialist era (1949–1989), due to its border position to contemporary Yugoslavia, the Drava floodplain was one of the most neglected regions of Hungary. After World War II, nationalizations fundamentally transformed the structure of agriculture, nevertheless under socialism, the majority of the population still lived from agricultural activities (Gyenizse and Lóczy 2010).

Traditional extensive cattle breeding survived till the 1990s and preserved valuable species-rich meadows (Dénes et al. 1998; Ortmann-Ajkai and Dénes 1999). Being a strictly protected (military) border zone, the active floodplain remained mostly undisturbed although some natural willow-poplar stands were changed into monodominant cultivated willow and poplar plantations. Seminatural oak forests were managed by a clearcut system with 100-year rotations. The intensification of agriculture resulted in severe nutrient pollution of non-point origin. Large oxbows on the protected floodplain have been used for extensive recreation angling by local people, which slowed down their terrestrialization. (This is evidenced from the comparison of aerial photographs from 1960s and 2010s.) In contrast, the decline of open water surfaces is apparent in the undisturbed oxbows of the Danube-Drava National Park (Ortmann-Ajkai et al. 2017b). At the same time, fish introduction and fish feeding posed threats.

13.3.3 Changes and Threats Since 1990

Diverse and valuable natural vegetation of the Drava floodplain basically have been preserved, but not without losses in the last two decades. The following changes were identified (Ortmann-Ajkai et al. 2017b):

1. increased frequency of shrub and forest habitats;
2. reduced frequency of some dystrophic and protected species (*Acorus calamus, Hottonia palustris, Ranunculus lingua, Salvinia natans, Thelypteris palustris, Urtica kioviensis, Utricularia vulgaris*) and
3. declining diversity of communities.

According to the most recent study on the threats for European Red List habitats, wetlands are endangered by 12 factors (Janssen et al. 2016). In the Drava floodplain, the main threatening factor is hydroseral succession, accelerated by dropping groundwater levels. Other threats are excess nutrient load, the spreading of invasive species, and the overpopulation of large game and herbivorous fish. Agriculture has not intensified further in the last 20 years; on the contrary, it rather decreased since the 1990s. Forestry management plays a minor part, as the economic value of small-sized softwood riparian forests is low.

Hydroseral succession is the key natural process in floodplains. High diversity is maintained in the successional series of numerous habitats (Ward and Stanford

1995). At the same time, since new plesio- and paleopotamon water bodies cannot come about, succession leads to the only climax stage, floodplain forests. As clearly indicated by the increase of woody habitats in the last 20 years, declining habitat diversity exerts deleterious effects on habitat-specific species. Hydroseral succession is accelerated by the dropping groundwater table (Lóczy et al. 2017). Desiccation has been the main factor of degradation since the 1990s (Ortmann-Ajkai and Dénes 1998; Ortmann-Ajkai et al. 2003). Water replenishment can slow down or reset succession.

Pollution is a foremost threat (Janssen et al. 2016). In the Drava floodplain it is manifested as *nutrient overload* from non-point agricultural sources in the neighbouring fields. Nutrient overload was regarded as serious in the 1990s (Dénes and Ortmann-Ajkai 1998), declined after the change of the socio-economic regime (1989). As a consequence of the dissolution of large co-operative farms (Gyenizse and Lóczy 2010), large arable areas were abandoned (Ortmann-Ajkai and Dénes 1999). Although there was 85% decrease in fertilizer use between the peak in the end of the 1980s and 1995 (Central Statistical Office 2017), it is doubtful, whether and when it will apparently be indicated by the biota, as nutrients accumulate in soils and in waters (Blackwell and Pilgrim 2011; Mitsch et al. 2012). Furthermore, fertilizer use has shown a slow but steady decrease in the past 10 years again.

Nutrient overload accelerates the hydroseral succession (Bornette and Puijalon 2011) and leads to reduced biodiversity, especially in the case of dystrophic communities. Nitrogen overload destroys dystrophic communities by extirpating dystrophic species. Its landscape-level effect is indicated by the strong decrease of dystrophic communities, e.g. *Caricetum elatae* (Ortmann-Ajkai 2002). The serious decrease from nine to three occurrences of *Lemno-Utricularietum vulgaris* since the 1990s is also a warning signal. The frequency of occurrences of other mire associations (*Thelypteridi-Typhaetum angustifoliae, Calamagrostio-Salicetum cinereae and Fraxino pannonicae-Alnetum*) did not decrease, but there are no comparable data on the change of their extent or quality (Ortmann-Ajkai et al. 2017b).

Excess nutrients can also facilitate the spreading of *invasive plant species*, which are considered one of the most dangerous conservational problems globally (Mölder and Schneider 2011). *Echinocystis lobata (E. echinata), Phytolacca americana* (Vöröss 1963), *Symphyotrichium (Aster) lanceolatus* (Vöröss 1964), *Ambrosia artemisiifolia* (Vöröss 1965) were documented in the sixties as rarities. In the 1990s, twenty-three invasive species were described from the Drava (Ortmann-Ajkai and Dénes 1999). Fourteen of them were spreading spontaneously in wetland habitats; three were reported only from agricultural fields; six planted species were regarded as potentially invasive. For most species, there were no changes on this level. *Acer negundo, Echinocystis lobata, Erigeron annuus, Parthenocissus quinquefoliae*, *Robinia pseudacacia, Solidago gigantea* and *Vitis riparia* occurred "in masses" that time (Ortmann-Ajkai and Dénes 1999). The growth of invasive plant species in the last 20 years is apparent. Since the 1990s the spontaneous spreading of *Juglans nigra* and *Morus alba* have been observed and also the appearance of a new invasive species group, North American *Aster* species. *Impatiens glandulifera* and *Asclepias syriaca* are widespread recently, although their presence was limited in the 1990s (Ortmann-Ajkai et al. under review). Interestingly, no invasive aquatic plant (according to EU and Hungarian lists: www.ec.europa.eu, www.termeszetvedelem.hu) has been found in oxbow areas. At the same time, *Salvinia natans* (presently protected in Hungary) can be regarded as potentially invasive (Netten et al. 2010), and more thermophilous species can expand due to the warming climate.

For profitable *angling and hunting* abundant fish and game populations are maintained, which, in turn, damage and even destroy natural habitats. Decreasing diversity of euhydrophyte habitats, especially of communities composed of small-sized species, e.g. *Salvinia natans* or *Lemna* spp., is apparent. Vegetation removal by anglers inevitably harmed the riparian reed belt, but at the same time it slowed down succession in oxbows used for angling (Ortmann-Ajkai et al. 2017b). Park-like management of angling sites prevented the spread of most invasive species, although not of invasive lianas (*Echinocystis lobata, Vitis riparia*). Negative effects of herbivorous fish species on aquatic macrophytes (Laguna et al. 2016) are also documented by Balázs and Erdős (2011) in our region. More potentially invasive fish species are present in the oxbows. *Ctenopharyngodon idella* and *Hypophthalmichthys* spp. were introduced in the sixties. *Ctenopharyngodon* is continually introduced since then and present now in all oxbows used for angling. *Hypophthalmichthys* spp. are less common, while *Ameiurus melas* is present in all oxbows and seems to suppress *A. nebulosus,* introduced around 1900 into Hungary. Now both *Ameiurus* species are listed as invasives (www.termeszetvedelem.hu).

Hunting as a threat, resulting extreme high game populations, has emerged since the 1990s. *Game damage* had not been mentioned in earlier publications, but in 2015 it was observed in many sites in the form of uprooting, trampling or browsing. Hunting tourism is an important source of income, as the study area is one of the most renowned large game (especially *Cervus elaphus*) hunting regions of Europe. Hunting-oriented large game management has led to excessive game populations and damage. Wild boar numbers had doubled between 1994 and 2002 and have remained stagnant since then. Red deer have risen in a similar way until 2009, but have shown 25% rise recently (www.ova.hu). The regeneration of floodplain forests in Central Europe is heavily suppressed by game species (Hobza 2007), as it was also documented in this area (Ortmann-Ajkai et al. 2017a). Although game prefer undisturbed wetlands for feeding, resting and hiding, few publications are concerned with the damage caused by them in such habitats. Wild boars are especially attracted to wetlands. They inflict serious damage on vegetation (Barrios-Garcia and Ballari 2012), often destroy of growing sites of valuable, rare communities, as sedge beds or the herb layer of remnants of near-natural floodplain forests (Ortmann-Ajkai et al. under review). Furthermore, game spread propagules of weeds and invasive species (Mölder and Schneider 2011), and are vectors for pathogen organisms, e.g. ticks (Barrios-Garcia and Ballari 2012). Game damage is an important threat and without proper regulations and management it will further increase.

13.3.4 Possible Conservation and Rehabilitation Measures

Passive conservation is not sufficient to preserve wetlands under hydroseral succession. Active conservation, on the other hand, had not been an issue till the early 21st century. Management plans prepared for the national park before the 2000s usually suggested 'water replenishment' in general. Planning and implementation of large-scale rehabilitation projects began around 2010, due to the opening of EU financial sources. Four side-arm rehabilitation projects were implemented along this Drava section in 2011–2012 (Purger 2013), but without any impact on the oxbows.

In recent years, the implementation of a complex landscape-level rehabilitation scheme, the Old Drava Programme aims to integrate conservational measures (wetland rehabilitation) and eco-tourism development (Lóczy et al. 2016; see also Chap. 20). Unfortunately, wetland rehabilitation is often simply equalled to water replenishment or retention (Lóczy et al. 2016; 'techno-arrogance'—Clare et al. 2011). From 621 case studies, Moreno-Mateos et al. (2012) concluded that ecosystem structure and functions do not recover fully even after 100 years, and the recovery of biological components is slower than that of hydrologic features. At the same time, multifunctional approaches to floodplain management provide new opportunities to restore degraded floodplain areas, as proven by a synthesis for six European states (Schindler et al. 2016). If projects aim only the improvement of water supply, they will reset hydroseral succession to a former stage, but will not automatically lead to better conservational state. Other above-mentioned landscape-scale problems have to be managed too.

The National Agro-Environmental Programme of Hungary, started in 2002 (Nemes and High 2011), offers financial support for farmers who voluntarily adapt environment-friendly management methods, particularly in agricultural areas with drought and/or flood hazard. These methods include, among others, no dredging, no irrigation, fallowing in every five years, no fertilizing with sludge, limited pesticide use, fertilizing plan based on soil analysis, 12 m buffer zone next to water bodies. Support for environment-friendly management of reedbeds is also available (NAK 2015). Nutrient infiltration can be alleviated by buffer zones between the oxbows and the fields, but they are often missing (Ortmann-Ajkai et al. 2003), represented only by a tree row (Fig. 13.2).

The issue of excess game and fish populations is even more difficult to handle. Key stakeholders often disregard this problem. Large game populations are needed for the profitable hunting tourism. Although monetary compensation for game damage in forests (especially for forest regeneration) is legally mandatory, there is no such obligation for wetlands.

The introduction of alien, even potentially invasive, fish species into oxbow lakes provides better catch possibilities, and even considered 'increasing biodiversity'. In addition to the native *Cyprinus carpio*, potentially invasive alien species, like *Ctenopharyngodon idella* and *Clarias gariepinus,* are introduced regularly in amounts of up to five tons per year into the intensively used oxbows, disregarding legal regulations and also the proven negative effects of herbivorous fish species on aquatic macrophytes (Laguna et al. 2016).

As far as invasive plant species are concerned, *Robinia pseudacacia* has been a preferred tree species for plantations. *Fraxinus pennsylvanica, Acer negundo* and *Ailanthus altissima* had been intentionally introduced too, but regarded as a nuisance now. Some other invasives, e.g. *Solidago gigantea* or *Amorpha fruticosa,* are important food sources for honeybees. There are no integrated, landscape-level projects for control of invasive plant species, except that in a nature conservation area a *Robinia* stand has been replaced by a *Populus alba* stand.

13.4 Conclusions

Despite the above threats and effects, recent changes in oxbow vegetation are relatively low. However, we have to raise attention to the possible extinction debt (Kuussaari et al. 2009): conservationally valuable habitats and species are probably elements of the ecological memory of the floodplain (Ortmann-Ajkai et al. 2014), relicts of the pre-regulation conditions. It is to be noted that, although floodplains respond rather rapidly to interventions like land drainage, sometimes they respond slowly to other kinds of human impact, and no information is available on resilience thresholds of the studied systems. The unidirectional evolution of a floodplain may finally lead to a substantial reduction in diversity and a major shift in community composition (Tockner and Stanford 2002). Delayed effects may be more severe than expected from short-term projections (Johnson et al. 2012). Future sustenance of floodplain wetlands, including oxbows, needs intensive, multi-sided and coordinated conservation measures with participation of all possible stakeholders. Raising the awareness of decision-makers and local people to environmental problems is indispensable. In addition to providing moderate income for local people, ecotourism development can be instrumental to this purpose.

References

Andrásfalvy B (1974) A Duna mente népének ártéri gazdálkodása Tolna és Baranya megyében az ármentesítések befejezéséig (Floodplain management along the Danube in Tolna and Baranya counties till the end of river regulation). In: Tanulmányok Tolna megye történetéből VII. 480 pp (in Hungarian)

Balázs A, Erdős L (2011) Botanical survey of Southern Transdanubian oxbow lakes and the changes of the last 10 years. Natura Somogyiensis 19:41–50

Barrios-Garcia MN, Ballari SA (2012) Impact of wild boar (*Sus scrofa*) in its introduced and native range: a review. Biol Invasions 14:2283–2300

Blackwell MSA, Pilgrim ES (2011) Ecosystem services delivered by small-scale wetlands. Hydrol Sci J 56:1467–1484

Borhidi A, Kevey B, Lendvai G (2012) Plant communities of Hungary. Akadémiai Kiadó, Budapest, 600 pp

Bornette G, Puijalon S (2011) Response of aquatic plants to abiotic factors: a review. Aquat Sci 73:1–14
Central Statistical Office (KSH) (2017) https://www.ksh.hu/docs/hun/xstadat/xstadat_eves/i_omf002.html. Accessed 06 Mar 2017
Clare S, Krogman N, Foote L, Lemphers N (2011) Where is the avoidance in the implementation of wetland law and policy? Wetlands Ecol Manag 19:165–182
Dénes A, Kevey B, Ortmann-Ajkai A, Pálfai L (1998) A Dráva-sík védelmet érdemlő területei. Areas worth for protection in the Drava floodplain. In Hungarian with English abstract. Janus Pannonius Múzeum Évkönyve 41–42: 5–12
Dénes A, Ortmann-Ajkai A (1998) A Dráva baranyai holtágai. Általános és botanikai jellemzés, természetvédelmi kérdések. Drava oxbows in Baranya county. General and botanical description, conservation issues. In Hungarian, with English abstract and with 6 maps. Janus Pannonius Múzeum Évkönyve 43:5–26
Dénes A, Ortmann-Ajkai A (2006) A Dráva baranyai holtágai 2: Vegetációtérképek (Drava oxbows in Baranya county 2: Vegetation maps). Natura Somogyiensis 9:27–38 (in Hungarian with English abstract and 10 vegetation maps)
European Commission (2007) Interpretation Manual of European Union Habitats. http://ec.europa.eu/environment/nature/legislation/habitatsdirective/docs/2007_07_im.pdf
Gyenizse P, Lóczy D (2010) The impact of microtopography and drainage on land use and settlement development in the Hungarian Drava Plain. Hrvatski Geografski Glasnik 72:5–20
Hobza P (ed) (2007) Forest management systems and regeneration of floodplain forest sites. Reviewed proceedings from the international conference. Mendel University of Agriculture and Forestry, Brno. ISBN 978-80-7375-089-3, 242 pp
Ihrig D (ed) (1973) A magyar vízszabályozás története (History of river regulation in Hungary). National Water Management Office (OVH), Budapest. 398 pp (in Hungarian)
Janssen JAM, Rodwell JS, Garcia Criado M, Gubbay S, Haynes T, Nieto A, Sanders N, Landucci F, Loidi J, Ssymank A, Tahvanainen T, Valderrabano M, AF, Bita-Nicolae C, Acosta A, Aronsson M, Arts G, Attorre F, Bergmeier E, Bijlsma R-J, Bioret, F, Bita-Nicole, C, Biurrun I, Calix M, Capelo J, Carni M, Chytry M, Dengler J, Dimopoulos P, Essl F, Gardjfell H, Gigante D, Giusso del Galdo G, Hajek M, Jansen F, Kapfer J, Mickolajczak A, Molina JA, Molnár Z, Paternoster D, Piernik A, Poulin B, Renaux B, Schaminee JHJ, Sumberova K, Toivonen H, Tonteri T, Tsiripidis I, Tzonev R, Valachovic M (2016) European Red List of Habitats, Part 2: Terrestrial and freshwater habitats. EU. 44 pp
Johnson WC, Dixon MD, Scott ML, Raabe L, Larson G, Volke M, Werner B (2012) Forty years of vegetation change on the Missouri River floodplain. Bioscience 62:123–135
Kovács M, Kárpáti I (1973) Untersuchungen über die Zonations- und Produktionsverhaltnisse im Überschwemmungsgebiet der Drau (Studies on zonation and production in the Drava floodplain). Acta Botanica Hungarica 18:323–353 (in German)
Kuussaari M, Bommarco R, Heikkinen RK, Helm A, Krauss J. Lindborg R, Öckinger E, Partel M, Oino J, Roda F, Stefanescu C, Teder T, Zobel M, Steffan-Dewenter I (2009) Extinction debt: a challenge for biodiversity conservation. Trends Ecol Evol 24:564–571
Laguna C, Lopez-Perea JJ, Vinuela J, Florin M, Feliu J, Chicote A, Cirujano S, Mateo R (2016) Effects of invasive fish and quality of water and sediment on macrophyte biomass, and their consequences for the waterbird community of a Mediterranean floodplain. Sci Total Environ 551–552:513–521
Lóczy D, Dezső J, Sz Czigány, Prokos H, Tóth G (2017) An environmental assessment of water replenishment to a floodplain lake. J Environ Manag 202(2):337–347. https://doi.org/10.1016/j.envman.2017.01.020
Lóczy D, Dezső J, Ronczyk L (2016) Floodplain rehabilitation projects in Hungary: case studies from the Danube, Tisza, Körös and Drava rivers. Bull Serbian Geograph Soc 96:1–10
Mitsch WJ, Zhang L, Stefanik KC, Nahlik AM, Anderson CJ, Bernal B, Hernandez M, Song K (2012) Creating wetlands: primary succession, water quality changes, and self-design over 15 years. Bioscience 62:237–250

Moreno-Mateos D, Power ME, Comin FA, Yockteng R (2012) Structural and functional loss in restored wetland ecosystems. PLoS Biol 10:1–8
Mölder A, Schneider E (2011) On the beautiful diverse Danube? Danubian floodplain forest vegetation and flora under the influence of river eutrophication. River Res Appl 27:881–894
NAK (2015) Agrár-környezetgazdálkodás (Agro-Environmental Management. Handbook on submission of request for support). National Chamber of Agrarian Economy. http://www.nak.hu/kiadvanyok/kiadvanyok/130-akg-kezikonyv?path=kiadvanyok (in Hungarian). Accessed 31 Aug 2017
National Game Management Database (Hungary) http://ova.info.hu/vgstat.html. Accessed 06 Apr 2017
Nemes G, High C (2011) Implementation of Agri-Environmental Programme in Hungary. Regional Studies Association Annual International Conference, 17–20 April 2011. Newcastle-upon-Tyne, UK. http://www.regionalstudies.org/uploads/conferences/presentations/international-conference-2011/nemes.pdf
Netten JJC, Arts GHP, Gylstr R, van Nees EH, Scheffer M, Roijackers RMM (2010) Effect of temperature and nutrients on the competition between free-floating *Salvinia natans* and submerged Elodea nuttallii in mesocosms. Fund Appl Limnol 177:125–132
Ortmann-Ajkai A (2002) Ökológiai mutató-párok alkalmazása ártéri társulások vizsgálatában. Applying pairs of ecological indicator values in study of floodplain communities. In: Salamon-Albert É (ed) Magyar botanikai kutatások az ezredfordulón. Tanulmányok Borhidi Attila 70.születésnapja tiszteletére. PTE TTK Növénytani Tanszék, Pécs. pp 485–493
Ortmann-Ajkai A, Csicsek G, Lukács M, Horváth F (2017a) Regeneration patterns in a pedunculate oak (*Quercus robur* L.) in a strict forest reserve in southern Hungary. Sumarski List 91:39–46
Ortmann-Ajkai A, Czirok A, Dénes A, Oldal I, Fehér G, Zs Gots, Kamarásné Buchberger E, Szabó E, Zs Vörös, Wágner L (2003) Dráva holtágak komplex állapotértékelése (Complex evaluation of Drava oxbows). Az EU Víz Keretirányelvének bevezetése a Dráva vízgyűjtőjén: A WWF Magyarország tudományos tanácskozásának összefoglalója (Introducing the Water Framework Directive on the Drava catchment). WWF Hungary, Budapest, pp 68–79
Ortmann-Ajkai A, Dénes A (1998) Baranyai Dráva holtágak fátlan növénytársulásai (Non-forest plant associations of Drava oxbows in Baranya county). In Hungarian with English abstract and 113 coenological releves. Janus Pannonius Múzeum Évkönyve 43:27–39
Ortmann-Ajkai A, Dénes A (1999) Changing floodplain ecosystems in the last 200 years in the plain of Drava. Publ Univ Horticulturae industriaque alimentariae 59:130–134
Ortmann-Ajkai A, Lóczy D, Gyenizse P, Pirkhoffer E (2014) Wetland habitat patches as ecological components of landscape memory in a highly modified floodplain. River Res Appl 30:874–886
Ortmann-Ajkai A, Kovács A, Lóczy D (2017b) Dráva holtágak tájtörténete az utolsó 50 évben (History of Drava oxbows in the last 50 years). Janus Pannonius Múzeum Évkönyve 54, Pécs, 105–128 (in Hungarian)
Purger J (ed) (2013) Biotas and rehabilitation of four Drava river side-branches in Hungary. Danube-Drava National Park Directorate, Pécs, 164 pp
Remenyik B. (2004) A Dráva szabályozása a XVIII. századtól a XIX. század végéig (Regulation of Drava from the 18th century till the end of the 19th century) In: Proceedings of the 2nd Hungarian Conference on Geography, Szeged. http://geography.hu/mfk2004/mfk2004/cikkek/remenyik_bulcsu.pdf. Accessed 31 Aug 2017 (in Hungarian)
Schindler S, O'Neill FH, Biró M, Damm C, Gasso V, Kanka R, van der Sluis T, Krug A, Lauwaars S, Sebesvári Z, Pusch M, Baranovsky B, Ehlert T, Neukirchen B, Martin JR, Euller K, Mauerhofer V, Wrbka T (2016) Multifunctional floodplain management and biodiversity effects: a knowledge synthesis for six European conuntries. Biodiversity Conserv 25:1349–1382
Tockner K, Stanford JA (2002) Riverine floodplains: present state and future trends. Environ Conserv 29:308–330

Vöröss LZs (1963) Adatok drávamenti holtágak cönológiai és florisztikai ismeretéhez (Floristic and coenological data from Drava oxbows). Pécsi Tanárképző Főiskola Tudományos Közleményei 7:271–279 (in Hungarian)

Vöröss LZs (1964) Újabb adatok a szaporcai holtágak cönológiai és florisztikai ismeretéhez (New floristic and coenological data from the Szaporca oxbows) Szegedi Tanárképző Főiskola Tudományos Közleményei 8:75–95 (in Hungarian)

Vöröss LZs (1965) Adatok a szaporcai Dráva hullámtér vizeinek cönológiai és florisztikai ismeretéhez (Floristic and coenological data from the active Drava floodplain near Szaporca). Pécsi Tanárképző Főiskola Tudományos Közleményei 9:123–145 (in Hungarian)

Ward JV, Stanford JA (1995) Ecological connectivity in alluvial river ecosystems and its disruption by flow regulation. Regulated Rivers: Res Manag 11:105–119

http://ec.europa.eu/environment/nature/invasivealien/list/index_en.htm. Accessed 14 Aug 2017

http://mapire.eu/hu/map/firstsurvey/. Accessed 14 Aug 2017

http://www.termeszetvedelem.hu/termeszetes-allat-es-novenyvilagra-veszelyt-jelento-idegenhonos-invazios-fajok-hazai-tudomanyos-alapu-jegyzekei. Accessed 14 Aug 2017

Chapter 14
Floodplain Connectivity

József Dezső, Dénes Lóczy, Ali Mohamed Salem and Gábor Nagy

Abstract Floodplains fulfil vital ecosystem services (supporting water management, biodiversity, agricultural production, ecotourism and others). Since a satisfactory water supply is indispensable for the provision of such services, in addition to longitudinal channel connectivity, lateral channel/floodplain hydrological connectivity is of primary importance. As a consequence of river regulation, however, floodplains shrunk considerably in area, 'protected' floodplains connection with the river channel which had produced them and became severely threatened ecosystems. In the Drava Plain, too, disconnected (or 'geographically isolated') oxbows became typical. With reduced surface connectivity, groundwater flow becomes the main driver of connecting processes (profundal type of oxbow). Effective porosity and hydraulic conductivity of alluvial deposits and seepage from an oxbow lake (the degree of clogging of floor deposits) were calculated to estimate groundwater movements and to reveal water exchange between oxbow lakes and the active river channel. Subsurface connectivity under drought conditions was simulated by hydrological modelling with the help of HYDRUS-1D and MODFLOW 6 packages. Planning rehabilitation efforts subsurface connectivity too should be considered.

Keywords Channel/floodplain connectivity · Land drainage · Groundwater flow Hydraulic conductivity · Seepage · Hydraulic modelling · Rehabilitation

---

J. Dezső (✉) · D. Lóczy
Department of Physical and Environmental Geography, Institute of Geography and Earth Sciences, Faculty of Sciences, University of Pécs, Ifjúság útja 6, Pécs 7624, Hungary
e-mail: dejozsi@gamma.ttk.pte.hu

D. Lóczy
e-mail: loczyd@gamma.ttk.pte.hu

A. M. Salem · G. Nagy
Earth Sciences Doctoral School, University of Pécs, Ifjúság útja 6, Pécs 7624, Hungary
e-mail: eng_ali_salem2010@mu.edu.eg

G. Nagy
e-mail: gnagy@gamma.ttk.pte.hu

D. Lóczy (ed.), *The Drava River*, Springer Geography,
https://doi.org/10.1007/978-3-319-92816-6_14

14.1 Introduction

In geomorphology and landscape ecology, the concept of connectivity is applied in three different meanings (Croke et al. 2013):

(1) *landscape (topographic) connectivity* denotes the physical coupling of landforms (e.g. hillslope to channel) within a drainage basin (e.g. Michaelides and Wainwright 2002; Brierley et al. 2006);
(2) *hydrological connectivity* refers to the communication of landforms manifested in surface and subsurface water flow (e.g. Ambroise 2004; Bracken and Croke 2007; Opperman et al. 2010) and
(3) *sedimentological connectivity* relates to the movement of sediments and attached pollutants between landscape units (e.g. Hooke 2003; Fryirs et al. 2007; Fryirs 2013).

Although the term is widely used, confusion remains on how to quantify connectivity at various spatial and temporal scales (Croke et al. 2013).

Even before the concept of hydrological connectivity became accepted, it had been pointed out that, nutrients, sediments and organic matter carried by water move laterally and are deposited on floodplain surfaces (Gregory et al. 1991; Zwoliński 1992). A central theme in large river ecosystem functioning has been the flood pulse concept (Junk et al. 1989; Sparks et al. 1990; Tockner et al. 2000), which holds that, during floods of proper level and duration, floodplain connectivity allows the exchange of nutrients and feeding and spawning for some fish species in the floodplain. In the often heavily altered flow regimes of regulated rivers, however, flood pulses, which had formerly ensured connectivity, are reduced in magnitude, frequency and duration (APFM-WMO 2017).

Landscape ecological studies underline that longitudinal hydrological connectivity is a key property for the survival of the whole system (Tockner and Stanford 2002; Brierley et al. 2006). The term does not only refer to the river channel, but to the riverine corridor, too, and it denotes a central concept of restoration/rehabilitation projects (Piégay et al. 2000; Sulc-Michalková and Sulc 2011).

In addition to surface connectivity, water may also connect with the floodplain and its wetlands through groundwater flow (Brinson et al. 1995; Jacobson et al. 2011). Unconsolidated and permeable floodplain deposits facilitate rapid and dynamic connections between river channels and wetlands adjacent to the channel (Amoros and Bornette 2002) or even at considerable distance from that (in 'geographically isolated' position—Ameli and Creed 2017). (It is also observed in the case of the Drava River—Dezső et al. 2017). The hyporheic corridor may extend some kilometres away from the river channel into the alluvium (Stanford and Ward 1993) and can be an important source of water supply to oxbow lakes (Tockner et al. 1999). Other sources of groundwater are neighbouring hills and uplands, where run-off maintains hillslope/floodplain links (Kelly 2001) and establish a connectivity chain.

In view of complex interactions among surface and groundwater, topography and alluvial deposits present a challenge to the assessment of rehabilitation potential

(Fryirs and Brierley 2000, 2016). For successful interventions, detailed historical geomorphic analyses at floodplain and catchment scales are needed (Brierley and Fryirs 2005, 2008; Kondolf et al. 2007; Hohensinner et al. 2011). Analysing a Danube restoration project, Tockner et al. (1998) claim that primarily fluvial dynamics, the associated connectivity gradients and a natural disturbance regime have to be reestablished. Then the ecosystem can hopefully be maintained with only minimum effort.

14.2 Types of Hydrological Connectivity

Floodplain hydrology, morphology and hydraulic connectivity are influenced by a wide range of factors (Shankman 1993; Amoros and Bornette 2002; Hudson 2010; Kupfer et al. 2010). The stage of infilling for abandoned channels (including oxbows) depends on the time elapsed since cut-off, avulsion or bifurcation and the rate of plugging processes. The density of vegetation controls the accumulation of organic fill. The links established with cross-floodplain tributaries help abandoned channels remain hydrologically active longer (Phillips 2009). Additional local processes can also affect the evolution of floodplain depressions.

Phillips (2013) identified six different modes of surface hydrological connectivity for floodplain wetlands (oxbow lakes, sloughs and paleochannels):

(1) *flow through* (regular river flow to and from the main channel);
(2) *floodchannel* (flow at high water stages, partly reaches the main channel);
(3) *fill and spill* (flow at high stages, fill to a threshold level and then overflow into flood basins);
(4) *fill and drain* (fills at high river discharge and returns to the main channel);
(5) *tributary occupation* of former river reaches and
(6) *disconnected* (no flow except during large floods).

He found that lateral distance from the active channel is poorly related to hydrological connectivity.

In their lake typology, Dawidek and Ferencz (2014) also claim that 'the hydrological (the degree of filling of the basin) and ecological state mostly depends on the type of connections that the lake has to the parent river'. Dawidek and Turczyński (2006) identify four types of connection between floodplain lakes and the main river (Fig. 14.1):

(1) *confluent* (the floodplain lake receives water from the river in the direction of general slope of the valley);
(2) *contrafluent–confluent* (the lake is fed by the river from both upstream and downstream directions);
(3) *contrafluent* (water supply during floods and backflow to lake at the same site) and
(4) *profundal–confluent* (the lake is fed primarily from groundwater and secondarily from the river channel).

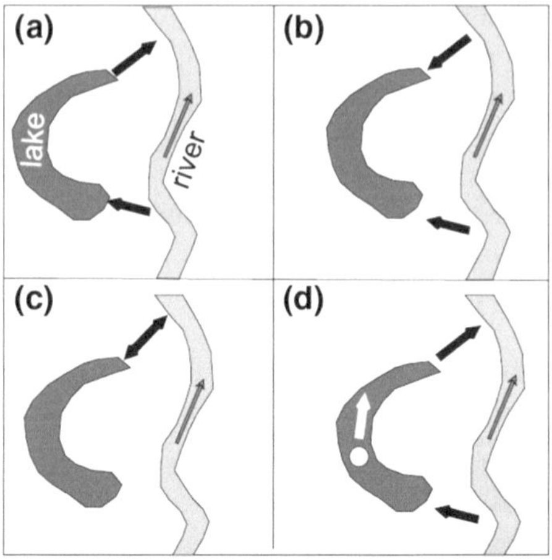

Fig. 14.1 Types of oxbow lake recharge from the main river channel (on the example of the Bug River, Poland—after Dawidek and Turczyński 2006). A, confluent; B, contrafluent–confluent; C, contrafluent; D, profundal–confluent

Naturally, there is an extreme profundal type, too, where groundwater flow is the sole supplier of water to the lake. Even for lakes where no profundal connection is predominant, groundwater flow is important (Dawidek and Ferencz 2014). As a matter of course, the rate of recharge is also dependent on these types. The interception and transpiration of riparian forests can significantly modify the water balance of floodplains (see Chap. 12 and Piégay et al. 1997).

The EU Water Framework Directive (WFD) (European Commission 2000) also acknowledges that surface/subsurface water interactions play a crucial role in the water budget of floodplains for the restoration of floodplain habitats (Downs and Thorne 2000). Major streams, like the Drava River, and their hyporheic zones maintain a hydraulic balance with groundwater. Among groundwater bodies and aquifers, the unconfined aquifer reacts most rapidly to rainfall events. Therefore, groundwater and surface water have to be conceived as components of a single system (Winter et al. 1999). Although the protected floodplain is only exceptionally covered by water along regulated rivers like the Drava, inundations in the perirheic zone (as defined by Mertes 1997) are regularly observed during wet spells.

Along the Drava River, flood-control structures also disrupted links on the surface and reduced connectivity to groundwater flow. Parafluvial and floodplain hyporheic flows also seem to be inhibited. Increasingly, severe drought conditions are indicated by dropping water levels recorded in groundwater observation wells even at 2–3 km distance from the channel.

14.3 Methods and Discussion

There are analytical and synthetical approaches to quantify floodplain/channel connectivity (also embracing flood risk). Studying surface connectivity, Croke et al. (2013) identified macrochannel width as a crucial factor. Macrochannels and associated landforms (within-channel benches, macrochannel banks and floodplains) were represented on fine-resolution digital elevation models based on LiDAR survey. Bankfull Average Recurrence Interval (ARI) was computed for channel reaches of expansion and contraction using the one-dimensional (1D) flow hydraulic model HEC-RAS. Valley floor width was established from the valley bottom flatness index (MrVBF—Gallant and Dowling 2003). Significant nonlinear changes in channel capacity were found to control the spatial pattern of hydrological connectivity. A network index (NI) has also been proposed to predict surface connectivity in agricultural catchments (Lane et al. 2004; Shore et al. 2013).

An index for river/floodplain connectivity, called the Land Capability Potential Index (LCPI), was suggested to assist regional-scale restoration planning of agricultural land along the Lower Missouri River (Jacobson et al. 2007, 2011). The LCPI integrates modelled water-surface elevations, floodplain topography and soils to index relative wetness of floodplain patches. Schwarz et al. (1996) assessed hydrological connectivity in terms of distance, elevation difference and channelized connections to the active channel. These approaches are directed at studying the fertility of agricultural land and not really appropriate for the investigation of the Drava floodplain.

To find suitable indicators for *subsurface connectivity* is even more problematic (Golden et al. 2014; McLaughlin et al. 2014). Most authors suggest the application of coupled surface–subsurface flow models, like MODFLOW (Brunner et al. 2010). The MODFLOW wetlands package and its recent modified versions (MODFLOW 6) are useful tools to depict subsurface hydrologic connectivity of wetlands, where groundwater is the dominant flow pathway. MIKE-SHE covers six key processes at the watershed scale: overland flow, channel flow, unsaturated and saturated flow, interception and evaporation, snowmelt and exchanges between aquifers and rivers. Ameli and Creed (2017) and Ameli and Craig (2014) used a 3D groundwater–surface water interaction model to reveal spatially variable recharge rates and groundwater depths. Supported by empirical groundwater table observations, these authors assumed a steady-state subsurface flow.

In the Drava research project, surface connectivity was found to be temporarily limited and of subordinate significance. Therefore, a central task was to search for hydrogeological parameters, which act as boundary conditions for groundwater flow in the floodplain (Dezső et al. 2017). In this approach, the highly changeable water transfer processes can be evaluated in the light of temporally much more stable sedimentological properties (such as grain-size distribution, effective porosity, hydraulic conductivity). Dynamic data on groundwater flow (depth to groundwater table, soil moisture content, rate of seepage from lake) are supplied by groundwater observation wells and soil moisture monitoring.

Table 14.1 Typical effective porosity data of the soil and sediment samples

Sediment	Maximum porosity (V/V%)	Effective porosity (V/V%)
Pale brown loam, root canals	46.52	22.14
Brown, subangular clayey loam	44.78	13.40
Pale brown loamy silt with ferruginous precipitations	43.19	17.15
Grey, single grained, fine sand	48.28	22.90
Pale grey, single grained sand	49.51	30.15

14.3.1 Effective Porosity

Effective porosity (Singhal and Gupta 2010) was measured by the gravimetric method on undisturbed samples. The samples were oven-dried at 105 °C for 24 h and filled with water to saturation. After saturation, samples were placed on a sand bed for 24 h to lose gravitational water.

Although microstratification varied with samples, effective porosity was equally low in all samples. As a consequence, relatively low infiltration rates were found. Effective porosity is reduced by various precipitations (calcareous, ferruginous and organic matter), while increased by root canals and outwash of sediments from the matrix (Table 14.1).

14.3.2 Hydraulic Conductivity

The alluvial sequences in floodplains are rather heterogeneous for physical and hydraulic properties (Wang et al. 2017). The preliminary soil survey, however, allows some typology and a simplified approach can be applied. The sediments of the studied oxbow floor can fundamentally be divided into two types: a clayey-silty and a calcareous sandy unit (Dezső et al. 2017). Hydraulic experiments were carried out on undisturbed sediment samples taken from the deepest part (the former channel thalweg) of the oxbow (Kp1) and parallel with the shoreline (Kp3). The modi of the PSD curves were markedly different, 80 μm along the shoreline and 10 μm in the deepest part of the lake (Fig. 14.2).

Based on hydraulic analyses of the undisturbed sediment samples in the laboratory, different hydraulic conductivity values were found in the middle and shoreline section of the oxbow (from 8.34×10^{-8} to 2.82×10^{-7} m s^{-1})—which is just the opposite to the corresponding pattern in river channels. It is explained by the presence of fine lacustrine silts in the deepest part and the dominance of fine fluvial sands (D_{med} = 80 μm) with high organic matter content in the offshore region.

At the beginning of the hydraulic conductivity experiments, a hydraulic head of 1.5 m was set on the top of the samples (Fig. 14.3). Subsequently, saturated hydraulic conductivity was calculated using the falling head method. The pressure

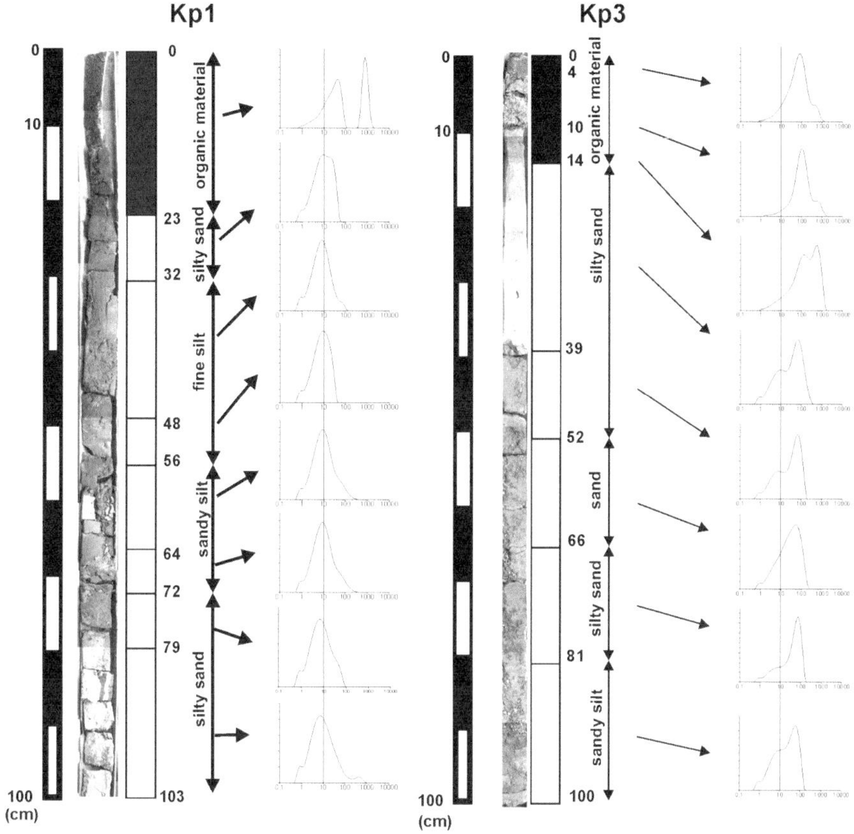

Fig. 14.2 Stratification and grain-size distribution of deposits in the deepest (Kp1) and offshore part (Kp3) of Lake Kisinc (Cún-Szaporca oxbow)

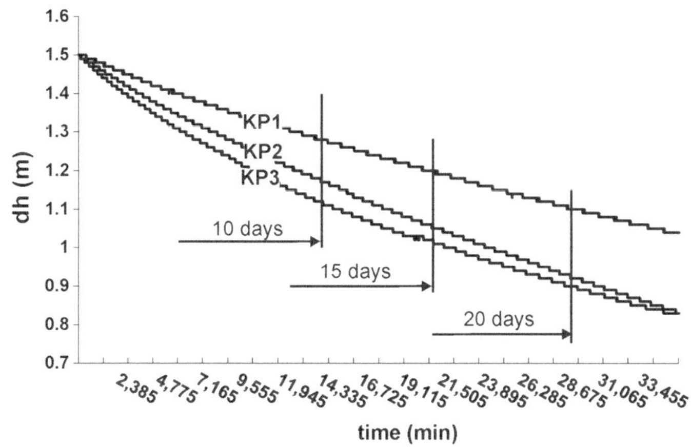

Fig. 14.3 Cumulative infiltration as a function of time for the undisturbed samples Kp1, Kp2 and Kp3

head is representative of the water column height in Lake Kisinc (Fig. 12.3) after the accomplishment of the water replenishment scheme. (Water levels will be raised from the current 90.5 m to an operational level of 91.5–92 m). Saturated hydraulic conductivity was calculated with the following formula (Reynolds and Elrick 1985):

$$k = \frac{L}{(t_2 - t_1)} * \ln\left(\frac{h_1}{h_2}\right)$$

where k is saturated hydraulic conductivity, L is the height of the soil core, t_1 and t_2 are initial and final times of the experiment, respectively, h_1 and h_2 are the corresponding pressure head heights.

Hydraulic conductivity was 8.3×10^{-8} m s^{-1} for Kp1 sediment samples and 2.82×10^{-7} m s^{-1} for the cores Kp2 and Kp3 (Fig. 14.3). When an initial head of 1.5 m was used during the laboratory experiments, water-level drop ranged between 0.38 and 0.60 m for the sediment samples taken from the clogging zones. However, since its cut-off, the oxbow has been functioning as a depositional basin with ever finer sedimentation. The sediments taken from the shoreline borehole originate and have been transported into the lake from the levee of the oxbow. In addition to the dissimilarity in PSD, variations in hydraulic conductivity may be caused by the development of biofilms. (This latter effect, however, cannot be proved from our measurements unambiguously).

14.3.3 Groundwater Flow Modelling During Drought

To study connectivity through groundwater flow, a drought period of 30 days (with no rainfall, 3 mm d^{-1} evaporation) was simulated using the HYDRUS-1D model (Nagy et al. 2017). The model was run for altogether 19 sampled sites with different soil textures under 3 hydrological boundary conditions; dry ($\psi < 15{,}000$ H_2O-cm), normal ($-15{,}000 < \psi < -10$ H_2O-cm) and saturated conditions or excess ponding (-10 H_2O-cm $< \psi$).

Multilayered sandy and sandy-silt loam soil profiles showed no capillary rise and groundwater recharge. With groundwater table 1 m below the average depth, permanent wilting point was reached at all sites after a 30-day drought. Water retention varied widely with soil texture type, and the best correlation coefficient ($r = 0.79$) between observed and measured volumetric water content was found for loam layers in the studied profile.

14.3.4 Seepage from Lake

The rate of seepage from the oxbow can be calculated from the water balance equation. The used input parameters included geomorphological data for the oxbow

Table 14.2 Input and output parameters for seepage calculations

Input data					
From geomorphological survey				From field and laboratory investigations	
Volume of oxbow	V_{oxbow}	(m^3)	Hydraulic conductivity (depending on sedimentological properties of the oxbow and newly flooded areas)	k	($m\ s^{-1}$)
Area of oxbow lake surface	A_{oxbow}	(m^2)			
(relative) Water level of oxbow	h_{oxbow}	(m)	Effective porosity	n_o	(-)
Output (calculation)					
Amount of seepage from the oxbow				Q_s	($m^3\ d^{-1}$)
Total seepage area				A_s	(m^2)
Change of oxbow lake surface				A_{om}	(m^2)
Calculated water level of oxbow				h_{cw}	(m a.s.l)
Calculated water storage capacity of oxbow				V_{cws}	(m^3)

and the measured hydrological parameters (Table 14.2). Since the planned replenishment rates for the months of March and June were 43,200 and 24,512 $m^3\ d^{-1}$, respectively (data obtained from DDKÖVÍZIG 2012), a mean 30,000 $m^3\ d^{-1}$ replenishment rate was used in our calculations. The thickness of the saturated zone was set to 4 m.

The amount of exchanged water between the oxbow lakes and the surrounding groundwater is proportional to the change of hydraulic head (*dh*), the surface area and the hydraulic conductivity (*k*) of the sediments and inversely proportional to the thickness of the clogging zone (*d*) as it is described by the modified Darcy equation (Brunner et al. 2010):

$$Q_s = \frac{k}{d}\left(h_{ox} - h_{grw}\right) * \Delta x \Delta y$$

where Q_s is total outflow from the oxbow; k is hydraulic conductivity ($m\ s^{-1}$); d is soil depth (m); h_{ox} is relative water level of the oxbow lakes (m); h_{grw} is the depth of the adjacent relative groundwater table (m) and $\Delta x \Delta y$ is the change of seepage surface area that corresponds to A_s.

The lakebed was divided into two zones, a shallow zone (less than 1.5 m deep) and a deeper (1.5–2.4 m) zone. The hydraulic conductivities of the recently flooded shoreline areas are similar to the relevant values obtained from the pumping tests. Very different hydraulic conductivity values were found: for the deeper zone (median particle size: D_{med} = 10 μm) an order of magnitude lower hydraulic conductivity ($k \sim 10^{-8}$ $m\ s^{-1}$) than in the shallow zone (D_{med} = 80 μm) ($k \sim 10^{-7}$ $m\ s^{-1}$). The additionally inundated areas have an even higher conductivity ($k \sim 10^{-5}$ $m\ s^{-1}$).

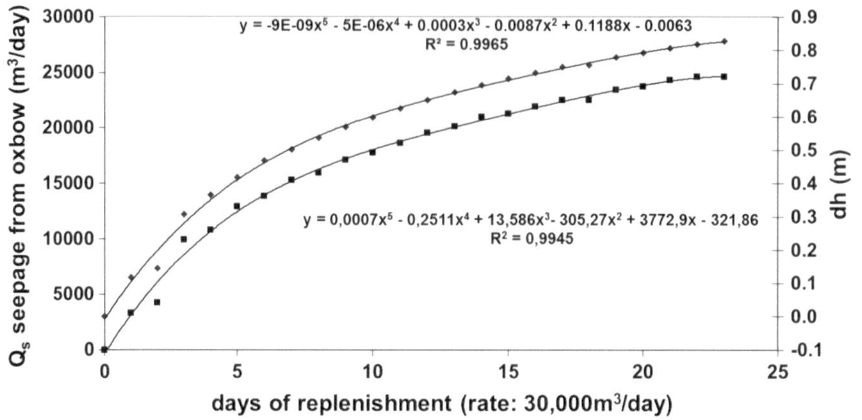

Fig. 14.4 Functional relationship between the hydraulic pressure head difference and seepage

The rising water level in the oxbow triggers an increasing hydraulic pressure difference compared to the adjacent areas. With the increasing volume in the oxbow, the potential contact surface and seepage area also increases. Due to the increasing total seepage area, more and more added water is lost (Fig. 14.4).

14.3.5 Groundwater Flow Modelling with MODFLOW-2005

Water transfer from the oxbow lakes to the main Drava channel was simulated using the MODFLOW 6 model (Langevin et al. 2017). The groundwater flow model includes calculations of groundwater flow (discretization, initial conditions, hydraulic conductance and storage), stress packages (constant heads, wells, recharge, rivers, general head boundaries, drains and evapotranspiration) and advanced stress packages (streamflow routing, lakes, multi-aquifer wells and unsaturated zone flow).

A 10-m resolution Digital Elevation Model (DEM) was built for the area between the Cún–Szaporca oxbow and the Drava main channel. The model is discretized with a finite-difference grid (60 rows, 14 columns and 10 layers; cell size: 10 m $\times$ 10 m) (Salem et al. 2017). The eastern and western boundaries are marked with constant head values from previous monitoring of precipitation, groundwater table, infiltration and groundwater recharge. The evapotranspiration is estimated at 13 mm d^{-1}.

The model was run using the replenishment scenarios identified in the Old Drava Programme (DDKÖVÍZIG 2012—see Chap. 21). In scenario 1 (lake level increases by 0.5–91 m above sea level), seepage from the lake was found as 1,298.8 m^3 d^{-1} and water loss through evapotranspiration rise was estimated at

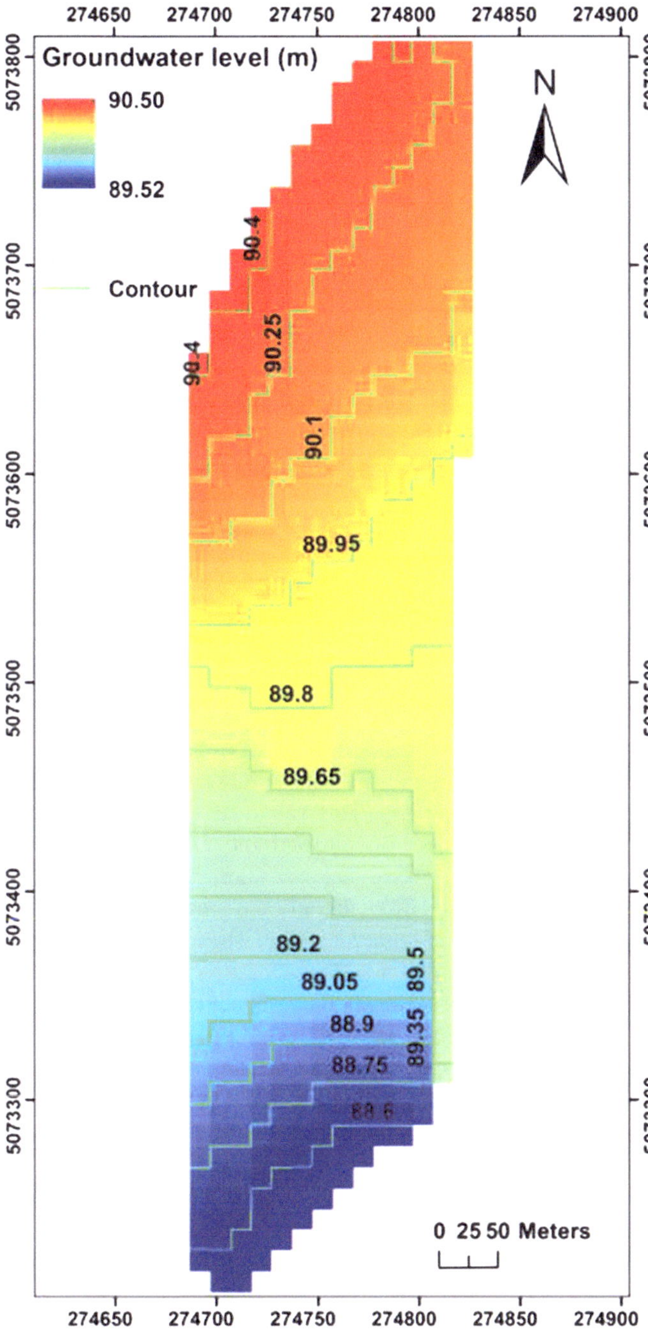

Fig. 14.5 Simulated groundwater levels between the oxbow lakes and the main Drava channel under modelling conditions (by Ali Mohamed Salem)

161.20 m^3 d^{-1} higher than the baseline situation. This equals 0.28 m average groundwater level rise. In scenario 2 (lake level raised by 1 m to 91.5 m), recharge rate from the lake to the aquifer grows by 745.59 m^3 d^{-1}. Consequently, the average water table rises by 0.77 m (Salem et al. 2017).

The advantage of this modelling approach is that it detects water transfer from the lake to inflow into the river. Great variations are revealed according to the grain-size distribution of the alluvial sequence. Calculating with a single uniform aquifer 344.29 m^3 d^{-1} seepage is found for a silty deposit (k = 60 m d^{-1}) and 1,468.95 m^3 d^{-1} for sand (k = 500 m d^{-1}). Recharge to the river is 176.34 m^3 d^{-1} for silt and 1,477.15 m^3 d^{-1} for sand. More realistic estimates of subsurface connectivity are made through the vertical discretization of the aquifer based on the geomorphological interpretation of satellite images supplemented with ground-penetrating radar surveys (Dezső et al. 2017). This way, six layers with different texture and conductivity were distinguished and incorporated in the model (Salem et al. 2017). In this case, water loss amounts to 347.79 m^3 d^{-1} and inflow into the channel is 707.74 m^3 d^{-1}. The figures underline a very high horizontal connectivity between water bodies (Fig. 14.5). At the same time, in the vadose and saturated zones 65% of the water leaked from the oxbow lake may be retained (Salem et al. 2017).

14.4 Conclusions

Surface connectivity of the protected floodplain with the main channel of the Drava is very difficult to provide. Groundwater flow, however, ensures some degree of subsurface hydrological connectivity. This has to be assessed in close association with the water retention capacities of soils and alluvial deposits.

Our research shows that the critical factor in water retention is the transmissivity of lakebed and adjacent deposits. It is not only the present lakeshore that has to be examined hydrodynamically but also the future shallow lakebed zone inundated after water replenishment. Based on the laboratory hydraulic analyses of the undisturbed sediment samples, highly different conductivity values were found for the middle and offshore parts of the oxbow lakes—a pattern just the opposite expected for active river channels. Relatively, coarse fraction (~80 μm) dominates the shoreline zone and allows higher seepage rate from the oxbow lake. Considerable losses to groundwater (and indirectly to the Drava River) are expected and may jeopardize the success of the replenishment scheme. Allowing for the hydraulic variability of the sediment sequence, HYDRUS-1D and MODFLOW 6 simulations showed to what extent the planned replenishment scenarios will raise the groundwater level and allowed the estimation of soil water retention capacity (Valentová et al. 2010).

Acknowledgements The authors are grateful for financial support from the Hungarian National Scientific Fund (OTKA, contract no. 104552) and for cooperation with the South-Transdanubian Water Management Directorate.

References

Ambroise B (2004) Variable 'active' versus 'contributing' areas or periods: a necessary distinction. Hydrol Process 18:1149–1155. https://doi.org/10.1002/hyp.5536

Ameli AA, Craig JR (2014) Semianalytical series solutions for three-dimensional groundwater-surface water interaction. Water Resour Res 50:3893–3906. https://doi.org/10.1002/2014WR015394

Ameli AA, Creed IF (2017) Quantifying hydrologic connectivity of wetlands to surface water systems. Hydrol Earth Syst Scient Discuss 21(3):1791–1808. https://doi.org/10.5194/hess-2016-404, 2016

Amoros C, Bornette G (2002) Connectivity and biocomplexity in water bodies of riverine floodplains. Freshw Biol 47:761–776. https://doi.org/10.1046/j.1365-2427.2002.00905.x

APFM-WMO (2017) Selecting measures and designing strategies for integrated flood management. Associated Programme on Flood Management. World Meteorological Organization, Geneva. 35 pp

Bracken LJ, Croke J (2007) The concept of hydrological connectivity and its contribution to understanding runoff-dominated geomorphic systems. Hydrol Process 21:1749–1763. https://doi.org/10.1002/hyp.6313

Brierley GJ, Fryirs KA (2005) Geomorphology and river management. Applications of the river styles framework. Blackwell Publishing, Carlton, Victoria, p 398

Brierley GJ, Fryirs KA, Jain V (2006) Landscape connectivity: the geographic basis of geomorphic applications. Area 38(2):165–174. https://doi.org/10.1111/j.1475-4762.2006.00671.x

Brierley GJ, Fryirs KA (2008) River futures. an integrative scientific approach to river repair. Island Press, Washington, DC. 328 pp (The Science and Practice of Ecological Restoration Series)

Brinson MM, Rheinhardt RD, Hauser FR, Lee LC, Nutter WL, Smith D, Whigham D (1995) A Guidebook for Application of Hydrogeomorphic Assessments to Riverine Wetlands. Wetlands Research Program, Technical Report WRP-DE-11: U.S. Army Corps of Engineers. http://www.water-research.net/Waterlibrary/geologicdata/riverinehydrodynamics.pdf

Brunner P, Simmons CT, Cook PG, Therrien R (2010) Modeling surface water-groundwater interaction with MODFLOW: some considerations. Groundwater 48(2):174–180. https://doi.org/10.1111/j.1745-6584.2009.00644.x

Croke J, Fryirs K, Thompson C (2013) Channel–floodplain connectivity during an extreme flood event: implications for sediment erosion, deposition, and delivery. Earth Surf Proc Land 38:1444–1456. https://doi.org/10.1002/esp.3430

Dawidek J, Ferencz B (2014) Water balance of selected floodplain lake basins in the Middle Bug River valley. Hydrol Earth Syst Sci 18:1457–1465. https://doi.org/10.5194/hess-18-1457-2014

Dawidek J, Turczyński M (2006) Recharge of lakes with river waters in the Middle Bug valley. Limnol Rev 7:65–72

DDKÖVÍZIG (2012) Revitalization of the Cún-Szaporca oxbow system. Final Master Plan. South-Transdanubian Environment and Water Directorate, Pécs. 100 pp

Dezső J, Halász A, Czigány Sz, Tóth G, Lóczy D (2017) Estimating seepage loss during water replenishment to a floodplain oxbow: a case study from the Drava Plain. Geografia física e dinamica quaternaria 40(1):61–75. https://doi.org/10.4461/GFDQ2017.40.6

Downs PW, Thorne CR (2000) Rehabilitation of a lowland river: reconciling flood defence with habitat diversity and geomorphological sustainability. J Environ Manage 58:249–268

Commission European (2000) Directive 2000/60/EEC. Establishing a framework for community action in the field of water policy. Off J Europ Commun, Luxemburg L327:1–71

Fryirs K (2013) (Dis)connectivity in catchment sediment cascades: a fresh look at the sediment delivery problem. Earth Surface Proc Landforms 38:30–46. https://doi.org/10.1002/esp.3242

Fryirs KA, Brierley GJ (2000) A geomorphic approach for the identification of river recovery potential. Phys Geogr 21:244–277

Fryirs KA, Brierley GJ (2016) Assessing the geomorphic recovery potential of rivers: forecasting future trajectories of adjustment for use in management. Water 3(5):727–748. https://doi.org/10.1002/wat2.1158

Fryirs KA, Brierley GJ, Preston NJ, Spencer J (2007) Catchment-scale (dis)connectivity in sediment flux in the upper Hunter catchment, New South Wales, Australia. Geomorphology 84:297–316. https://doi.org/10.1016/j.geomorph.2006.01.044

Gallant J, Dowling T (2003) A multi-resolution index of valley bottom flatness for mapping depositional areas. Water Resour Res 39:1347–1360

Golden HE, Lane CR, Amatya DM, Bandilla KW, Kiperwas HR, Knightes CD, Ssegane H (2014) Hydrologic connectivity between geographically isolated wetlands and surface water systems: a review of select modeling methods. Environ Model Softw 53:190–206

Gregory SV, Swanson FJ, McKee WA, Cummins KW (1991) An ecosystem perspective ofriparian zones. Bioscience 41:540–551

Hohensinner S, Jungwirth M, Muhar S, Schmutz S (2011) Spatio-temporal habitat dynamics in a changing Danube River landscape, 1812–2006. River Res Appl 27:939–955. https://doi.org/10.1002/rra.1407

Hooke J (2003) Coarse sediment connectivity in river channel systems: a conceptual framework and methodology. Geomorphology 56:79–94. https://doi.org/10.1016/S0169-555X(03)00047-3

Hudson PF (2010) Floodplain lake formation and dynamics in the lower reaches of the large texas coastal plain rivers: Brazos, Guadalupe, and San Antonio Rivers. Austin: Texas Water Development Board, Report no. 0600010583. http://www.twdb.state.tx.us/RWPG/rpgm_rpts/IndividualReportPages/0600010583.asp

Jacobson RB, Chojnacki KA, Reuter JM (2007) Land capability potential index (LCPI) for the Lower Missouri River valley. U.S. Geological Survey, Reston, VA19 pp (Scientific Investigations Report 2007-5256)

Jacobson RB, Janke TP, Skold JJ (2011) Hydrologic and geomorphic considerations in restoration of river-floodplain connectivity in a highly altered river system, Lower Missouri River, USA. Wetlands Ecol Manag 19:295–316. https://doi.org/10.1007/s11273-011-9217-3

Junk WJ, Bayley PB, Sparks RE (1989) The flood pulse concept in river floodplain systems. In: Dodge DP (ed) Proceedings of the international large river symposium fisheries and oceans. Ottawa. 110–127

Kelly BP (2001) Relations among river stage, rainfall, ground-water levels, and stage at two missouri river flood-plain wetlands. USGS water-resources investigations report 01–4123, 18 pp. https://mo.water.usgs.gov/Reports/01-4123-kelly/index.htm

Kondolf GM, Anderson S, Lave R, Pagano L, Merenlender A, Bernhardt ES (2007) Two decades of river restoration in california: what can we learn? Restor Ecol 15(3):516–523

Kupfer JA, Meitzen KM, Pipkin AR (2010) Hydrogeomorphic controls of early post-logging successional pathways in a southern floodplain forest. For Ecol Manage 259:1880–1889. https://doi.org/10.1016/j.foreco.2010.01.050

Lane SN, Brookes CJ, Kirkby MJ, Holden J (2004) A network-index-based version of TOPMODEL for use with high resolution digital topographic data. Hydrol Process 18:191–201. https://doi.org/10.1002/hyp.5208

Langevin CD, Hughes JD, Banta ER, Niswonger RG, Panday S, Provost AM (2017) Documentation for the MODFLOW 6 Groundwater Flow Model. US Geological Survey Techniques and Methods 6. Chapter A55. 197 pp. https://doi.org/10.3133/tm6a55

McLaughlin DL, Kaplan DA, Cohen MJ (2014) A significant nexus: geographically isolated wetlands influence landscape hydrology. Water Resour Res 50:7153–7166. https://doi.org/10.1002/2013WR015002

Mertes LAK (1997) Documentation and significance of the perirheic zone on inundated floodplains. Water Resour Res 33:1749–1762

Michaelides K, Wainwright J (2002) Modelling the effects of hill slope channel coupling on catchment hydrological response. Earth Surf Proc Land 27:1441–1457. https://doi.org/10.1002/esp.440

Nagy G, Dezső J, Czigány Sz, Pirkhoffer E, Lóczy D (2017) Modeling of soil moisture dynamics in multilayered Entisols under drought: a case study in the Drava valley. In: Conference paper. "Degradation and Revitalisation of Soil and Landscape", Olomouc, 10–13 September 2017

Opperman JJ, Laster R, McKenney BA, Roberts M, Meadows AW (2010) Ecologically functional floodplains: connectivity, flow regime and scale. J Am Water Resour Assoc 46:211–226. https://doi.org/10.1111/j.1752-1688.2010.00426.x

Phillips JD (2009) Avulsion regimes in southeast Texas rivers. Earth Surf Proc Land 34:75–87. https://doi.org/10.1002/esp.1692

Phillips JD (2013) Hydrological connectivity of abandoned channel water bodies on a coastal plain river. River Res Appl 29:149–160. https://doi.org/10.1002/rra

Piégay H (1997) Interactions between floodplain forests and overbank flows: data from three piedmont rivers of southeastern France. Glocal Ecol Biogeogr Lett 6:187–196

Piégay H, Bornette G, Citterio A, Herouin E, Moulin B (2000) Channel instability as a control on silting dynamics and vegetation patterns within the perifluvial aquatic zones. Hydrol Process 14(16–17):3011–3029. https://doi.org/10.1002/1099-1085(200011/12)14:16/17<3011:AID-HYP132>3.0.CO;2-B

Reynolds WD, Elrick DE (1985) In-situ measurement of field-saturated hydraulic conductivity, sorptivity and α parameter using the Guelph permeameter. Soil Sci 140(4):292–302

Salem A, Dezső J, Lóczy D, Słowik M (2017) Modeling surface water-groundwater interaction in an oxbow of the Drava floodplain. Extended abstract. In: HIC 2018 International Conference on Hydroinformatics, Palermo, 1–6 July 2018. 4 pp

Schwarz WL, Malanson GP, Weirich FH (1996) Effect of landscape position on the sediment chemistry of abandoned-channel wetlands. Landscape Ecol 11:27–38

Shankman D (1993) Channel migration and vegetation patterns in the southeastern coastal plain. Conserv Biol 7:176–185

Shore M, Murphy PNC, Jordan P, Mellander P-E, Kelly-Quinn M, Cushen M, Mechan S, Shine O, Melland AR (2013) Evaluation of a surface hydrological connectivity index in agricultural catchments. Environ Model Softw 47:7–15. https://doi.org/10.1016/j.envsoft.2013.04.003

Singhal BBS, Gupta RP (2010) Applied hydrogeology of fractured rocks. Springer, Dordrecht, 408 pp

Sparks RE, Bayley PB, Kohler SL, Osborne LL (1990) Disturbance and recovery of large floodplain rivers. Environ Manage 14(5):699–709

Stanford JA, Ward JV (1993) An ecosystem perspective of alluvial rivers: connectivity and the hyporheic corridor. J North American Bentholog Soc 12:48–60

Šulc-Michalková M, Šulc V (2011) Floodplain lakes, river regulation and restoration strategies—what do we know? J Landscape Manag 2(2):31–40

Tockner K, Malard F, Ward JV (2000) An extension of the flood pulse concept. Hydrol Process 14:2861–2883

Tockner K, Pennetzdorfer D, Reiner R, Schiemer F, Ward JV (1999) Hydrological connectivity and the exchange of organic matter and nutrients in a dynamic river-floodplain system (Danube, Austria). Freshw Biol 41:521–535

Tockner K, Schiemer F, Ward JV (1998) Conservation by restoration: the management concept for river floodplain system on the Danube River in Austria. Aquatic Conserv: Marine Freshw Ecosyst 8:71–86

Tockner K, Stanford JA (2002) Riverine floodplains: present state and future trends. Environ Conserv 29:308–330. https://doi.org/10.1017/S037689290200022X

Valentová J, Valenta P, Weyskrabová L (2010) Assessing the retention capacity of a floodplain using a 2D numerical model. J Hydrol Hydromech 58(4):221–232. https://doi.org/10.2478/v10098-010-0021-1

Wang YL, Yeh TCJ, Wen JC, Huang SY, Zha Y, Tsai JP, Hao YH, Liang Y (2017) Characterizing subsurface hydraulic heterogeneity of alluvial fan using riverstage fluctuations. J Hydrol 547:650–663. https://doi.org/10.1016/j.jhydrol.2017.02.032

Winter TC, Harvey JW, Franke OL, Alley WM (1999) Ground water and surface water, a single resource. US Geol Survey, Circular 1139. Denver, CO. 79 pp

Zwoliński Z (1992) Sedimentology and geomorphology of overbank flows on meandering river floodplains. Geomorphology 4:367–379

Chapter 15
Water Quality of the Lower Drava River

Anita Dolgosné Kovács, Gabriella Tóth and Dénes Lóczy

Abstract The tendencies of change in river water quality are difficult to establish since the evaluation systems has changed on several occasions in the past decades. At present, according to the comprehensive assessment system of the EU Water Framework Directive (WFD), on the upper and middle sections (in Austria and Slovenia), the Drava has good and the Mura somewhat worse water quality. In 2010, a campaign found the Croatian section of the Drava River in excellent condition based on the Water Quality Index. Regular data collection on water quality of the Hungarian Drava section began in the 1960s at three sampling sites (Őrtilos, Barcs and Drávaszabolcs) and soon continued in international cooperation with Yugoslavian authorities. The monitoring and evaluation systems changed in 1981 and again in 1994. In 2001, an automatic Drava Monitor Station (DAM) began to operate at Barcs and complex (hydromorphological, physico-chemical, biological, and biochemical) monitoring according to the WFD guidelines was introduced. The influence of tributaries on the quality of the Drava water is represented on a map series. A case study illustrates the water quality problems of oxbow lakes in the Drava floodplain.

Keywords Water quality monitoring · Chemical status · Research history National standards · Water Framework Directive

A. Dolgosné Kovács
Department of Environmental Engineering, Faculty of Engineering and Informatics, University of Pécs, Boszorkány út 2, Pécs 7624, Hungary
e-mail: kanita@mik.pte.hu

G. Tóth · D. Lóczy (✉)
Department of Physical and Environmental Geography, Institute of Geography and Earth Sciences, Faculty of Sciences, University of Pécs, Ifjúság útja 6, Pécs 7624, Hungary
e-mail: loczyd@gamma.ttk.pte.hu

G. Tóth
e-mail: toth.gabi90@gmail.com

D. Lóczy (ed.), *The Drava River*, Springer Geography,
https://doi.org/10.1007/978-3-319-92816-6_15

15.1 Introduction

Water pollution disturbs the natural renewal of water resources. Since the 1950s diffuse contamination of agricultural origin (from wastes, fertilizers and pesticides —Novotny 2005) in Hungary has reached the same dimensions as point-source pollution (HAS 2017). For instance, two-thirds of diffuse phosphorus load comes from soil erosion (both agricultural and natural) and the rest from municipal wastewater treatment plants. Sampling campaigns measuring pesticides in Hungarian surface waters including streams, rivers and lakes have also been performed (e.g. Maloschik et al. 2007).

The pressures are manifested in water quality regularly monitored on the Drava since the 1960s (Dolgosné Kovács 2008). The comparability of data is made difficult by the fact that in the meantime the official system of assessment has changed on several occasions (Kovács 2002). In 1994 the Hungarian Standard 12749 (Surface water quality, quality indicators and assessment), based on chemical indices and less concerned with biological and ecological conditions, was introduced. After Hungary's accession to the European Union the EU Water Framework Directive (WFD—EC 2000) has been followed. In addition to chemical status, the WFD assesses ecological conditions and is meant to ensure sustainable water use. Instead of point sampling, it strives to collect information on entire water bodies within the framework of river basin management plans.

15.2 Water Quality of the Upper and Middle Drava

Naturally, the water quality of Hungarian rivers strongly depends on the quality of river inflow into Hungary, a country in the centre of the Carpathian Basin. Since the adoption of the EU WFD, which allows comprehensive ecological assessment, it is easier to compare the status of water bodies in the member countries of the European Union, including those in the upper Drava catchment.

In *Austria* river water quality is assessed by the Federal Office for Environment (Umweltbundesamt) in six-year cycles (Bundesministerium 2006). Temporal trends can be reconstructed from the previous cycle, the years 2007, 2010 and 2013 (Bundesministerium 2015). Although there are no fixed quality classes, for the Drava and Mura rivers the chemical parameters (monitored on a yearly basis) show good status (Table 15.1). The hydromorphological characteristics, however, did not present an improvement for 40% of Austrian rivers by the end of the cycle in 2013.

In *Slovenia,* the Environmental Agency of the Republic of Slovenia is responsible for water quality monitoring and evaluation of water quality status (Urbanić 2011). Monitoring programmes are designed in accordance with European directives and include assessment and pressure analysis for each individual water body (rivers, lakes, sea, groundwater and water bodies in protected areas—ARSO 2008). Priority substances were monitored monthly, while nationally relevant substances

Table 15.1 Chemical status of the Drava and Mura Rivers in Austria, 2013 (Bundesministerium 2015)

River	*Monitoring Site*	*Metals*	*Organic solvent*	*PAHs*	*Pesticide*	*AOX*	*Other*
Drava	Lavamünd	good	good	good	good	good	good
Drava	Rosegger Schleife	good	good	good	good	good	good
Mura	Leoben	good	good	good	good	moderate	good
Mura	Kalsdorf	good	good	good	good	moderate	good
Mura	Spielfeld	good	good	good	good	poor	moderate

	poor chemical status	AOX, Adsorbable Organic Halogens
	moderate chemical status	PAH, Polycyclic Aromatic Hydrocarbons
	good chemical status	in sed.: upward trend in sediment
	very good chemical status	FS, phenol substances
	the monitoring site was not included in the monitoring programme	

four times a year. The findings show that the water quality of the Mura and Drava meets the standards proposed by the European Commission and they are in good chemical status (Table 15.1), but some parameter values are worse than in Austria (Table 15.2).

Surface waters are monitored by the Hydrometeorological Institute at more than 100 sampling sites along the main rivers. The main parameters which indicate pollution are dissolved oxygen, chemical oxygen demand (COD), biochemical oxygen demand (BOD), phenols, nitrogen compounds, detergents, formaldehyde and mineral oil. Surface water quality is referred into four classes: Class I can be

Table 15.2 Chemical status of the Mura and Drava Rivers in Slovenia, 2002–2006 (after ARSO 2008) (for legend see Table 15.1)

River	*Monitoring site*	*2002*	*2003*	*2004*	*2005*	*2006*
Mura	Ceršak	AOX	Cd in sed.	good	AOX	AOX
Mura	Petanjci	good	good	good	good	good
Mura	Mota	good	good	AOX	AOX, FS	AOX
Drava	Dravograd	good	good	good	good	good
Drava	Brezno	good	good	good	good	n.a.
Drava	Mariborski otok	good	good	good	good	good
Drava	Duplek	good	good	good	good	n.a.
Drava	Ptuj	good	good	n.a.	n.a.	n.a.
Drava	Borl	good	good	good	good	n.a.
Drava	Ormož	Hg, Cd in sed.	Hg in sed.	good	good	good

used as drinking water; Class II water needs pre-treatment before used as drinking water; Class III includes water polluted with degradable compounds from domestic sewage, which does not necessarily preclude its use in agriculture or as industrial cooling water; Class IV is polluted water not suitable for any direct use.

The monitoring of groundwater quality is carried out by the Hydrometeorological Institute of the Slovenian Republic. Along the Mura and Drava, agricultural activities directly influence water quality. NO_3 concentrations are between 31 and 242 mg L^{-1}, and some pesticides also exceed EU drinking water standards. Potassium and zinc concentrations are increasing along the Drava.

In 2010, an investigation using the Water Quality Index (CCME 2001), calculated from a mathematical formula, found the *Croatian* section of the Drava River in excellent condition (Tomas et al. 2013). In the Water Quality Index six main parameters (pH, dissolved oxygen, biochemical oxygen demand, nitrate, total nitrogen and total phosphorus) are combined. Sampling was conducted mainly in the Croatian-Hungarian border at seven monitoring sites.

The Croatian surface water regulations identify five quality classes depending on the usage of water (Gvozdić et al. 2011). Their goal is to achieve that the Drava River should remain in Class II on the long run. This means that after purification river water can be used as either drinking or industrial water. Authors emphasize that for the consideration of water quality of the Drava River the largest tributary, the Mura, which is classified as Class IV exerts a huge effect on the Drava.

According to the most recent data, the physico-chemical properties of the Drava in Croatia were invariably in the 'very good' category in 2012 (Table 15.3—Republika Hrvatska 2013).

Table 15.3 Physico-chemical properties of the Drava River in Croatia, 2012. *Source* Republika Hrvatska (2013)

Monitoring site (Drava)	Chemical and physico-chemical properties									
	Conductivity	Alkalinity	pH	dissolved oxygen	COD	BOD	NH_4+	NO_3-	N	P
Donji Miholjac	very good	very good	very good	very good	very good	very good	very good	very good	very good	good
Belišće	very good	very good	very good	very good	very good	very good	very good	very good	very good	good
upstream Osijek	very good	very good	very good	very good	very good	very good	very good	very good	very good	good
before the Danube	very good	very good	very good	very good	very good	very good	very good	very good	very good	good
boundary profile	very good	very good	very good	very good	very good	very good	very good	moder-ate	very good	good

15.3 History of Water Quality Research in Hungary

15.3.1 Water Quality Assessment Before 2000

In Hungary, water pollution was first prohibited in a government decree of 1952. In 1963, an observation network was established by the National Water Management Directorate (OVF 1965) and in 1965 regional water quality planning started in the most polluted regions. Water quality was primarily studied from the aspect of public health. The assessment for the Drava in 1960 was based on oxygen budget and coli number and claimed that the Drava water had a low sodium concentration; bacteriologically it was slightly polluted, but after proper treatment and disinfection suitable for drinking water or utilization in food industry, bathing and water sports (OVF 1965).

The basic network of sampling (of constant components and with defined frequency) was established in 1968 (Katona 1984). Sites for weekly sampling for the Drava were appointed at Őrtilos, Barcs and Drávaszabolcs (Fig. 15.1). (In 1982 sampling at Őrtilos was stopped, but resumed in 1994.) From 1990 heavy metal content was also measured in every three month by the South Transdanubian Environmental Protection Directorate. Although Kiss (1974) described the Pécsi-víz stream as the most polluted in the catchment, he claimed that the Pécsi-víz and Fekete-víz streams only have a minimal influence, but the Mura River, under serious pressure upstream, has a serious impact on the water quality of the Drava. Uherkovich (1974) proposed an evaluation system which, along with physical and chemical techniques, also uses hydrological methods to describe water contamination. The international coordination of measurements became necessary and

Fig. 15.1 Sampling sites on the Hungarian Drava section 1, sampling sites

between 1966 and 1977, Croatia and Hungary conducted analysis separately four times a year at the Letenye–Goričan road bridge on the Mura and at the Drávaszabolcs–Donji Miholjac road bridge on the Drava (Stundl 1976). The findings were jointly evaluated twice a year. However, biological properties (such saprobity index available since 1970) were not included in water quality assessment yet. Between 1978 and 1989 sampling frequency was raised to 10 occasions per year at the same locations and once a year joint sampling was also performed.

In 1981, a new system was introduced with a reduced number of sampling sites (from 300 to 120) in Hungary but with increased frequency. Criticizing the system, Katona (1982) claimed that the assessment of natural waters only shows the conditions during the sampling period. To follow the changes in quality tendency and

Table 15.4 Parameters to measure (after Hungarian Standard 12749, by Dolgosné Kovács 1993)

Water quality properties		
Group A: oxygen properties		Dissolved oxygen; Oxygen saturation; Biological Oxygen Demand (BOD_5); Chemical Oxygen Demand (COD_{ps}), (COD_k); Total Organic Carbon (TOC); Saprobity (Pantle-Buck) index
Group B: nitrogen and phosphorus properties		Ammonium-nitrogen (NH_4–N); Nitrite-nitrogen (NO_2–N); Nitrate-nitrogen (NO_3–N); Organic nitrogen; Total phosphorus, Orthophosphate-phosphorus (PO_4–P); a-chlorophyll
Group C: microbiological properties		Coliform number per 1 ml; Fecal coliform number in 1 ml; Fecal streptococcus in 1 ml; Salmonella in 1 L; Colony count at 37 °C; Colony count at 22 °C
Group D: micropollutants and toxicity	Group D_1: inorganic micropollutants	Aluminum; Arsenic; Boron; Cyanide; Zinc; Mercury; Cadmium; Chromium; Chromium (VI); Nickel; Lead; Copper
	Group D_2: organic micropollutants	Phenols; Detergents (anion active, non-ionic); Hydrocarbon derivates (mineral oil and its products, PAHs, benzo(a)pyrene); Volatile chlorinated hydrocarbons (chloroform, carbon-tetrachloride, trichloroethylene, tetrachloroethylene); Pesticides; PCB; PCP
	Group D_3: toxicity	Daphnia test; Seedling plant test; Static fish test
	Group D_4: radioactive material	Total β-activity; $Cesium^{137}$; $Strontium^{90}$; Tritium
Group E: other properties		pH; Specific conductivity (20 °C); Iron; Manganese
		Water temperature; Air temperature; Total suspended matter; Turbidity; Alkalinity; Hardness; Sodium; Potassium; Calcium; Magnesium; Carbonate; Bicarbonate; Sulphate, Chloride
		Colour; Odour; Transparence

pollution load tests should be made. In international comparison, Hungary has the least strict duration values (80%). Somlyódi et al. (1990) suggested that it should be raised to 95%.

In 1979, heavy metal measurements began and in 1986 became regular. Total nitrogen and phosphorus were also covered. Because of political issues, measurements were only performed on the Hungarian side between 1989 and 1992. Biological parameters (like Liebmann's saprobity system) were also included in water quality assessment in 1994, when the Hungarian Standard 12749 (1993) came into force (see below). In 1992, joint sampling resumed and in 1995, a joint Hungarian-Croatian committee was formed and confirmed the protocols of analyses. Organic micro-pollutants, residues, and hydrobiological data were collected 36 times a year (Table 15.4). In the case of river sediments chemical tests, organic micro-pollutants, residuals were studied with variable sample numbers. For the Drava oxbows and five major gravel pits chemical and hydrobiological tests were carried out. From 2000, the studied parameters were extended to additional biological properties. In 2001, an automatic Drava Monitor Station (DAM) began to operate at Barcs, which measures chemical and hydrometeorological parameters (Kovács 2002; Dolgosné Kovács 2003).

15.4 Assessment According to the Water Framework Directive

In the EU member countries water management has been basically transformed after the adoption of the European Water Framework Directive (WFD, Directive 2000/60/EC) (Urbanić 2011). The most important change is that the emphasis placed on the ecological status of waterbodies. Previously, physico-chemical properties were the main basis of determining water quality, supplemented with only a few biological parameters. Since plant and animal communities respond sensitively to the change of main aquatic components, the focus on ecological status results in a more complex and reliable overall picture of river water quality.

After Hungary's accession to the European Union, the WFD (EC 2000) was introduced in Hungary too. The WFD ensured the protection of all kinds of water bodies with the purpose to reach a 'good state' for them by 2015. It is nature-oriented and ecologically based. Another important difference from the system previously applied in Hungary (based on Hungarian Standard 12749) is that the entire water body should be monitored for water quality instead of defining the status at individual monitoring sites (Dworak et al. 2005; Allan et al. 2006).

In order to establish the ecological status of waters, the definition of biological, hydromorphological, physico-chemical and biochemical parameters (Table 15.5) is necessary in five quality classes. Considering the ecological state, the worst biological and physico-chemical values have to be taken into account as an essential determinant of water quality. The determination of chemical status is based on

Table 15.5 Components of the chemical assessment of surface waters according to the WFD (EC 2000)

	Conditions	Parameters
Physico-chemical properties	Thermal	Temperature
	Oxygen balance	Dissolved oxygen
	Salinity	Conductivity
	Acidity	pH, alkalinity
	Nutrients	Total P, ORP, total N, nitrate and nitrite, ammonium
	Clarity	Secchi disk depth, turbidity, colour (for lakes)
	Other	Suspended matter, turbidity (for rivers)
Specific synthetic pollutants	All substances on the WFD priority list	
	Other substances dependent on the pressures affecting the drainage basin	
Specific non-synthetic substances	All substances on the WFD priority list	
	Other substances dependent on the pressures affecting the drainage basin	

further chemical parameters. In the final assessment, the 'one bad, all bad' principle prevails, i.e. the worst condition decides the status of the water body (Szilágyi 2005; Ijjas 2005; Clement 2005).

An important principle of the WFD is that the status of waters should be compared to undisturbed (reference) conditions. The member states should set reference criteria and ecological class limits for each type of surface water and all relevant quality elements. In 2005 16 and in 2006 84 water bodies were studied in the Drava subcatchment of the Danube basin. In 2007 the Integrated Drava Monitoring programme was launched and the use of the Hungarian Standard was officially terminated. However, due to cross-border conventions and the Government Decree 2066/1999 (31 March), at some sampling sites data collection and evaluation on the Drava and its region still has to follow the Hungarian Standard. Moreover, joining the Transnational Monitoring Network (TNMN) at national level resulted in additional tasks in the design of the complex monitoring system.

15.5 Results and Discussion

From the data series of monitoring at three sampling sites (Őrtilos, Barcs and Drávaszabolcs), some trends in the water quality of the Hungarian Drava section can be revealed.

For both dissolved *oxygen* and oxygen saturation favorable results have been detected. Drava water was found 'excellent' on the basis of 90% duration from the 1990s (Dolgosné Kovács 2008). The values were not significantly influenced by the tributaries, their different origin, hydromorphology and discharge, reflecting the amount of organic matter in the water, oxygen concentrations were the highest in the 1970s, usually at Őrtilos. Until the mid-1980s, based on chemical and biological oxygen demands, the water was only of 'tolerable' quality at all three sampling sites. From the 1990s, there was some improvement, but higher organic matter loads were found at Drávaszabolcs, due to the loads arriving with the Fekete-víz Stream and the lowest at Barcs. Apart from the beginning of the study period, the Drava had no direct untreated wastewater inflow from Hungarian territory. The amount of treated sewage was also minimal before its inflow was stopped. The tributaries usually had high organic matter contents which occasionally influenced the samples from the Drava at Barcs and Drávaszabolcs. From 2000, the oxygen balance of water was referred to Class III.

The parameters of *nitrogen* balance in the 1970s showed high pressure for ammonium and nitrogen mostly at Őrtilos (Dolgosné Kovács 2008). Water quality from this respect was Class III or IV. Since the 1990s, more severe water pollution in the section of Drávaszabolcs was mainly due to the strongly contaminated Pécsi-víz Stream. Overall, the indicators of nitrogen balance often showed more favorable values for the Drava than for the tributaries.

The poorest *phosphorous* balance values were found at Drávaszabolcs, attributable to contaminants arriving with the Fekete-víz, and at Őrtilos, where the upstream Croatian section was polluted at the beginning of the monitored period (Dolgosné Kovács 2008). Based on orthophosphate and phosphorus concentrations, from the mid-1990s water was mostly assessed as 'good' at sampling sites. In addition to the tributaries, nutrient balance was also affected by livestock ranches (mostly because of improper strawy manure storage and liquid manure use on fields). From 2000 onwards, in terms of nutrient balance the water at the sampling sites was referred to class II or III.

The assessment of physico-chemical properties shows favorable water quality for most of the tributary streams (data extended to represent entire subbasins) of the Hungarian Drava catchment (Fig. 15.2). As far as the chemical status is concerned, 'poor' category was only recorded for the Pécsi-víz unit (Fig. 15.3). (It has to be noted that only 69% of the water bodies were assessed in the Drava catchment for the period 2009–2012, while extensive areas, mainly in Zala County, could not be evaluated.)

Microbiological contamination, mostly due to shortcomings in disinfection at municipal sewage treatment plants, was generally the most favorable at Őrtilos and the most unfavorable in the Drávaszabolcs section (Dolgosné Kovács 2008). Outstanding values at Drávaszabolcs indicate fresh fecal contamination (probably

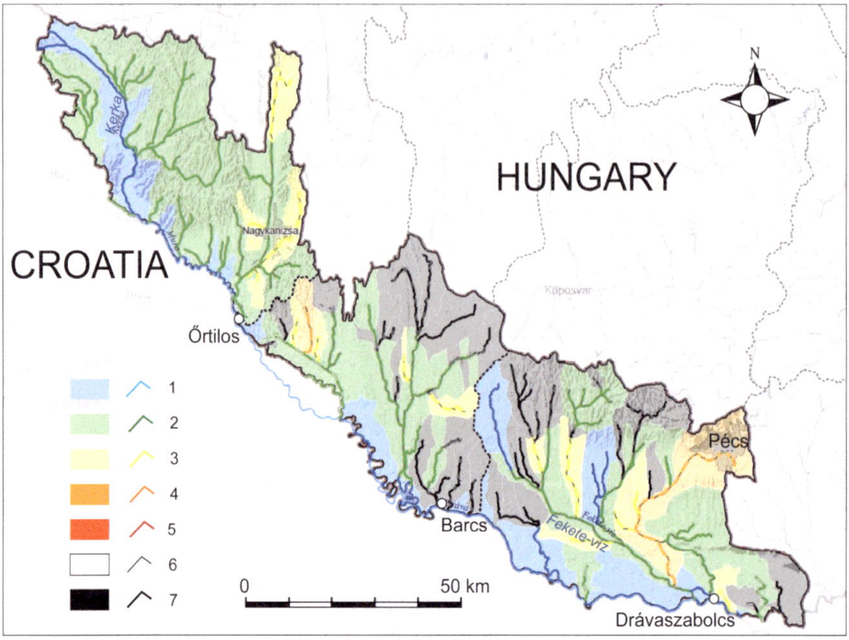

Fig. 15.2 Assessment of physico-chemical status for subbasins. *Source* STWMD (2015). 1, excellent; 2, good; 3, moderate; 4, poor; 5, very poor; 6, not assessed; 7, no data

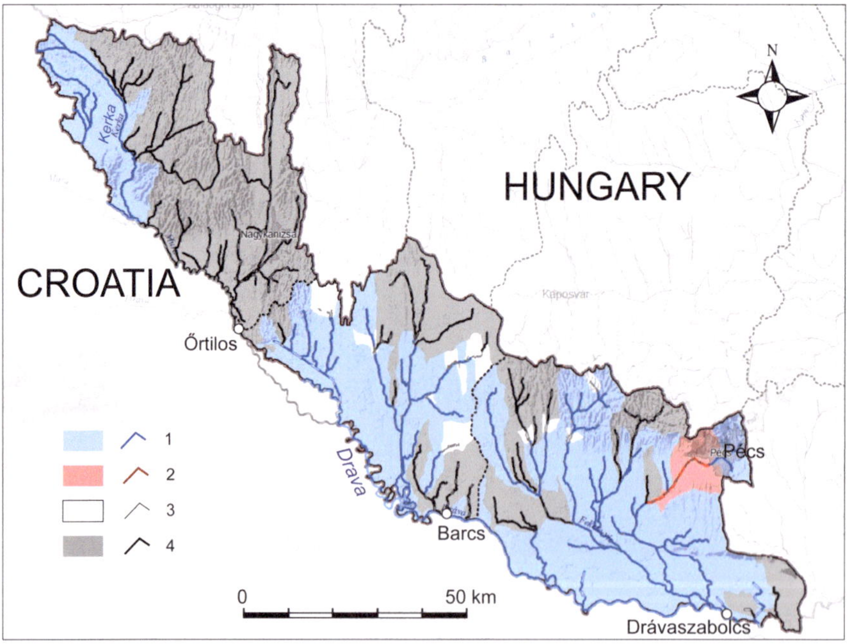

Fig. 15.3 Assessment of the chemical status for subbasins. *Source* STWMD (2015). 1, good; 2, poor; 3, not assessed; 4, no data

again through the Pécsi-víz). From 2000, the microbiological parameters at the sampling points were in Class III–V and Class II–III.

Based on 90% duration values, for *inorganic micro-pollutants* the Drava water was mostly assessed 'excellent' and in the case of aluminium 'good'. Data from Barcs were generally the most favorable. Although somewhat higher values were measured at both Őrtilos and Drávaszabolcs, the assessment stayed in the same class. Among *organic micro-pollutants*, crude oil and its products were classifiable in the study period. Until 1996, apart from a few years, the sampling sites were 'highly contaminated'. At Őrtilos pollutants came from outside study area, at Barcs the values are ascribed to hydrocarbon extraction, sediment dredging and renewed shipping (although at low intensity), while at Drávaszabolcs petrol products carried in the Drava by the Fekete-víz also contributed to the results. The values of other parameters show a favorable water quality without negative impacts from tributaries.

Even though the physico-chemical status was mostly 'excellent', the ecological assessment presents 'excellent' or 'good' values only over 6% of the catchment for the same period (Figs. 15.4 and 15.5).

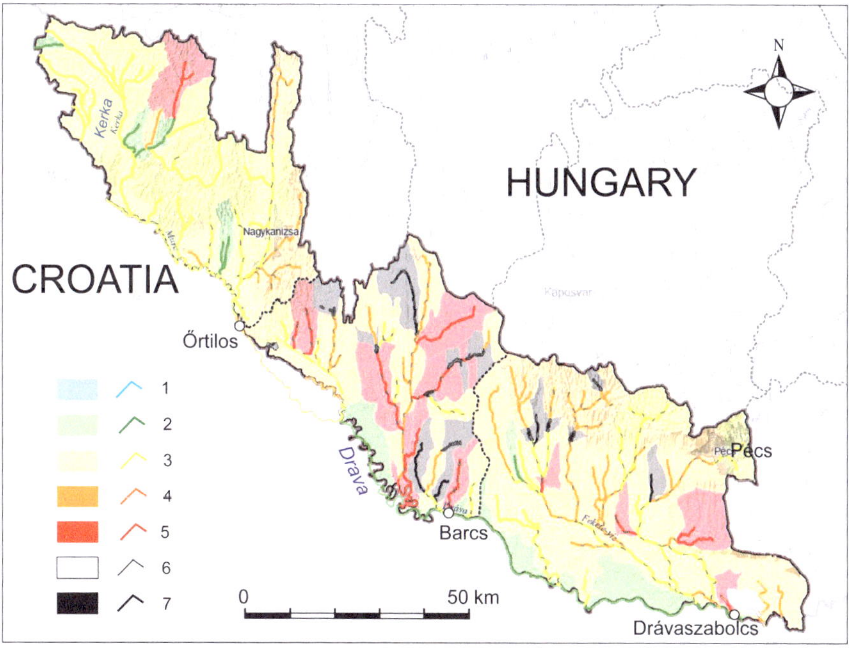

Fig. 15.4 Ecological assessment of surface waters in the Hungarian Drava catchment. For legend see Fig. 5.2. *Source* STWMD (2015)

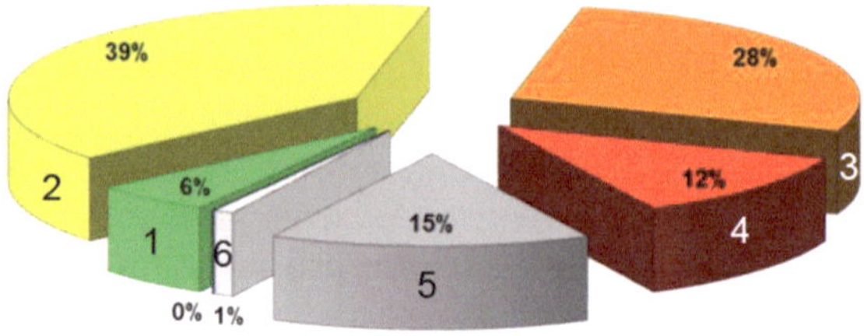

Fig. 15.5 Ecological status of subbasins. For the meaning of colours see Fig. 5.2

Table 15.6 Chemical status of lake Kisinc in 2014 (Tóth et al. 2015)

Monitoring site	Depth	Chemical and physico-chemical properties				
		dissolved oxygen	ammonium	nitrate	COD	phosphate
Kisinc oxbow lake	30 cm	poor	very good	very good	poor	good
	210 cm	poor	moderate	very good	poor	moderate

15.6 Case Study: Water Quality of an Oxbow Lake

After a rainy summer period in 2014, a water sampling campaign was organized at Lake Kisinc, the largest water body of the Cún-Szaporca oxbow (for description see Chap. 12). The aim of the investigation was to discover the daily periodicity in the concentration of chemical substances in the lake. Thus, every hour samples from −30 to −210 cm water depths (30 cm above the surface of lake sediments) were collected. The variable properties such as pH, redox potential, temperature, dissolved oxygen, were measured in the field. Other components like ammonium, nitrate, phosphorus forms (total and ortho-phosphorus) and COD were analyzed in the soil and water laboratory of the Institute of Geography, University of Pécs.

Evaluating the findings, the conclusion can be drawn that most of the chemical parameters do not follow the daily fluctuation of environmental parameters (with the exception of dissolved oxygen in the pondweed, next to the shore). Due the accumulation of organic matter in the sediment, the COD values are extremely high. We believe that after artificial water replenishment, which raised the water level by 1 m, COD values have increased (Table 15.6).

15.7 Conclusions

In the previous assessment systems, the locations, dates and frequency of water quality monitoring was not properly coordinated. Between 1980 and 1994, sampling sites were selected with the purpose of protecting the water quality of the main recipient, the Drava River. On tributaries sampling points were close to confluences. Therefore, pressures (sources of pollution) upstream could not be precisely located. By the time the monitoring network was built, large-scale contaminations had already affected the rivers. Moreover, water quality modelling lacked hydromorphological parameters, including water discharge, and ecological indicators.

A novelty of the EU WFD is that it purports integrated water basin management in a complex hierarchical system, instead of focusing on individual water bodies. Thus, the water quality of minor streams acquires the same importance as that of big water bodies. This attitude requires a new approach to the allocation of sampling sites too. However, the efficiency and reliability of the new monitoring system can only be judged after the evaluation of 8–10 years of experience.

As far as the current situation is concerned, compared to major rivers in Hungary, like the Danube and the Tisza, our results confirm that at present the Hungarian Drava section is assessed as one of the cleanest water bodies. The reason for the relatively higher concentration of contaminants at the Drávaszabolcs sampling site is the polluting effect of the Fekete-víz stream, which is the recipient of sewage from Pécs through the Pécsi-víz Stream.

References

Allan IJ, Vrana B, Greenwood R, Mills GA, Roig B, Gonzalez C (2006) A "toolbox" for biological and chemical monitoring requirements for the European Union's water framework directive. Talanta 69:302–322

ARSO (2008) Water quality in Slovenia. Environmental Agency of the Republic of Slovenia, Ljubljana, 74 pp

Bundesministerium (2006) Wassergüte in Österreich. Jahresbericht 2006. Bundesministerium für Land- und Forstwirtschaft, Umwelt und Wasserwirtschaft, Wien–Umweltbundesamt, Wien. 207 pp

Bundesministerium (2015) Wassergüte in Österreich. Jahresbericht 2015. Bundesministerium für Land- und Forstwirtschaft, Umwelt und Wasserwirtschaft, Wien. 184 pp

CCME (2001) Canadian water quality guidelines for the protection of aquatic life: Canadian Water Quality Index 1.0 Technical Report. In: Canadian environmental quality guidelines 1999. Winnipeg, MA

Clement A (2005) Felszíni vizek monitoring rendszere a VKI követelményei szerint (Surface water monitoring system according to WFD requirements) Manuscript educational material for the WFD. Department of Water Utilities and Environmental Engineering, Budapest (in Hungarian)

Dolgosné Kovács A (2003) Analitikai módszerek alkalmazása a Dráva-folyó vízminőségének monitorozása során (Analytical methods in monitoring the water quality of the Drava River) In: Börcsök E, Albert L (eds), A Magyar Tudomány Napja 2002 (Day of Hungarian Science). Scientific Conference of the Institute of Chemistry, Faculty of Forestry, University of West-Hungary, Sopron. 7 November 2002. pp 22–25 (in Hungarian)

Dolgosné Kovács A (2008) Antropogén tevékenységek hatásainak megjelenése a Dráva és főbb hazai mellékvízfolyásainak vízminőségén (Anthropogenic impacts on the water quality of the Drava River and its main tributaries in Hungary. Manuscript PhD Thesis. University of Pécs, Pécs. 220 pp (in Hungarian)

Dworak T, Gonzalez C, Laaser C, Interwies E (2005) The need for new monitoring tools to implement the WFD. Environ Sci Policy 8:301–306

Commission European (2000) Directive 2000/60/EEC. Establishing a framework for community action in the field of water policy. Off J European Communities, Luxemburg. L327:1–71

Gvozdić V, Brana J, Puntarić D, Vidosavljević D, Roland D (2011) Changes in the lower Drava River water quality parameters over 24 years. Arhiv za higijenu rada i toksikologiju/Arch Indus Hygiene and Toxicol 62:325–333

HAS (2017) Water in Hungary: Status overview for the National Water Programme of the Hungarian Academy of Sciences. Water Research Coordination Group, Centre for Ecology, Hungarian Academy of Sciences, Budapest. 97 pp. http://mta.hu/data/dokumentumok/Viztudomanyi%20Program/Water_in_Hungary_2017_07_20.pdf

Ijjas I (2005) Környezeti célkitűzések a Víz Keretirányelv (VKI) szerint (Environmental objectives according to the Water Framework Directive). Manuscript educational material for the WFD. Department of Water Utilities and Environmental Engineering, Budapest (in Hungarian)

Katona E (1982) A vizek minőségének védelme (Water quality protection). In: Wisnovszky I (ed) Water resource management, water quality protection, industrial water management.) VIZDOK, Budapest. pp 28–56 (in Hungarian)

Katona E (ed) (1984) Vízminőségi kárelhárítás kézikönyve (Handbook of water quality damage prevention). VIZDOK, Budapest. 323 pp (in Hungarian)

Kiss Gy (1974) Vízgazdálkodás és környezetvédelem Baranyában. (Water management and environmental protection in Baranya). In: Koltai J (ed) Környezetvédelem Baranyában és Pécsett (Environmental protection in Baranya and Pécs). Proceedings of the meeting in the frame of the Engineering and Economics Propaganda Month, Pécs, 19–20 March 1974. pp 95–103 (in Hungarian)

Kovács A (2002) Experiences on the water-quality analysis of the River Drava and the possible ways of the data-processing in the future. In: Jancskárné Anweiler I, Orbán F (eds) 40th Anniversary of Pollack Mihály College of Engineering International Symposium. Proceedings volume I. PTE PMMFK, Pécs, pp 387–394

Maloschik E, Ernst A, Gy Hegedűs, Darvas B, Székács A (2007) Monitoring water-polluting pesticides in Hungary. Microchem J 85(1):88–97. https://doi.org/10.1016/j.microc.2006.05.002

Novotny V (2005) Diffuse pollution from agriculture in the world. Proceedings of the European Commission Workshop "Where do the fertilizers go?" Ispra, Italy, 28–29 June 2005

OVF (1965) Dél-Dunántúl vízgazdálkodási keretterve (Water management framework plan for Southern Transdanubia). Manuscript. National Water Management Directorate (OVF), Budapest. 391 pp (in Hungarian)

Republika Hrvatska (2013) Informacija o stanju I kvaliteti voda, te izvorima onečišćenja voda na području Osječko-Baranjske Županije. Republika Hrvatska Osječko-Baranjska Županja Skupština, Osijek. (in Croatian) http://www.obz.hr/hr/pdf/2013/4_sjednica/07-Informacija_o_stanju_i_kvaliteti_voda_te_izvorima_oneciscenja_voda_na_porucju_Osjecko-baranjske_zupanije.pdf

Somlyódi L, Hock B, Gorzó Gy (1990) Felszíni vizek minőségének értékelése: javaslat a korszerűsítésre. (Surface water quality assessment: proposal for modernization). Vízügyi Közlemények 72(2):121–142 (in Hungarian)

Stundl K (1976) A Mura és a Dráva vízminősége (Water quality of the Mura and Drava Rivers) Hidrológiai Közlöny 6:268–272 (in Hungarian)

STWMD (2015) River basin management plan—Drava subcatchment. Southern Transdanubian Water Management Directorate, Pécs http://www.vizeink.hu/

Szilágyi F (2005) Felszíni vizek referencia állapota, referencia helyek (Reference status of surface waters, reference sites). Manuscript educational material for the WFD. Department of Water Utilities and Environmental Engineering, Budapest (in Hungarian)

Tomas D, Ćuk R, Senta Marić A, Mijatović I (2013) Assessment of the Drava River (Croatia) by Water Quality Index method. Fresenius Environ Bull 22(3a):904–913

Tóth G, Dezső J, Halász A, Miřijovský J, Lóczy D (2015) Deterioration of water quality in protected oxbows due to organic matter accumulation and sedimentation. In: Mentlík P, Stacke V (eds), Role of fieldwork in geomorphology. Abstracts. In: Annual conference of the Czech Association of Geomorphologists, Plzeň, 2015. 108 pp

Uherkovich G (1974) Megyénk hidrobiológiai problémái (Hydrobiological problems in Baranya). In: Koltai J (ed) Környezetvédelem Baranyában és Pécsett (Environmental protection in Baranya and Pécs). In: Proceedings of the meeting in the frame of the Engineering and Economics Propaganda Month, Pécs, 19–20 March 1974. pp 107–113. (in Hungarian)

Urbanić G (2011) Ecological status assessment of rivers in Slovenia—an overview. Natura Sloveniae 13(2):5–16

Chapter 16
Aquatic Macroinvertebrates of the Drava River and Its Floodplain

Arnold Móra and Zoltán Csabai

Abstract Due to the untouched and almost pristine conditions of the Drava region, to date 438 species belonging to various aquatic macroinvertebrate animal groups have been recorded from the Drava and its floodplain. Therefore the aquatic fauna of this area is among the richest in Hungary. The most unique species of the Drava is the caddisfly *Platyphylax frauenfeldi*, since its population along the river might be the last in the world. The occurrence of many rare, Natura 2000, protected or strictly protected species enhances the nature conservation value of the region. Unfortunately, the large number of non-indigenous species indicates the vulnerability of natural assemblages devastated by aquatic invasions. Moreover, the aquatic macroinvertebrate assemblages are vulnerable, and might be threatened by any unconsidered human action, especially measures of river regulation, since the change or loss of habitats might cause serious damage to the populations of rare species.

Keywords Diversity · Natura 2000 · Protected species · Invasive species *Platyphylax frauenfeldi* · *Graphoderus bilineatus*

16.1 Introduction

The lower Drava River has been a national border for a millennium. Access to the river and its floodplain was particularly strictly restricted in the second half of the 20th century, during the time of the Cold War. This might be the reason why this area remained nearly untouched and almost pristine with high habitat heterogeneity and preserved one of the most diverse biota in Europe. On the other hand, the area

A. Móra (✉) · Z. Csabai (✉)
Department of Hydrobiology, Institute of Biology, Faculty of Sciences, University of Pécs, Ifjúság útja 6, Pécs 7624, Hungary
e-mail: marnold@gamma.ttk.pte.hu

Z. Csabai
e-mail: csabai@gamma.ttk.pte.hu

D. Lóczy (ed.), *The Drava River*, Springer Geography,
https://doi.org/10.1007/978-3-319-92816-6_16

was not accessible for scientists either in this period and, consequently, its diverse biota remained unknown for a long time. Modern studies on the fauna of Drava and its floodplain started in the 1990s, in connection with the establishment of the Danube–Drava National Park in Hungary (Uherkovich 1995, 1998). During these works, nearly all aquatic invertebrate groups were taken into consideration, however, in many cases only sporadic data were provided. Later monitoring continued (Ábrahám 2005; Purger 2008) and many further collections were carried out for certain groups (see below). These studies have revealed a very rich macroinvertebrate fauna for the Drava and its floodplain water bodies.

16.2 Diversity of Aquatic Macroinvertebrates

To date 438 species belonging to various aquatic macroinvertebrate animal groups (e.g. molluscs, leeches, crustaceans, and insects) have been recorded from the Drava and its floodplain (Table 16.1). Due to the high number of species, the aquatic fauna of this area is among the richest in Hungary.

The first compilation of mollusc fauna of Drava and its floodplain area was published by Varga (1995). Based on former literature data, materials deposited in public and private collections and recent field works, he mentioned some 30 aquatic species to occur along the Drava. The extension of this work provided further data (Varga and Uherkovich 1998). Unfortunately, the intensive collections ended at that time, and later only sporadic data were published from the river itself (e.g., Héra 2005; Juhász et al. 2006a; Czirok et al. 2008; Horvai et al. 2012) and from its floodplain area (Purger 2013; Csabai et al. 2015a). Based on these results, the mollusc fauna proves to be very diverse with 36 aquatic snails and 16 bivalves, especially that of the Drava from where 47 species (31 snails and 16 bivalves) are known. The mollusc fauna of the floodplains is poorer or lesser known with 29 species (25 snails and 4 bivalves). The majority of the species are common in Hungary, but species valuable for nature conservation are remarkably represented too, as many rare species (e.g. *Amphimelania holandrii, Anisus vorticulus, Fagotia daudebartii, Theodoxus danubialis, Sphaerium rivicola, Pseudanodonta complanata, Unio crassus*) have been found in the Drava (Fig. 16.1).

The Annelida fauna of the Drava and its floodplain is poorly studied, and only the leeches (Hirudinea) were included in the investigations. Based on the sporadic data (Juhász et al. 2006b; Nesemann 1998; Czirok et al. 2008; Horvai et al. 2012) only 16 leech species are known from the area, 11 species from the river and 8 species from the floodplains.

Only sporadic data are known on the macroscopic crustacean fauna of both the Drava (Juhász et al. 2006c; Czirok et al. 2008; Borza 2011; Borza et al. 2011; Horvai et al. 2012) and its floodplain (Forró and Meisch 1998; Purger 2013; Csabai et al. 2015a). However, the fauna is relatively diverse with 16 species. The majority of the species were found in the Drava, but the two Branchiopoda species (the fairy shrimp *Branchipus schaefferi* and the tadpole shrimp *Triops cancriformis*) occur only in the

Table 16.1 List of aquatic macroinvertebrate species known from the Drava River and its floodplain

	Drava	Floodplain
MOLLUSCA		
GASTROPODA		
Acroloxidae		
Acroloxus lacustris (Linnaeus, 1758)	x	x
Bithyniidae		
Bithynia tentaculata (Linnaeus, 1758)	x	x
Hydrobiidae		
Potamopyrgus antipodarum (Gray, 1843)[Ni]	x	
Lithoglyphidae		
Lithoglyphus naticoides (C. Pfeiffer, 1828)[Ni]	x	
Lymnaeidae		
Galba truncatula (O.F. Müller, 1774)	x	x
Lymnaea stagnalis (Linnaeus, 1758)	x	x
Radix auricularia (Linnaeus, 1758)	x	x
Radix balthica (Linnaeus, 1758)	x	x
Radix labiata (Rossmassler, 1835)		x
Stagnicola corvus (Gmelin, 1791)	x	
Stagnicola palustris (O.F. Müller, 1774)	x	x
Stagnicola palustris complex		x
Melanopsidae		
Amphimelania holandrii (C. Pfeiffer, 1828)[a]	x	
Fagotia daudebartii (Prevost, 1821)[a]		x
Neritidae		
Theodoxus danubialis (C. Pfeiffer, 1828)[a]	x	
Theodoxus fluviatilis (Linnaeus, 1758)[Ni]	x	
Theodoxus transversalis (C. Pfeiffer, 1828)[a, N2000]	x	
Physidae		
Haitia acuta (Draparnaud, 1805)[Ni]	x	x
Physa fontinalis (Linnaeus, 1758)	x	x
Planorbidae		
Ancylus fluviatilis (O.F. Müller, 1774)	x	
Anisus septemgyratus (Rossmassler, 1835)	x	x
Anisus spirorbis (Linnaeus, 1758)	x	
Anisus vortex (Linnaeus, 1758)	x	x
Anisus vorticulus (Troschel, 1834)[a, N2000]	x	x
Ferrissia fragilis (Tryon, 1863)[Ni]		x
Gyraulus albus (O.F. Müller, 1774)	x	x
Gyraulus laevis (Alder, 1838)	x	
Hippeutis complanatus (Linnaeus, 1758)	x	x

(continued)

Table 16.1 (continued)

	Drava	Floodplain
Planorbarius corneus (Linnaeus, 1758)	x	x
Planorbis carinatus (O.F. Müller, 1774)		x
Planorbis planorbis (Linnaeus, 1758)	x	x
Segmentina nitida (O.F. Müller, 1774)	x	
Valvatidae		
Borysthenia naticina (Menke, 1845)[a]	x	
Valvata cristata (O.F. Müller, 1774)	x	x
Valvata piscinalis (O.F. Müller, 1774)		x
Viviparidae		
Viviparus acerosus (Bourguignat, 1862)	x	x
Viviparus contectus (Millet, 1813)	x	x
BIVALVIA		
Corbiculidae		
Corbicula fluminea (O.F. Müller, 1774)[Ni]	x	
Dreissenidae		
Dreissena polymorpha (Pallas, 1771)[Ni]	x	
Sphaeriidae		
Musculium lacustre (O.F. Müller, 1774)	x	x
Pisidium amnicum (O.F. Müller, 1774)	x	
Pisidium henslowanum (Sheppard, 1823)	x	
Pisidium subtruncatum (Malm, 1855)	x	
Pisidium supinum (A. Schmidt, 1851)	x	
Sphaerium corneum (Linnaeus, 1758)	x	x
Sphaerium rivicola (Lamarck, 1818)	x	
Unionidae		
Anodonta anatina (Linnaeus, 1758)	x	x
Anodonta cygnea (Linnaeus, 1758)	x	x
Anodonta woodiana (Lea, 1834)[Ni]	x	
Pseudanodonta complanata (Rossmassler, 1835)[a]	x	
Unio crassus (Philipson, 1788)[a, N2000]	x	
Unio pictorum (Linnaeus, 1758)	x	
Unio tumidus (Philipson, 1788)	x	
ANNELIDA		
HIRUDINEA		
Glossiphoniidae		
Alboglossiphonia heteroclita (Linnaeus, 1761)		x
Alboglossiphonia hyalina (O.F. Müller, 1774)		x
Glossiphonia complanata (Linnaeus, 1758)	x	x
Glossiphonia nebulosa (Kalbe, 1964)		x
Glossiphinia paludosa (Carena, 1824)	x	

(continued)

Table 16.1 (continued)

	Drava	Floodplain
Helobdella stagnalis (Linnaeus, 1758)	x	x
Hemiclepsis marginata (O.F. Müller, 1774)		x
Piscicolidae		
Caspiobdella fadejewi (Epshtein, 1961)	x	
Piscicola geometra (Linnaeus, 1758)	x	
Piscicola haranti (Jarry, 1960)	x	
Haemopidae		
Haemopis sanguisuga (Linnaeus, 1758)		x
Erpobdellidae		
Dina punctata (Johansson, 1927)	x	
Erpobdella octoculata (Linnaeus, 1758)	x	x
Erpobdella nigricollis (Brandes, 1900)	x	
Erpobdella vilnensis (Liskiewicz, 1925)	x	
Trocheta cylindrica (Örley, 1886)	x	
CRUSTACEA		
AMPHIPODA		
Corophiidae		
Chelicorophium curvispinum (G.O. Sars, 1895)[Ni]	x	
Chelicorophium sowinskyi (Martynov, 1924)[Ni]	x	
Gammaridae		
Dikerogammarus bispinosus (Martynov, 1925)[Ni]	x	
Dikerogammarus haemobaphes (Eichwald, 1841)[Ni]	x	
Dikerogammarus villosus (Sowinsky, 1894)[Ni]	x	
Gammarus fossarum (Koch, 1835)	x	
Gammarus roeselii (Gervais, 1835)	x	x
Niphargus valachicus (Dobreanu & Manolache, 1933)	x	x
Synurella ambulans (Müller, 1846)[Ni]	x	
DECAPODA		
Astacidae		
Astacus leptodactylus (Eschscholtz, 1823)[a]	x	
Pacifastacus leniusculus (Dana, 1852)[Ni]	x	
ISOPODA		
Asellidae		
Asellus aquaticus (Linnaeus, 1758)	x	x
Janiridae		
Jaera sarsi (Valkanov, 1936)[Ni]	x	
MYSIDA		
Mysidae		
Limnomysis benedeni (Czerniavsky, 1882)[Ni]	x	
ANOSTRACA		

(continued)

Table 16.1 (continued)

	Drava	Floodplain
Branchipodidae		
Branchipus schaefferi (Fischer, 1834)		x
NOTOSTRACA		
Triopsidae		
Triops cancriformis (Bosc, 1801)		x
CHELICERATA		
ARANEAE		
Cybaeidae		
Argyroneta aquatica (Clerck, 1757)[a]		x
INSECTA		
EPHEMEROPTERA		
Ametropodidae		
Ametropus fragilis (Albarda, 1878)[a]	x^L	
Baetidae		
Baetis alpinus (Pictet, 1843)	x	
Baetis buceratus (Eaton, 1870)	x^L	
Baetis fuscatus (Linnaeus, 1761)	x^L	
Baetis pentaphlebodes (Ujhelyi, 1966)	x^L	
Baetis rhodani (Pictet, 1843)	x^L	
Baetis tracheatus (Keffermüller & Machel, 1967)	x^L	
Baetis vardarensis (Ikomonov, 1962)	x^L	
Baetis vernus (Curtis, 1834)	x^L	
Centroptilum luteolum (Müller, 1776)	x^L	
Cloeon dipterum (Linnaeus, 1761)	x^L	x^L
Procloeon bifidum (Bengtsson, 1912)	x^L	
Procloeon macronyx (Kluge & Novikova, 1992)	x^L	
Raptobaetopus tenellus (Albarda, 1878)	x^L	
Caenidae		
Cercobrachys minutus (Tshernova, 1952)	x^L	
Caenis horaria (Linnaeus, 1758)	x^L	x
Caenis luctuosa (Burmeister, 1839)	x^L	
Caenis macrura (Stephens, 1835)	x^L	
Caenis pseudorivulorum (Keffermüller, 1960)	x^L	
Caenis robusta (Eaton, 1883)	x	x^L
Ephemerellidae		
Ephemerella ignita (Poda, 1761)	x^L	
Ephemerella notata (Eaton, 1887)	x	
Ephemeridae		
Ephemera danica (Müller, 1764)	x^L	
Ephemera glaucops (Pictet, 1843)	x	

(continued)

Table 16.1 (continued)

	Drava	Floodplain
Ephemera vulgata (Linnaeus, 1758)	x^L	
Heptageniidae		
Epeorus assimilis (Eaton, 1885)	x^L	
Heptagenia coerulans (Rostock, 1877)	x^L	
Heptagenia flava (Rostock, 1977)	x^L	
Heptagenia sulphurea (O.F. Müller, 1776)	x^L	x^L
Leptophlebiidae		
Habrophlebia fusca (Curtis, 1834)	x^L	
Paraleptophlebia submarginata (Stephens, 1835)	x^L	
Oligoneuriidae		
Oligoneuriella pallida (Hagen, 1855)[a]	x^L	
Oligoneuriella rhenana (Imhoff, 1852)[a]	x	x
Potamanthidae		
Potamanthus luteus (Linnaeus, 1767)	x^L	
Siphlonuridae		
Siphlonurus lacustris (Eaton, 1870)	x^L	
ODONATA		
Lestidae		
Chalcolestes viridis (Vander Linden, 1825)		x
Lestes barbarus (Fabricius, 1798)	x^L	x
Lestes dryas (Kirby, 1890)[a]		x
Lestes sponsa (Hansemann, 1823)	x	x^L
Lestes virens (Charpentier, 1825)	x	x^L
Sympecma fusca (Vander Linden, 1820)	x	x^L
Calopterygidae		
Calopteryx splendens (Harris, 1782)	x^L	x^L
Calopteryx virgo (Linnaeus, 1758)[a]	x^L	x
Platycnemididae		
Platycnemis pennipes (Pallas, 1771)	x^L	x^L
Coenagrionidae		
Coenagrion ornatum (Selys, 1850)[a, N2000]	x^L	x^L
Coenagrion puella (Linnaeus, 1758)	x^L	x^L
Coenagrion pulchellum (Vander Linden, 1825)	x	x^L
Coenagrion scitulum (Rambur, 1842)[a]		x
Enallagma cyathigerum (Charpentier, 1840)	x^L	x
Erythromma najas (Hansemann, 1823)	x	x^L
Erythromma viridulum (Charpentier, 1840)	x	x^L
Ischnura elegans (Vander Linden, 1820)	x^L	x^L
Ischnura pumilio (Charpentier, 1840)		x
Pyrrhosoma nymphula (Sulzer, 1776)		x^L

(continued)

Table 16.1 (continued)

	Drava	Floodplain
Aeshnidae		
Aeshna affinis Vander (Linden, 1820)		x^L
Aeshna cyanea (Müller, 1764)		x
Aeshna grandis (Linnaeus, 1758)		x
Aeshna isoceles (Müller, 1767)[a]		x^L
Aeshna mixta (Latreille, 1805)	x	x^L
Aeshna viridis (Eversmann, 1836)[b, N2000]		x^L
Anax imperator (Leach, 1815)	x^L	x^L
Anax parthenope (Selys, 1839)		x
Brachytron pratense (Müller, 1764)		x^L
Gomphidae		
Gomphus flavipes (Charpentier, 1825)[a, N2000]	x^L	x^L
Gomphus vulgatissimus (Linnaeus, 1758)[a]	x^L	x
Onychogomphus forcipatus (Linnaeus, 1758)[a]	x^L	
Ophiogomphus cecilia (Fourcroy, 1785)[a, N2000]	x^L	
Corduliidae		
Cordulia aenea (Linnaeus, 1758)	x^L	x^L
Epitheca bimaculata (Charpentier, 1825)[a]		x^L
Somatochlora flavomaculata (Vander Linden, 1825)[a]	x^L	x^L
Somatochlora metallica /meridionalis		x^L
Libellulidae		
Crocothemis erythraea (Brullé, 1832)		x^L
Leucorrhinia caudalis (Charpentier, 1840)[b, N2000]		x^L
Leucorrhinia pectoralis (Charpentier, 1825)[b, N2000]		x^L
Libellula depressa (Linnaeus, 1758)	x	x^L
Libellula fulva (Müller, 1764)[a]		x^L
Libellula quadrimaculata (Linnaeus, 1758)	x	x^L
Orthetrum albistylum (Selys, 1848)		x^L
Orthetrum brunneum (Fonscolombe, 1837)[a]	x	x
Orthetrum cancellatum (Linnaeus, 1758)	x	x
Orthetrum coerulescens (Fabricius, 1798)		x^L
Sympetrum depressiusculum (Selys, 1841)[a]		x
Sympetrum flaveolum (Linnaeus, 1758)	x	x^L
Sympetrum fonscolombii (Selys, 1840)		x
Sympetrum meridionale (Selys, 1841)	x	x
Sympetrum sanguineum (Müller, 1764)	x^L	x
Sympetrum striolatum (Charpentier, 1840)		x
Sympetrum vulgatum (Linnaeus, 1758)	x^L	x^L
PLECOPTERA		
Chloroperlidae		

(continued)

Table 16.1 (continued)

	Drava	Floodplain
Isoptena serricornis (Pictet, 1841)[a]	x[L]	
Xanthoperla apicalis (Newman, 1836)	x[L]	x[L]
Leuctridae		
Leuctra nigra (Olivier, 1811)	x[L]	
Nemouridae		
Nemoura cinerea (Retzius, 1783)	x[L]	
HETEROPTERA		
Aphelocheiridae		
Aphelocheirus aestivalis (Fabricius, 1794)	x	
Nepidae		
Nepa cinerea (Linnaeus, 1758)	x	x
Ranatra linearis (Linnaeus, 1758)	x	x
Naucoridae		
Ilyocoris cimicoides (Linnaeus, 1758)	x	x
Notonectidae		
Notonecta glauca (Linnaeus, 1758)	x	x
Notonecta lutea (Müller, 1776)[a]		x
Notonecta viridis (Delcourt, 1909)		x
Pleidae		
Plea minutissima (Leach, 1817)	x	x
Corixidae		
Callicorixa praeusta (Fieber, 1848)		x
Corixa affinis (Leach, 1817)		x
Corixa punctata (Illiger, 1807)		x
Hesperocorixa linnaei (Fieber, 1848)		x
Hesperocorixa sahlbergi (Fieber, 1848)		x
Micronecta griseola (Horváth, 1899)	x	x
Micronecta scholtzi (Fieber, 1860)	x	
Sigara falleni (Fieber, 1848)	x	x
Sigara fossarum (Leach, 1817)		x
Sigara lateralis (Leach, 1818)		x
Sigara nigrolineata (Fieber, 1848)		x
Sigara striata (Linnaeus, 1775)	x	x
Hydrometridae		
Hydrometra gracilenta (Horváth, 1899)		x
Hydrometra stagnorum (Linnaeus, 1758)	x	x
Hebridae		
Hebrus pusillus (Fallén, 1807)	x	x
Mesoveliidae		
Mesovelia furcata (Mulsant & Rey, 1852)		x

(continued)

Table 16.1 (continued)

	Drava	Floodplain
Mesovelia thermalis (Horváth, 1915)		x
Veliidae		
Microvelia buenoi (Drake, 1920)		x
Microvelia pygmaea (Dufour, 1833)		x
Microvelia reticulata (Burmeister, 1835)		x
Gerridae		
Aquarius najas (De Geer, 1773)[a]	x	
Aquarius paludum (Fabricius, 1794)	x	x
Gerris argentatus (Schummel, 1832)	x	x
Gerris asper (Fieber, 1861)	x	
Gerris lacustris (Linnaeus, 1758)	x	x
Gerris odontogaster (Zetterstedt, 1838)	x	x
Gerris thoracicus (Schummel, 1832)		x
COLEOPTERA		
Haliplidae		
Haliplus flavicollis (Sturm, 1834)		x
Haliplus fluviatilis (Aubé, 1836)		x
Haliplus furcatus (Seidlitz, 1887)		x
Haliplus heydeni (Wehncke, 1875)		x
Haliplus immaculatus (Gerhardt, 1877)		x
Haliplus obliquus (Fabricius, 1787)	x	
Haliplus ruficollis (De Geer, 1774)		x
Peltodytes caesus (Duftschmid, 1805)		x
Noteridae		
Noterus clavicornis (De Geer, 1774)		x
Noterus crassicornis (Müller, 1776)	x	x
Dytiscidae		
Acilius canaliculatus (Nicolai, 1822)		x
Acilius sulcatus (Linnaeus, 1758)		x
Agabus bipustulatus (Linnaeus, 1767)		x
Agabus undulatus (Schrank, 1776)		x
Bidessus nasutus (Sharp, 1887)		x
Clemnius decoratus (Gyllenhal, 1808)		x
Colymbetes fuscus (Linnaeus, 1758)		x
Cybister lateralimarginalis (De Geer, 1774)		x
Deronectes latus (Stephens, 1829)	x	
Dytiscus circumflexus (Fabricius, 1801)		x
Dytiscus dimidiatus (Bergsträsser, 1778)		x
Dytiscus marginalis (Linnaeus, 1758)		x
Graphoderus austriacus (Sturm, 1834)		x

(continued)

Table 16.1 (continued)

	Drava	Floodplain
Graphoderus bilineatus (De Geer, 1774)[b, N2000]		x
Graphoderus cinereus (Linnaeus, 1758)		x
Graptodytes granularis (Linnaeus, 1767)		x
Graptodytes pictus (Fabricius, 1787)		x
Hydaticus grammicus (Germar, 1830)		x
Hydaticus seminiger (De Geer, 1774)		x
Hydaticus transversalis (Pontoppidan, 1763)		x
Hydroglyphus geminus (Fabricius, 1792)	x	x
Hydroporus angustatus (Sturm, 1835)	x	x
Hydroporus dorsalis (Fabricius, 1787)		x
Hydroporus palustris (Linnaeus, 1761)		x
Hydroporus planus (Fabricius, 1781)		x
Hydroporus striola (Gyllenhal, 1827)	x	x
Hygrotus impressopunctatus (Schaller, 1783)		x
Hygrotus inaequalis (Fabricius, 1776)		x
Hygrotus versicolor (Schaller, 1783)		x
Hyphydrus ovatus (Linnaeus, 1761)	x	x
Ilybius ater (De Geer, 1774)	x	x
Ilybius fenestratus (Fabricius, 1781)		x
Ilybius fuliginosus (Fabricius, 1792)		x
Ilybius quadriguttatus (Lacordaire, 1835)		x
Laccophilus hyalinus (De Geer, 1774)	x	
Laccophilus minutus (Linnaeus, 1758)		x
Laccophilus poecilus (Klug, 1834)		x
Liopterus haemorrhoidalis (Fabricius, 1787)		x
Platambus maculatus (Linnaeus, 1758)	x	x
Porhydrus lineatus (Fabricius, 1775)		x
Rhantus bistriatus (Bergsträsser, 1778)		x
Rhantus consputus (Sturm, 1834)		x
Rhantus exsoletus (Forster, 1771)		x
Rhantus frontalis (Marsham, 1802)		x
Rhantus grapii (Gyllenhal, 1808)		x
Rhantus latitans (Sharp, 1882)		x
Rhantus suturalis (MacLeay, 1825)		x
Gyrinidae		
Gyrinus colymbus (Erichson, 1837)		x
Gyrinus substriatus (Stephens, 1828)		x
Orectochilus villosus (Müller, 1776)	x	
Hydraenidae		
Hydraena palustris (Erichson, 1837)	x	

(continued)

Table 16.1 (continued)

	Drava	Floodplain
Hydraena riparius species group		x
Limnebius atomus (Duftschmid, 1805)		x
Limnebius papposus (Mulsant, 1844)		x
Ochthebius minimus (Fabricius, 1792)		x
Spercheidae		
Spercheus emarginatus (Schaller, 1783)		x
Hydrochidae		
Hydrochus crenatus (Fabricius, 1792)		x
Hydrochus elongatus (Schaller, 1783)		x
Hydrochus flavipennis (Küster, 1852)		x
Hydrochus megaphallus (van Berge Henegouwen, 1988)		x
Helophoridae		
Helophorus aquaticus (Linnaeus, 1758)		x
Helophorus brevipalpis (Bedel, 1881)		x
Helophorus granularis (Linnaeus, 1761)		x
Helophorus griseus (Herbst, 1793)		x
Helophorus minutus (Fabricius, 1775)		x
Helophorus montenegrinus (Kuwert, 1886)		x
Helophorus nubilis (Fabricius, 1776)		x
Helophorus paraminutus (Angus, 1986)		x
Hydrophilidae		
Anacaena globulus (Paykull, 1798)		x
Anacaena limbata (Fabricius, 1792)		x
Anacaena lutescens (Stephens, 1829)		x
Berosus frontifoveatus (Kuwert, 1890)		x
Berosus luridus (Linnaeus, 1761)		x
Berosus signaticollis (Charpentier, 1825)		x
Cercyon convexiculus (Stephens, 1829)	x	x
Coelostoma orbiculare (Fabricius, 1775)		x
Cymbiodyta marginella (Fabricius, 1792)		x
Enochrus affinis (Thunberg, 1794)		x
Enochrus bicolor (Fabricius, 1792)		x
Enochrus coarctatus (Gredler, 1863)	x	x
Enochrus melanocephalus (Olivier, 1792)		x
Enochrus ochropterus (Marsham, 1802)		x
Enochrus quadripunctatus (Herbst, 1797)		x
Enochrus testaceus (Fabricius, 1801)		x
Helochares lividus (Forster, 1771)		x
Helochares obscurus (Müller, 1776)		x
Hydrobius fuscipes (Linnaeus, 1758)		x

(continued)

Table 16.1 (continued)

	Drava	Floodplain
Hydrochara caraboides (Linnaeus, 1758)	x	x
Hydrochara flavipes (Steven, 1808)		x
Hydrophilus aterrimus (Eschscholz, 1822)		x
Hydrophilus piceus (Linnaeus, 1758)		x
Laccobius bipunctatus (Fabricius, 1775)		x
Laccobius gracilis (Motschulski, 1855)		x
Laccobius minutus (Linnaeus, 1758)		x
Laccobius simulatrix (Orchymont, 1932)		x
Limnoxenus niger (Zschach, 1788)		x
Dryopidae		
Dryops anglicanus (Edwards, 1909)		x
Elmidae		
Macronychus quadrituberculatus (Müller, 1806)[a]	x	
Limnius volckmari (Panzer, 1793)	x	
MEGALOPTERA		
Sialidae		
Sialis lutaria (Linnaeus, 1758)	x^L	x^L
NEUROPTERA		
Sisyridae		
Sisyra fuscata (Fabricius, 1793)	x	x
Sisyra terminalis (Curtis, 1854)	x	x
TRICHOPTERA		
Rhyacophilidae		
Rhyacophila dorsalis (Curtis, 1834)	x	
Glossosomatidae		
Agapetus laniger (Pictet, 1834)	x^L	
Glossosoma boltoni (Curtis, 1834)	x	
Hydroptilidae		
Agraylea sexmaculata (Curtis, 1834)	x	x
Hydroptila angustata (Mosely, 1939)	x	
Hydroptila forcipata (Eaton, 1873)	x	x
Hydroptila lotensis (Mosely, 1930)	x	
Hydroptila sparsa (Curtis, 1834)	x	x
Hydroptila vectis (Curtis, 1834)	x	
Ithytrichia lamellaris (Eaton, 1873)	x	
Orthotrichia angustella (McLachlan, 1865)	x	
Orthotrichia costalis (Curtis, 1834)	x	
Orthotrichia tragetti (Mosely, 1930)	x	x
Oxyethira falcata (Morton, 1893)	x	
Oxyethira flavicornis (Pictet, 1834)	x	x

(continued)

Table 16.1 (continued)

	Drava	Floodplain
Hydropsychidae		
Hydropsyche angustipennis (Curtis, 1834)	x^L	x
Hydropsyche bulbifera (McLachlan, 1878)	x^L	
Hydropsyche bulgaromanorum (Malicky, 1977)	x^L	x
Hydropsyche contubernalis (McLachlan, 1865)	x^L	x
Hydropsyche incognita (Pitsch, 1993)	x^L	
Hydropsyche modesta (Navás, 1925)	x^L	x
Hydropsyche ornatula (McLachlan, 1878)	x^L	x
Hydropsyche pellucidula (Curtis, 1834)	x^L	x
Hydropsyche siltalai (Döhler, 1963)	x	
Polycentropodidae		
Cyrnus crenaticornis (Kolenati, 1859)	x	x
Cyrnus trimaculatus (Curtis, 1834)	x	x
Holocentropus dubius (Rambur, 1842)	x	x
Holocentropus picicornis (Stephens, 1836)	x	
Neureclipsis bimaculata (Linnaeus, 1758)	x^L	x
Plectrocnemia conspersa (Curtis, 1834)	x	
Polycentropus irroratus (Curtis, 1835)	x	
Psychomyiidae		
Lype phaeopa (Stephens, 1836)	x	x
Lype reducta (Hagen, 1868)	x^L	x
Psychomyia pusilla (Fabricius, 1781)	x^L	x
Ecnomidae		
Ecnomus tenellus (Rambur, 1842)	x	x
Phryganeidae		
Agrypnia varia (Fabricius, 1793)	x	x
Phryganea bipunctata (Retzius, 1783)		x
Phryganea grandis (Linnaeus, 1758)	x	x
Trichostegia minor (Curtis, 1834)	x	x
Brachycentridae		
Brachycentrus subnubilus (Curtis, 1834)	x^L	
Limnephilidae		
Anabolia furcata (Brauer, 1857)	x^L	x
Chaetopteryx fusca (Brauer, 1857)	x	
Chaetopteryx major (McLachlan, 1876)	x	
Glyphotaelius pellucidus (Retzius, 1783)	x	x
Grammotaulius nigropunctatus (Retzius, 1783)	x	x
Halesus tesselatus (Rambur, 1842)	x^L	x
Limnephilus affinis (Curtis, 1834)	x	x
Limnephilus auricula (Curtis, 1834)	x	x

(continued)

Table 16.1 (continued)

	Drava	Floodplain
Limnephilus bipunctatus (Curtis, 1834)	x	x
Limnephilus decipiens (Kolenati, 1848)	x	
Limnephilus extricatus (McLachlan, 1865)	x	
Limnephilus flavicornis (Fabricius, 1787)	x	x
Limnephilus griseus (Linnaeus, 1758)	x	x
Limnephilus incisus (Curtis, 1834)	x	
Limnephilus lunatus (Curtis, 1834)	x	x
Limnephilus rhombicus (Linnaeus, 1758)	x	x
Limnephilus vittatus (Fabricius, 1798)	x	x
Micropterna lateralis (Stephens, 1837)	x	
Platyphylax frauenfeldi (Brauer, 1857)[b]	x	
Potamophylax luctuosus (Piller & Mitterpacher, 1783)	x	
Potamophylax rotundipennis (Brauer, 1857)	x[L]	x
Stenophylax meridiorientalis (Malicky, 1980)		x
Stenophylax permistus (McLachlan, 1895)	x	x
Lepidostomatidae		
Crunoecia irrorata (Curtis, 1834)	x	
Goeridae		
Goera pilosa (Fabricius, 1775)	x[L]	x
Silo nigricornis (Pictet, 1834)	x	
Silo piceus (Brauer, 1857)	x[L]	
Leptoceridae		
Adicella syriaca (Ulmer, 1907)	x	
Athripsodes albifrons (Linnaeus, 1758)	x	
Athripsodes aterrimus (Stephens, 1836)	x	x
Athripsodes cinereus (Curtis, 1834)	x	
Ceraclea alboguttata (Hagen, 1860)	x	x
Ceraclea annulicornis (Stephens, 1836)	x	
Ceraclea aurea (Pictet, 1834)	x	
Ceraclea dissimilis (Stephens, 1836)	x[L]	x
Ceraclea riparia (Albarda, 1874)	x	
Ceraclea senilis (Burmeister, 1838)	x	x
Leptocerus tineiformis (Curtis, 1834)	x	xL
Mystacides azureus (Linnaeus, 1761)	x[L]	x
Mystacides longicornis (Linnaeus, 1758)	x	
Mystacides niger (Linnaeus, 1758)	x	
Oecetis furva (Rambur, 1842)	x[L]	x
Oecetis lacustris (Pictet, 1834)	x	x
Oecetis notata (Rambur, 1842)	x[L]	x
Oecetis ochracea (Curtis, 1825)	x	x

(continued)

Table 16.1 (continued)

	Drava	Floodplain
Oecetis tripunctata (Fabricius, 1793)	x	
Setodes punctatus (Fabricius, 1793)	x[L]	x
Setodes viridis (Fourcroy, 1785)		x
Triaenodes bicolor (Curtis, 1834)		x
Triaenodes simulans (Tjeder, 1929)	x	
Sericostomatidae		
Notidobia ciliaris (Linnaeus, 1761)		x
DIPTERA		
Culicidae		
Aedes cinereus (Meigen, 1818)		x
Aedes rossicus (Dolbeskin et al., 1930)		x
Aedes vexans (Meigen, 1830)		x
Anopheles claviger (Meigen, 1804)		x
Anopheles hyrcanus (Pallas, 1771)		x
Anopheles maculipennis (Meigen, 1818)		x
Anopheles messeae (Falleroni, 1926)		x
Coquillettidia richiardii (Ficalbi, 1889)		x
Culex modestus (Ficalbi, 1890)		x
Culex pipiens (Linnaeus, 1758)		x
Culex territans (Walker, 1856)		x
Culiseta annulata (Schrank, 1776)		x
Ochlerotatus annulipes (Meigen, 1830)		x
Ochlerotatus cantans (Meigen, 1818)		x
Ochlerotatus caspius (Pallas, 1771)		x
Ochlerotatus excrucians (Walker, 1856)		x
Ochlerotatus flavescens (Müller, 1764)		x
Ochlerotatus geniculatus (Olivier, 1791)		x
Ochlerotatus sticticus (Meigen, 1838)		x
Uranotaenia unguiculata (Edwards, 1913)		x
Simuliidae		
Simulium erythrocephalum (De Geer, 1776)	x[L]	
Simulium ornatum species group	x[L]	
Simulium reptans (Linnaeus, 1758)	x[L]	
Chironomidae		
Chernovskiia macrocera (Saether, 1977)	x[L]	
Chernovskiia orbicus (Townes, 1945)	x[L]	
Chironomus acutiventris (Wülker, Ryser et Scholl, 1983)	x[L]	
Chironomus nuditarsis (Keyl, 1961)		x[L]
Chironomus plumosus agg.		x[L]
Chironomus tentans (Fabricius, 1804)		x[L]

(continued)

Table 16.1 (continued)

	Drava	Floodplain
Cladotanytarsus vanderwulpi (Edwards, 1929)	x[L]	
Corynoneura scutellata (Winnertz, 1846)		x[L]
Cricotopus vierriensis (Goetghebuer, 1935)	x[L]	
Cryptochironomus rostratus (Kieffer, 1921)	x[L]	
Guttipelopia guttipennis (van der Wulp, 1861)		x[L]
Macropelopia nebulosa (Meigen, 1804)		x[L]
Monopelopia tenuicalcar (Kieffer, 1918)		x[L]
Orthocladius glabripennis (Goetghebuer, 1921)	x[L]	
Orthocladius oblidens (Walker, 1856)	x[L]	
Parachironomus gracilior (Kieffer, 1918)		x[L]
Paratendipes albimanus (Meigen, 1818)	x[L]	
Potthastia longimanus (Kieffer, 1922)	x[L]	
Rheocricotopus chalybeatus (Edwards, 1929)	x[L]	
Tanypus kraatzi (Kieffer, 1912)		x[L]

L collected also as larvae; *E* collected also as exuviae; [a]protected in Hungary; [b]strictly protected in Hungary; N2000, Natura 2000 species; Ni, Non-indigenous species in Hungary

Fig. 16.1 Typical native large-sized unionid mussels (Bivalvia: Unionidae) of the Drava. **a** *Anodonta anatina*, **b** *Unio crassus*, **c** *Unio pictorum*, **d** *Unio tumidus* (photos by Arnold Móra)

floodplain area, and both of them develop in temporary small water bodies. Other crustaceans occurring in the Drava and the floodplains belong to Malacostraca. Unfortunately, only 5 out of 14 malacostracan species are native (*Asellus aquaticus, Gammarus fossarum, G. roeselii, Niphargus valachicus, Astacus leptodactylus*), indicating the vulnerability of the natural assemblages of the river.

The only aquatic spider species, *Argyroneta aquatica*, can be found in the vegetated water bodies of the Drava floodplain (e.g., Csabai et al. 2015a). The species is regarded as an indicator of good water quality, partly this is the reason why it is protected for many years.

Insects represent the most diverse group along the Drava. In some groups (e.g. Odonata, Ephemeroptera, Trichoptera) only the larvae live in the water and adults may move for a long distance from their breeding places. Accordingly, when only adults of these groups were collected, we do not know if the specimens developed in the Drava or in its floodplains. In these cases the fauna of the river and the floodplain cannot be separated from each other. It is only possible if the ecological requirements of the species are well known and clearly classify them to either fast-flowing or standing waters.

The first information on the mayfly (Ephemeroptera) fauna of the area can be found in Sziráki (1995). His work contained some data based on investigation of adults collected in the early 1990s. Further adult data were published in Sziráki (1998) and Bauerfeind et al. (2005). Later the studies focused on larvae, and despite the sporadic collections, a high number of species was found in the Drava (Kovács et al. 1998, 1998–99, 1999a, 2003; Kovács 2005, 2006a, 2009a, 2011; Czirok et al. 2008; Horvai et al. 2012). Until now 35 mayfly species have been found in the river, but four of them (*Baetis alpinus, Ephemera glaucops, Ephemerella notata, Oligoneuriella rhenana*) were collected only as adults in the 1990s (Sziráki 1995) and no further data are available. Most of the species are common and widespread in Hungary, but some interesting species also occur. For example, *Ametropus fragilis, Procloeon macronyx* and *Cercobrachys minutus* are rare, but typical psammophilous species in larger rivers (Kovács 2011). *Oligoneuriella pallida* and *O. rhenana* are among the rare mayfly species in Hungary (Kovács et al. 1999a).

Although dragonflies and damselflies are among the most attractive and, thus, the most investigated insect groups, the first studies on the Odonata fauna along the Drava were completed only in the 1990s (Tóth 1995a, 1998). The results of these studies and a later monitoring work (Tóth 2005) were summarized in Tóth (2010), along with many new data based on the investigation of larvae, exuviae and adults. In addition, larvae and exuviae of Odonata were sporadically collected both in the river (Kovács et al. 2004; Müller et al. 2006; Czirok et al. 2008; Horvai et al. 2012) and in its floodplain (Purger 2013; Csabai et al. 2015a). Based on these works, the Odonata fauna of the Drava region is well known. However, since these studies covered a limited geographical area along the river, the distribution of Odonata species in the region is not fully revealed. Similarly to other larger rivers in Hungary, the most typical dragonflies in the Drava are riverine dragonflies (Gomphidae) and demoiselles (Calopterygidae). Since these species are intolerant to pollution and indicators of good water quality, the occurrence of all four Hungarian species of

Fig. 16.2 All four Hungarian riverine dragonfly (Odonata: Gomphidae) species can be found in the Drava (**a**–**b** *Gomphus flavipes*, **c**–**d** *Gomphus vulgatissimus*, **e**–**f** *Onychogomphus forcipatus*, **g**–**h** *Ophiogomphus cecilia*; **a**, **c**, **e**, **g** adult males, **b**, **d**, **a**, **h** exuviae; photos by Arnold Móra)

Gomphidae (Fig. 16.2) in the river is remarkably important for nature conservation (see Jakab and Dévai 2008). Both Hungarian demoiselles occur in the Dráva, but the Beautiful Demoiselle (*Calopteryx virgo*) is much less common than its widely distributed sister species Banded Demoiselle (*C. splendens*). Other species can occasionally be found in rivers too, especially in slower-flowing and vegetated reaches. Accordingly, altogether 30 dragonfly and damselfly species were recorded from the Drava, but 13 of them were observed only as adults and might not develop in the river. The other 17 species were collected as larvae and/or exuviae too. Besides the ubiquitous species (e.g. *Platycnemis pennipes, Ischnura elegans, Anax imperator, Cordulia aenaea, Sympetrum sanguineum*), some rare species (e.g. *Coenagrion ornatum, Somatochlora flavomaculata*) also occur in the river. With 51 species, many are rare and valuable (e.g. *Lestes dryas, Coenagrion scitulum, Aeshna viridis, Epitheca bimaculata, Leucorrhinia caudalis, L. pectoralis, Sympetrum depressiusculum*), the Odonata fauna of the floodplains is much more diverse.

The Hungarian section of the Drava mainly crosses lowlands, unfavourable environments for many stonefly (Plecoptera) species. Moreover, only very sporadic collections of stoneflies were performed along the river (Kovács et al. 2002; Kovács 2006b, c, 2009b, 2011; Horvai et al. 2012). Accordingly, only four Plecoptera species were recorded from the area. Three of them (*Leuctra nigra, Nemoura cinerea, Xanthoperla apicalis*) are common in Hungary, while the fourth, *Isoptena serricornis*, is a very rare species with only three known localities in the country (Kovács 2011).

Despite the sporadic collections, the aquatic and semi-aquatic Heteroptera (Nepomorpha, Gerromorpha) fauna of the area proved to be relatively diverse with 35 species (18 from Drava and 31 from the floodplain). The majority of the species are widespread and common in both running and standing waters of Hungary, including the Drava (e.g., Kondorosy and Földessy 1998; Soós et al. 2009a; Kálmán et al. 2011), but some interesting and rare species also occur in the area. *Aphelocheirus aestivalis* occurs exclusively in running waters and was found both in the main channel (Kiss et al. 2006; Czirok et al. 2008; Horvai et al. 2012) and its side-arms (Purger 2013). The occurrence of three rare species, *Notonecta lutea, Sigara fossarum* and *Mesovelia thermalis*, is important from a faunistical point of view. All these species were collected in oxbow lakes, where *M. thermalis* is apparently common (Kálmán et al. 2011; Csabai et al. 2015a), while other two species are very rare: *N. lutea* was collected in two oxbows (Kálmán et al. 2011) and *S. fossarum* was only found at a single locality (Csabai et al. 2015a).

Generally, with few exceptions, aquatic Coleoptera species are related to standing rather than running waters. This fact is also reflected in the local fauna: only 17 Coleoptera species were found in the Drava, while 104 species were recovered from floodplain oxbows, ponds and marshes. The aquatic beetle fauna of the area, however, is relatively poorly known: since the first compilation (Gidó and Szél 1998), no studies focused on aquatic beetles and only a single sampling campaign was directed at beetles in the floodplains (Kálmán et al. 2011). Additional information is available on the occurrence of species of nature conservation value, i.e. *Graphoderus bilineatus* (Csabai et al. 2015b) and *Macronychus quadrituberculatus* (Kovács et al. 1999b; Kovács and Merkl 2005). Sporadic data have been published from studies concerning a larger area of Hungary (Ködöböcz et al. 2006; Csabai et al. 2009) or arose from various assessment studies of ecological status

(Czirok et al. 2008; Horvai et al. 2012; Purger 2013; Csabai et al. 2015a). Most of the species known from the Drava are common in Hungary, but some interesting species also occur in the river. The Hairy Whirligig Beetle, *Orectochilus villosus*, and the riffle beetle *Macronychus quadrituberculatus* were considered as very rare species for a long time, but they both are relatively widespread in the running waters of Hungary. One of the most important faunistical results is the occurrence of the diving beetle *Deronectes latus* in the Drava, since this is the second Hungarian locality for this species (see Csabai et al. 2009).

Alderflies (Megaloptera: Sialidae) and spongeflies (Neuroptera: Sysiridae) are among the smallest insect families containing aquatic species. A single sialid species, *Sialis lutaria*, was recorded from the Drava and its floodplains, where the larvae can be found in standing and running water bodies (Ábrahám 1995, 1998; Horvai et al. 2012; Csabai et al. 2015a), in sediments with high organic matter content. The feeding of spongeflies larvae is special as they parasitize freshwater sponges. From this family, two species, *Sysira fuscata* and *Sysira terminalis*, are known to occur along the Drava (Ábrahám 1995, 1998).

The first data on the occurrence of caddisflies (Trichoptera) along the Drava came from a collection of adults by light trap at the very end of the 1980s (Uherkovich and Nógrádi 1992). Due to continuous collections of adults (Nógrádi and Uherkovich 1995, 1998; Nógrádi 2001; Uherkovich 2005), the Trichoptera fauna of the area became one of the best known in Hungary with 91 species. However, a large part of the collected species typically live in standing waters, and most probably developed in different water bodies of the floodplain instead of the river, e.g. larvae of *Leptocerus tineiformis* were collected only in oxbow lakes (Csabai et al. 2015a). However, based on sporadic collections of larvae, 23 species are known to occur in the Drava (Móra et al. 2006; Czirok et al. 2008; Szitta et al. 2009; Horvai et al. 2012). In the river, the caddisfly assemblages are dominated by species typical of larger rivers, like *Agapetus laniger, Hydropsyche* spp., *Neureclipsis bimaculata, Psychomyia pusilla, Brachycentrus subnubilus, Ceraclea dissimilis, Oecetis notata, Setodes punctatus*. Besides them, rare species in Hungary also occur along the Drava (e.g. *Hydroptila vectis, Platyphylax frauenfeldi, Silo nigricornis, Ylodes simulans*). Some of the rare species, e.g. *Silo piceus* and *Adicella syriaca*, became relatively common in the region in the early 2000s (Uherkovich and Nógrádi 2005).

One of the most diverse orders of insects is Diptera, with a plenty of species developing in aquatic habitats. Unfortunately, the aquatic Diptera were almost neglected during the studies on the fauna of the Drava region. To date, larvae of three black fly taxa (Simuliidae: *Simulium erythrocephalum, S. reptans, S. ornatum* species group) were only recorded from the river (Horvai et al. 2012). In a recent study, larvae and exuviae of 11 non-biting midge (Chironomidae) taxa were collected from the river, among them typical species that are characteristic for larger sandbed rivers, like *Chernovskiia macrocera, C. orbicus, Chironomus acutiventris, Cladotanytarsus vanderwulpi, Cricotopus vierriensis, Cryptochironomus rostratus, Orthocladius oblidens, O. glabripennis, Paratendipes albimanus, Potthastia longimanus, Rheocricotopus chalybeatus* (own unpublished data). However, the aquatic Diptera fauna (including these two families) of the Drava remained nearly completely unknown. Similarly to the river, the aquatic Diptera fauna of the floodplains is very poorly described, and mainly the families of medical and veterinary importance were studied, i.e. mosquitoes (Culicidae) and horseflies (Tabanidae). Tóth (1995b) reported the occurrence of 21 mosquito species based on collections of adults. Horseflies were studied extensively in both Croatian and Hungarian parts of the floodplain, with a result of the occurrence of 42 species (summarized in Majer and Krčmar 2006), but there are only some species whose larvae develop in water, so they are considered as aquatic macroinvertebrates only to some extent. It is not clearly known which species prefer the water against moist soil for early stage development. Therefore, the Tabanid species have not been listed in the appendix. Additionally, nine chironomid taxa were collected in an oxbow of the Drava (Csabai et al. 2015a), among them species characteristic for standing waters with dense vegetation (e.g., *Chironomus tentans, Corynoneura scutellata, Guttipelopia guttipennis, Monopelopia tenuicalcar, Parachironomus gracilior, Tanypus kraatzi*).

16.3 Species of Nature Conservation Interest

The most unique and valuable species of the Drava is the caddisfly *Platyphylax frauenfeldi*. The population of this large-sized species along the Drava might be the last one in the world (Malicky et al. 2002). Some time ago *P. frauenfeldi* was present in a relatively large part of Europe from France to the Danube region—although always rare and in very scattered distribution. The species exclusively live/lived in unregulated and non-polluted larger rivers, like the Aare, Danube, Drava, Enns, Inn, Mura and Rhône (Malicky et al. 2002), but nearly all habitats of *P. frauenfeldi* have been lost in Europe. Although in Hungary there are older records from many sites along the rivers Kerka, Mura and Drava (Uherkovich and Nógrádi 1997; Uherkovich 2004), the species has not been recovered for 15 years. All information on its recent distribution suggests that the remnant Hungarian population of *P. frauenfeldi* is small and therefore vulnerable. Because of its vulnerability, the species is strictly protected in Hungary, but not listed in either the Habitat Directive (Council of the European Union 2013) or the IUCN Red List (IUCN 2017). The future of *P. frauenfeldi* is uncertain, and any unconsidered human action (e.g. any measures of river regulation, heavy industrial activities) might lead to its extinction. In lack of knowledge on the life cycle of the species, the possibilities of its protection are very limited. The flight season lasts from October to November, with a peak in late October (Uherkovich and Nógrádi 1997). The larval biology is completely unknown, since no larvae have been collected in the nature; however, larvae were described based on reared specimens (Malicky et al. 2002).

The dytiscid beetle *Graphoderus bilineatus* (Fig. 16.3a) is a charismatic species for conservation. It has been granted special conservation status, is a Natura 2000 species of community interest, and strictly protected in Hungary. According to the

IUCN Red List of Threatened Species (IUCN 2017), its status is 'Vulnerable' and it also has been included in the Berne Convention, annexes II and IV of the Habitats Directive and in CORINE lists. In Hungary, the species has two restricted distribution areas: 1. along the Upper Tisza River and in the Bodrog River floodplain (Northeast-Hungary) and 2. along the Lower Danube and in the eastern Drava region (Southwest-Hungary). The latter units merge into a single contiguous area at the confluence of these rivers, at Kopački Rit, Croatia (Haraszthy 2014; Csabai 2015b). Along the eastern Drava, six different localities of the species became known from a long section (133–70 rkm) within the period 2012–2014 (Csabai et al. 2015b). Summing up our knowledge on its ecology, in Hungary the species is exclusively linked to small water bodies, gravel pits, ponds, and oxbow lakes within the floodplains of medium-sized and large rivers. While in the floodplains of Tisza, Bodrog and Danube, it mainly occurs in densely vegetated eu- or hypertrophic water bodies, along the Drava it prefers active side-arms and small ponds with less vegetation (Haraszthy 2014). All known localities are in the active floodplain, occasionally or regularly connected to the rivers for a while, inundated, and refreshed during floods.

In addition to *Graphoderus bilineatus*, many further aquatic macroinvertebrate species occurring along the Dráva can be found among the 'species of community interest whose conservation requires the designation of special areas of conservation' (Annex II) and 'species of community interest in need of strict protection' (Annex IV) that are listed in the latest consolidated version of the Habitat Directive (Council of the European Union 2013). These species are considered important for nature conservation, and all are protected or strictly protected in Hungary. The Lesser Ramshorn Snail (*Anisus vorticulus*), although widely distributed in Hungary, is threatened by loss of its habitats (Haraszthy 2014). Fortunately, a viable population can be found in the standing waters of the Drava floodplain (Csabai et al. 2015a; Varga and Uherkovich 1998). Only an old reference from times before 1933 is known (Varga and Uherkovich 1998) for the Striped Nerite (*Theodoxus transversalis*) from the Drava, and according to recent studies, this species disappeared from the river (see Haraszthy 2014). Similarly to other larger Hungarian rivers, a strong population of Thick Shelled River Mussel (*Unio crassus*) (Fig. 16.1b) lives in the Drava. *Unio crassus* is among the endangered (EN) species (IUCN 2017), and the Hungarian populations are important in its conservation. Most of the Hungarian Natura 2000 species belong to Odonata. The Ornate Bluet (*Coenagrion ornatum*) (Fig. 16.3b) is a near threatened (NT) damselfly showing decreasing population trend in Europe (Boudot and Kalkman 2015). Fortunately, it is still moderately frequent in Hungary. Larvae of the species were collected in both the Drava and the floodplain (see Tóth 2010), suggesting that there is a viable population along the river. The occurrence of Green Hawker (*Aeshna viridis*) is almost completely confined to large fields of Water Soldier (*Stratiotes aliodes*) since the larvae develop among the leaves of this plant. The populations show a decreasing trend, and the species is regarded as near threatened in Europe (Boudot and Kalkman 2015). The survival of the population found in the floodplains of the Drava (Tóth 2010) largely depends on the presence of *S. aloides*, which is strongly related to the good ecological state of the oxbows. The

Fig. 16.3 Natura 2000 species occurring in the Drava floodplain. **a** *Graphoderus bilineatus*, **b** *Coenagrion ornatum*, **c** *Leucorrhinia pectoralis* (photos: **a** by Nataša Turić, **b**, **c** by Arnold Móra)

River Clubtail (*Gomphus flavipes*) (Fig. 16.2a, b) is a characteristic species for larger lowland rivers. It has strong populations in Hungarian rivers, and it occurs all along the Hungarian Drava section (Jakab and Dévai 2008). Although *G. flavipes* suffered a very severe decline in the 19th and 20th centuries, and was regarded as one of the most threatened species in Europe, at present the populations show an increasing trend and the species is not threatened at present. However, declining water quality and improper river management can negatively influence the populations. The Green

Snaketail (*Ophiogomphus cecilia*) (Fig. 16.2g, h) also suffered a severe decline in the past, but a recovery started in the 1990s, and now the species is in the least concern (LC) category (Boudot and Kalkman 2015). In contrast, the Hungarian populations of *O. cecilia* are regarded as vulnerable, due to loss of habitat (Haraszthy 2014). In the Drava, a strong population of this species can apparently be found, suggested by the fact that larvae were collected during all samplings carried out on the river (e.g. Kovács et al. 2004; Müller et al. 2006; Czirok et al. 2008; Tóth 2010; Horvai et al. 2012). The European populations of Lilypad Whiteface (*Leucorrhinia caudalis*) showed a remarkable decline in the 20th century and the species became extinct in many countries of Europe. A recovery started in the 2000s, and now the species shows stable population trends (Boudot and Kalkman 2015). In Hungary *L. caudalis* is a rare species which is threatened by the eutrophication of its habitats. Along the Drava it prefers mesotrophic small standing waters (Tóth 2010). Although the Yellow-spotted Whiteface (*Leucorrhinia pectoralis*) (Fig. 16.3c) is in the least concern (LC) category, its populations are declining in Europe (Boudot and Kalkman 2015) and probably in Hungary too. This species can be found in some oxbows of the Drava (see Tóth 2010), but the size of the population, varying remarkably from year to year, can hardly be assessed in lack of long-term monitoring.

Besides the Natura 2000 species, a relatively large number of other nationally protected species belonging to various taxonomical groups occur along the Dráva. Protected aquatic snails (*Amphimelania holandrii, Borysthenia naticina, Fagotia daudebartii, Theodoxus danubialis*) are not frequent but characteristic species in the running waters of Hungary. The Depressed River Mussel (*Pseudanodonta complanata*) lives in both running and standing waters. It is threatened by pollution and eutrophication of waters, and listed among the vulnerable species (IUCN 2017). The Narrow-clawed Crayfish (*Astacus leptodactylus*) used to be very common in Hungarian waters once, but, mainly because of water pollution and the appearance of invasive crayfish species, it shows a decreasing population trend. The Water Spider (*Argyroneta aquatica*) is not rare in Hungarian water bodies with a dense aquatic vegetation. The protected mayflies (*Ametropus fragilis, Oligoneuriella pallida, O. rhenana*) and stoneflies (*Isoptena serricornis*) occurring in the Drava are seldom recovered riverine species in Hungary. Among the protected dragonflies and damselflies rare (*Lestes dryas, Coenagrion scitulum, Epitheca bimaculata, Somatochlora flavomaculata, Sympetrum depressiusculum*) and rather common (*Aeshna isoceles, Libellula fulva, Orthetrum brunneum*) species are equally found. Two aquatic and semi-aquatic Heteroptera species are protected in Hungary. *Aquarius najas* is relatively distributed mainly in slow-flowing streams and rivers of Hungary, and was collected only from the Drava (Czirok et al. 2008; Horvai et al. 2012). The distribution of *Notonecta lutea* is sporadic in Hungary (Soós et al. 2009b). It has been recovered from two oxbow lakes along the Drava (Kálmán et al. 2011). The riffle beetle *Macronychus quadrituberculatus* used to be regarded as threatened species near extinction. Therefore, it became protected after its rediscovery in Hungary, but recently it is known from numerous sites along medium-sized and larger Hungarian rivers (Kovács and Merkl 2005).

16.4 Non-indigenous Species

One of the major threats in natural water bodies is the invasion of non-native species, which can affect all levels of aquatic ecosystems. However, the ecological impacts of these species are difficult to evaluate, since both negative and positive effects have been reported (Simberloff et al. 2013). Nevertheless, the invasive species are often more compatitive than native ones, forcing the latter to abandon their natural habitats. Furthermore, new pathogens are often introduced along with the invasive species, against which the native species are defenseless. The strong competition and the diseases can cause the extinction of populations of native species. These problems may exist in the case of the Drava, where the non-indigenous species are represented mainly by molluscs and crustaceans.

Some invasive aquatic mollusc species, e.g. New Zealand Mud Snail (*Potamopyrgus antipodarum*) from New Zealand and Fragile Ancylid (*Ferrissia fragilis*) from North America, have well established populations in the Drava, but no direct negative effects on native species or on the river ecosystem were observed. The North American Acute Bladder Snail (*Haitia acuta*) and the Asian Chinese Pond Mussel (*Anodonta woodiana*) (Fig. 16.4a) have become widely distributed in Europe, including Hungary. They can be found in a wide range of aquatic habitats (e.g. in both running and standing waters along the Drava), where they can replace the related native species, i.e. Common Bladder Snail (*Physa fontinalis*) and large unionid mussels. Two bivalves, the Golden Freshwater Clam (*Corbicula fluminea*) from Asia and the Zebra Mussel (*Dreissena polymorpha*) from the Ponto-Caspian region (Fig. 16.4b), often occur in high densities in the Drava, but their impacts can hardly be assessed. The large number of specimens and empty shells can provide substrate and shelter, increasing the density of other macroinvertebrates, but they can strongly modify the quality of the sediment (e.g. chemistry, grain size, organic matter content). They can decrease the quantity of planktonic organisms and increase light penetration by their filter feeding (Sousa et al. 2009). The Zebra Mussels often attach themselves to shells of larger mussels by byssal threads

Fig. 16.4 Non-indigenous bivalves of the Drava. **a** Chinese Pond Mussel (*Anodonta woodiana*), **b** Zebra Mussels (*Dreissena polymorpha*) attached to the shell of a Golden Freshwater Clam (*Corbicula fluminea*) (photos by Arnold Móra)

(Fig. 16.4b), reducing their ability to move, feed, and breed, and eventually leading to their deaths. This way it can be a threat for native species, especially the large-sized Unionidae (Fig. 16.1), among them the protected *Pseudanodonta complanata* and *Unio crassus*. The River Nerite (*Theodoxus fluviatilis*) is a widely distributed aquatic snail in northern, central and eastern Europe, expanding its area towards the Carpathian Basin, where this species is invasive. It became the most frequent nerite species in the larger rivers of Hungary (in the Drava as well), replacing the native species due to its higher tolerance to pollution (IUCN 2017).

Among the Malacostraca occuring along the Dráva there are more non-native (*Chelicorophium* spp., *Dikerogammarus* spp., *Synurella ambulans, Jaera sarsi, Limnomysis benedeni, Pacifastacus leniusculus*) (Fig. 16.5) than native species (9 out of 14), which indicates that the problems with aquatic invasions are getting more serious in the Dráva. All but one (*P. leniusculus*) species are of Ponto-Caspian origin, extending their range in the Carpathian basin and, in many cases, all over Europe. The very high densities of *Chelicorophium curvispinum* (Fig. 16.5b) and *C. sowinskyi* might have an impact on the ecosystem of the Drava by changing food webs. For example, along with other invasive macroinvertebrate species (e.g. *Dikerogammarus* spp., *Jaera sarsi, Limnomysis benedeni, Corbicula fluminea, Dreissena polymorpha, Theodoxus fluviatilis*), they became the main food items in

Fig. 16.5 Invasive crustacean species of the Drava. **a** *Jaera sarsi*, **b** *Chelicorophium curvispinum*, **c** demon shrimp (*Dikerogammarus haemobaphes*), **d** killer shrimp (*Dikerogammarus villosus*) (photos by Michał Grabowski)

both native and invasive fish diet (Kelleher et al. 1998; Borza et al. 2009). The *Dikerogammarus* species (Fig. 16.5c, d) are strong predators preying on a wide range of aquatic organisms, accordingly they can pose a major threat to native species, as it is suggested by their common names (e.g. killer shrimp for *D. villosus* and demon shrimp for *D. haemobaphes*). It is well documented that after their invasion native and other non-native species retreated from their original habitats to less favourable ones or completely disappeared due to the high predation and competition pressure (Rewicz et al. 2014; Bovy et al. 2015). However, in the case of high habitat heterogeneity, like in the Drava, the co-existence of *Dikerogammarus* species with native and other non-native species is possible by niche partitioning (Kley and Maier 2005). The North American Signal Crayfish (*Pacifastacus leniusculus*) is spreading downstream from the Mura river (Hudina et al. 2009) and appeared in the Hungarian section in the last few years (András Weiperth pers. comm.). This invader is a carrier of crayfish plague fungus (*Aphanomyces astaci*). The Signal Crayfish is resistant to this disease, which is responsible for widespread mortality in native European crayfish populations (Edgerton et al. 2004). Due to the vulnerability of native species to crayfish plague and the high competitive abilities of Signal Crayfish, the population of Narrow-Clawed Crayfish (*Astacus leptodactylus*) is especially threatened in the Drava. Another North American invasive species, the Spiny-Cheek Crayfish (*Orconectes limosus*), spreads upwards from the Danube throughout the Drava (Hudina et al. 2009), but it has not been found in the Hungarian section yet. Although the *A. leptodactylus* and *O. limosus* can co-exist, the latter may be a threat since it is also possibly a vector of crayfish plague (Kozubíková et al. 2009).

16.5 Conclusions

Due to the heterogeneous habitat complex hosting diverse aquatic macroinvertebrate assemblages and a high number (39) of protected and strictly protected species, the Drava and its floodplains are among the most valuable and most important Hungarian regions from the point of view of nature conservation. Both the river and its floodplain deserve protection, for which the Danube–Drava National Park provides good frames. However, in case of all aquatic macroinvertebrate groups further studies are needed to achieve a thorough and satisfactory knowledge on the fauna of the river and its surroundings, especially in case of floodplain water bodies, and to explore the exact distribution of the species.

At the same time, the valuable aquatic macroinvertebrate assemblages of the Drava Region are vulnerable, and might be threatened by any unconsidered human action, especially measures of river regulation. The change or loss of habitats might cause serious damage to populations of rare species, or, in extreme cases like that of *Platyphylax frauenfeldi*, might lead to extinction.

Acknowledgements Authors were supported by the EU-funded project EFOP-3.6.1.-16-2016-00004 and by the Higher Education Institutional Excellence Program of the Ministry of Human Capacities in the framework of the research project 20765-3/2018/FEKUTSTRAT at University of Pécs.

References

Ábrahám L (1995) A tervezett Duna-Dráva Nemzeti Park recésszárnyú-alkatú (Megaloptera, Raphidioptera, Neuroptera) faunájának természetvédelmi vizsgálata, I. (Nature protection studies on the neuropteroids [Megaloptera, Raphidioptera, Neuroptera] fauna of the projected Duna-Drava National Park I] Dunántúli Dolgozatok. Természettudományi sorozat 8:53–70 (in Hungarian)

Ábrahám L (1998) Natural protection studies on the neuropteroids (Megaloptera, Raphidioptera, Neuroptera) fauna of the projected Duna-Drava National Park, II. Dunántúli Dolgozatok, Természettudományi sorozat 9:269–289

Ábrahám L (ed) (2005) Biomonitoring along the river Drava in Hungary, 2000–2004. Natura Somogyiensis 7:207 pp

Bauerfeind E, Kovács T, Ambrus A (2005) Collection of adult mayflies (Ephemeroptera) of the Mátra Museum, Hungary. Folia Historico-naturalia Musei Matraensis 29:91–94

Borza P (2011) Revision of invasion history, distributional patterns, and new records of Corophiidae (Crustacea: Amphipoda) in Hungary. Acta Zoologica Academiae Scientiarum Hungaricae 57:75–84

Borza P, Czirok A, Deák Cs, Ficsór M, Horvai V, Horváth Zs, Juhász P, Kovács K, Szabó T, Vad CsF (2011) Invasive mysids (Crustacea: Malacostraca: Mysida) in Hungary: distributions and dispersal mechanisms. North-Western J Zool 7:222–228

Borza P, Erős T, Oertel N (2009) Food resource partitioning between two invasive gobiid species (Pisces, Gobiidae) in the littoral zone of the River Danube, Hungary. Int Rev Hydrobiol 94:609–621

Boudot J-P, Kalkman VJ (eds) (2015) Atlas of the European dragonflies and damselflies. KNNV Publishing, The Netherlands. 381 pp

Bovy HC, Barrios-O'Neill D, Emmerson MC, Aldridge DC, Dick JTA (2015) Predicting the predatory impacts of the "demon shrimp" *Dikerogammarus haemobaphes*, on native and previously introduced species. Biol Invasions 17:597–607

Council of the European Union (2013) Council Directive 2013/17/EU of 13 May 2013 adapting certain directives in the field of environment, by reason of the accession of the Republic of Croatia. Official J European Union L158(2013):193–229

Csabai Z, Horvai V, Kálmán Z, Czirok A (2009) Contribution to the aquatic beetle fauna of the southern part of Transdanubian Region, Hungary (Coleoptera: Hydradephaga,Hydrophiloidea). Acta Biologica Debrecina Oecologica Hungarica 20:41–55

Csabai Z, Boda R, Boda P, Deák Cs, Mauchart P, Móra A, Pernecker B (2015a) A bentikus fauna felmérése és ökológiai értékelése, kiemelve a jelölő fajokat (Survey and ecological evaluation of the benthic fauna of the Old-Drava, with special emphasis on Natura 2000 indicator species). In: Purger J (ed) Az Ó-Dráva ökológiai alapállapot felmérése (Ecological baseline survey of the Old Drava). Scientific report, BioRes Kft., Pécs. 33–55 (in Hungarian)

Csabai Z, Kálmán Z, Mauchart P, Pernecker B, Soós N (2015b) New localities of *Graphoderus bilineatus* along Drava River (SW Hungary). Acta Biologica Debrecina Oecologica Hungarica 33:83–87

Czirok A, Horvai V, Sárfi N (2008) Adatok a magyar Dráva szakasz litorális zónájának makroszkopikus gerinctelen faunájáról (Faunistic data from the littoral zone of the Hungarian reach of River Drava). Acta Biologica Debrecina Oecologica Hungarica 18:27–36 (in Hungarian)

Edgerton BF, Henttonen P, Jussila J, Mannonen A, Paasonen P, Taugbøl T, Edsman L, Souty-Grosset C (2004) Understanding the causes of disease in European crayfish. Conserv Biol 18:1466–1474

Forró L, Meisch C (1998) A Duna-Dráva Nemzeti Park Dráva menti vizeinek rákfaunája (Crustacea) (Aquatic crustacean fauna of the Dráva region in the Duna-Dráva National Park, Hungary). Dunántúli Dolgozatok, Természettudományi sorozat 9:111–122 (in Hungarian)

Gidó Zs, Szél Gy (1998) Adatok a Duna–Dráva Nemzeti Park Dráva menti részének vízibogár (Coleoptera: Hydradephaga, Palpicornia, Dryopidae, Elmidae) faunájáról (Data to the aquatic beetle fauna [Coleoptera: Hydradephaga, Palpicornia, Dryopidae, Elmidae] of Dráva Region in Duna–Dráva National Park, South Hungary]. Dunántúli Dolgozatok, Természettudományi sorozat 9:189–202 (in Hungarian)

Haraszthy L (ed) (2014) Natura 200 fajok és élőhelyek Magyarországon (Natura 2000 species and habitats in Hungary). Pro Vértes Közalapítvány, Csákvár. 955 pp (in Hungarian)

Héra Z (2005) On experiences in monitoring molluscs (Mollusca) in the area of Duna-Dráva National Park. Natura Somogyiensis 7:25–34

Horvai V, Czirok A, Lőkkös A, Borza P, Bódis E, Deák Cs (2012) Újabb adatok a magyar-horvát Dráva szakasz ripális régiójának makroszkopikus gerinctelen faunájáról (New faunistic data from the riparian zone of the Hungarian-Croatian reach of Drava river). Acta Biologica Debrecina Oecologica Hungarica 28:109–120 (in Hungarian)

Hudina S, Faller M, Lucić A, Klobučar G, Maguire I (2009) Distribution and dispersal of two invasive crayfish species in the Drava River basin, Croatia. Knowl Manag Aquatic Ecosyst 394–395:09

IUCN (2017) The IUCN Red List of threatened species. Version 2017-1. http://www.iucnredlist.org (downloaded on 10 August 2017)

Jakab T, Dévai Gy (2008) A folyami szitakötők (Odonata: Gomphidae) előfordulása Magyarországon a lárva-és exuviumadatok alapján (The occurrence of the riverine dragonfly-species [Odonata: Gomphidae] in Hungary according to the data of larvae and exuviae). Acta Biologica Debrecina Oecologica Hungarica 18:53–65 (in Hungarian)

Juhász P, Kiss B, Müller Z (2006a) Faunistical results of the Hirudinea investigations carried out in the frames of the ecological survey of the surface waters of Hungary (ECOSURV) in 2005. Folia Historico-naturalia Musei Matraensis 30:315–318

Juhász P, Kovács K, Szabó T, Csipkés R, Kiss B, Müller Z (2006b) Faunistical results of the Malacostraca investigations carried out in the frames of the ecological survey of the surface waters of Hungary (ECOSURV) in 2005. Folia Historico-naturalia Musei Matraensis 30:319–323

Juhász P, Varga A, Kiss B, Müller Z (2006c) Faunistical results of the Mollusca investigations carried out in the frames of the ecological survey of the surface waters of Hungary (ECOSURV) in 2005. Folia Historico-naturalia Musei Matraensis 30:305–314

Kiss B, Juhász P, Müller Z (2006) Faunistical results of the Heteroptera (Gerromorpha et Nepomorpha) investigations carried out in the frames of the ecological survey of the surface waters of Hungary (ECOSURV) in 2005. Folia Historico-naturalia Musei Matraensis 30:343–348

Kálmán Z, Boda R, Kálmán A, Ortmann-Ajkai A, Soós N, Csabai Z (2011) Contribution to the aquatic Coleoptera (Hydradephaga, Hydrophiloidea, Dryopidae) and Heteroptera (Gerromorpha, Nepomorpha) fauna of Dráva Plain, SW Hungary. Acta Biologica Debrecina Oecologica Hungarica 26:117–134

Kelleher B, Bergers PJM, van den Bink FWB, Giller PS, van der Welde G, de Vaate AB (1998) Effects of exotic amphipod invasions on fish diet in the Lower Rhine. Archiv für Hydrobiologie 143:363–382

Kley A, Maier G (2005) An example of niche partitioning between *Dikerogammarus villosus* and other invasive and native gammarids: a field study. J Limnol 64:85–88

Ködöböcz V, Juhász P, Kiss B, Müller Z (2006) Faunistical results of the Coleoptera investigations carried out in the frames of the ecological survey of the surface waters of Hungary (ECOSURV) in 2005. Folia Historico-naturalia Musei Matraensis 30:349–355

Kondorosy E, Földessy M (1998) Adatok a Duna-Dráva Nemzeti Park Dráva menti területei poloska (Heteroptera) faunájához (Data to the Heteroptera fauna of areas along river Dráva in the Danube-Dráva National Park). Dunántúli Dolgozatok, Természettudományi sorozat 9:159–176 (in Hungarian)

Kovács T (2005) Data to the distribution of four species of *Baetis* in Hungary, based on larvae (Ephemeroptera: Baetidae). Folia Historico-naturalia Musei Matraensis 29:95–100

Kovács T (2006a) Faunistical results of the Ephemeroptera investigations carried out in the frames of the ecological survey of the surface waters of Hungary (ECOSURV) in 2005. Folia Historico-naturalia Musei Matraensis 30:325–331

Kovács T (2006b) Data to the Hungarian distribution of Plecoptera. Folia Historico-naturalia Musei Matraensis 30:181–188

Kovács T (2006c) Faunistical results of the Plecoptera investigations carried out in the frames of the ecological survey of the surface waters of Hungary (ECOSURV) in 2005. Folia Historico-naturalia Musei Matraensis 30:339–341

Kovács T (2009a) Data to the Hungarian mayfly (Ephemeroptera) fauna arising from collectings of larvae V. Folia Historico-naturalia Musei Matraensis 33:73–85

Kovács T (2009b) Data to the Hungarian distribution of Plecoptera II. Folia Historico-naturalia Musei Matraensis 33:103–108

Kovács T (2011) Psammophilous Ephemeroptera and Plecoptera in Hungary. Folia Historico-naturalia Musei Matraensis 35:33–38

Kovács T, Merkl O (2005) Data to the Hungarian distribution of some aquatic beetles, with notes on an extralimital species (Coleoptera: Gyrinidae, Haliplidae, Elmidae, Dryopidae). Folia Entomologica Hungarica 66:81–94

Kovács T, Ambrus A, Bánkuti K (1998–99) Data to the Hungarian mayfly (Ephemeroptera) fauna arising from collectings of larvae. Folia Historico-naturalia Musei Matraensis 23:157–170

Kovács T, Ambrus A, Bánkuti K (1999a) Data to the distribution of *Oligoneuriella* larvae in Hungary (Ephemeroptera: Oligoneuriidae). Folia Entomologica Hungarica 60:349–354

Kovács T, Ambrus A, Merkl O (1999b) Potamophilus acuminatus (Fabricius, 1792) and Macronychus quadrituberculatus P.W.J. Müller, 1806: new records from Hungary (Coleoptera: Elmidae). Folia Entomologica Hungarica 60:187–194

Kovács T, Ambrus A, Bánkuti K, Juhász P (1998) New Hungarian mayfly (Ephemeroptera) species arising from collectings of larvae. Miscellanea Zoologica Hungarica 12:55–60

Kovács T, Ambrus A, Juhász P (2003) Data to the Hungarian mayfly (Ephemeroptera) fauna arising from collectings of larvae II. Folia Historico-naturalia Musei Matraensis 25:59–72

Kovács T, Ambrus A, Juhász P, Bánkuti K (2004) Lárva és exuvium adatok Magyarország Odonata faunájához (Larval and exuvial data to the Odonata fauna of Hungary). Folia Historico-naturalia Musei Matraensis 28:97–110 (in Hungarian)

Kovács T, Weinzierl A, Ambrus A (2002) New and rare stoneflies (Plecoptera) from Hungary. Folia Entomologica Hungarica 63:43–48

Kozubíková E, Filipová L, Kozák P, Ďuriš Z, Martín MP, Diéguez-Uribeondo J, Oidtmann B, Petrusek A (2009) Prevalence of the crayfish plague pathogen *Aphanomyces astaci* in invasive American crayfishes in the Czech Republic. Conserv Biol 23:1204–1213

Majer J, Krčmar S (2006) Geographical distribution of Tabanids (Diptera: Tabanidae) of the Drava river along the Somogy County (Hungary), the Koprivničko-križevačka and one part Virovitičko-podravska County (Croatia). Somogyi Múzeumok Közleményei 17:183–198

Malicky H, Waringer G, Uherkovich Á (2002) Ein Beitrag zur Bionomie und Ökologie von *Platyphylax frauenfeldi* Brauer, 1857 (Trichoptera, Limnephilidae) mit Beschreibung der Larve (A contribution to bionomics and ecology of *Platyphylax frauenfeldi* Brauer, 1857 [Trichoptera, Limnephilidae], with description of the larva). Entomologische Nachrichten und Berichte 46:73–80 (in German)

Móra A, Juhász P, Kiss B, Müller Z (2006) Faunistical results of the Trichoptera investigations carried out in the frames of the ecological survey of the surface waters of Hungary (ECOSURV) in 2005. Folia Historico-naturalia Musei Matraensis 30:359–367

Müller Z, Juhász P, Kiss B (2006) Faunistical results of the Odonata investigations carried out in the frames of the ecological survey of the surface waters of Hungary (ECOSURV) in 2005. Folia Historico-naturalia Musei Matraensis 30:333–338

Nesemann H (1998) Beiträge zur Kenntnis der Egelfauna (Hirudinea) des Draugebietes in Südungarn (The leech (Hirudinea) fauna of the Dráva Region, South Hungary). Dunántúli Dolgozatok, Természettudományi sorozat 9:69–72 (in German)

Nógrádi S (2001) Further data to the caddisflies (Trichoptera) of Hungary. Folia Historico-naturalia Musei Matraensis 25:83–90

Nógrádi S, Uherkovich Á (1995) A Dráva magyarországi szakaszának tegzes (Trichoptera) faunája (The caddisfly (Trichoptera) fauna of the Hungarian reach of Dráva river). Dunántúli Dolgozatok, Természettudományi sorozat 8:117–137 (in Hungarian)

Nógrádi S, Uherkovich Á (1998) Újabb eredmények a Duna-Dráva Nemzeti Park Dráva menti területei tegzes (Trichoptera) faunájának kutatásában (Further results of the studies on the caddisflies [Trichoptera] of Dráva river and environments in the Duna-Dráva National Park, Southwest Hungary). Dunántúli Dolgozatok, Természettudományi sorozat 9:331–358 (in Hungarian)

Purger J (ed) (2008) Biodiversity studies along the Drava river. University of Pécs, Pécs, p 328

Purger J (ed) (2013) A Dráva négy magyarországi mellékágának élővilága és rehabilitációja/Biotas and rehabilitation of four Drava river side-branches in Hungary. BioRes Bt, Pécs, p 164

Rewicz T, Grabowski M, MacNeil C, Bącela-Spychalska K (2014) The profile of a 'perfect' invader – the case of killer shrimp, *Dikerogammarus villosus*. Aquat Invasions 9:267–288

Simberloff D, Martin JL, Genovesi P, Maris V, Wardle DA, Aronson J, Courchamp F, Galil B, García-Berthou E, Pascal M, Pyšek P, Sousa R, Tabachii E, Vilà M (2013) Impacts of biological invasions: what's what and the way forward. Trends Ecol Evol 28:58–66

Soós N, Boda P, Csabai Z (2009a) First confirmed occurrences of *Notonecta maculata* and *N. meridionalis* (Heteroptera: Notonectidae) in Hungary with notes, maps, and a key to the *Notonecta* species of Hungary. Folia Entomologica Hungarica 70:67–78

Soós N, Horvai V, Czirok A, Csabai Z (2009b) Contribution to the aquatic and semiaquatic Heteroptera (Nepomorpha, Gerromorpha) fauna of the Transdanubian Region, Hungary. Acta Biologica Debrecina Oecologica Hungarica 20:193–208

Sousa R, Gutiérrez JL, Aldridge DC (2009) Non-indigenous invasive bivalves as ecosystem engineers. Biol Invasions 11:2367–2385

Sziráki Gy (1995) Adatok a Dráva-mellék kérész (Ephemeroptera) faunájának ismeretéhez (Contribution to the knowledge of mayflies [Ephemeroptera] of the Dráva Region, Hungary). Dunántúli Dolgozatok, Természettudományi sorozat 8:37–39 (in Hungarian)

Sziráki Gy (1998) A Dráva mente kérész (Ephemeroptera) faunája (The mayfly [Ephemeroptera] fauna of the Dráva region, Hungary). Dunántúli Dolgozatok, Természettudományi sorozat 9:131–134 (in Hungarian)

Szitta E, Juhász P, Kiss B, Müller Z (2009) Contribution to the Hungarian caddisfly (Trichoptera) fauna, based on the nationwide surveys in 2007. Folia Historico-naturalia Musei Matraensis 33:205–210

Tóth S (1995a) A Dráva mente szitakötő (Odonata) faunájának előzetes vizsgálata (Preliminary report on the dragonfly [Odonata] fauna of the Dráva Region, Hungary). Dunántúli Dolgozatok, Természettudományi sorozat 8:41–52 (in Hungarian)

Tóth S (1995b) Adatok a Dráva mente kétszárnyú (Diptera) faunájához (Data to the knowledge of the dipterous fauna [Diptera] of the Dráva region, South Hungary). Dunántúli Dolgozatok, Természettudományi sorozat 8:161–172 (in Hungarian)

Tóth S (1998) A Duna-Dráva Nemzeti Park Dráva menti területének szitakötő (Odonata) faunája (Data to the dragonfly fauna of the Duna-Dráva National Park, South Hungary [Odonata]). Dunántúli Dolgozatok, Természettudományi sorozat 9:135–150 (in Hungarian)

Tóth S (2005) Monitoring dragonflies on the section of the Dráva between Őrtilos and Vízvár (Insceta: Odonata). Natura Somogyiensis 7:35–48

Tóth S (2010) A Dunántúli-dombság és környéke szitakötő faunája (Dragonfly fauna of the Transdanubian Hills and surroundings). Natura Somogyiensis 16:1–188 (in Hungarian)

Uherkovich Á (ed) (1995) A Dráva mente állatvilága, I (The fauna of the Drava region I). Dunántúli Dolgozatok, Természettudományi sorozat 8:210 pp (in Hungarian)
Uherkovich Á (ed) (1998) A Dráva mente állatvilága, II (The fauna of the Drava region, II). Dunántúli Dolgozatok, Természettudományi sorozat 9:509 pp (in Hungarian)
Uherkovich Á (2004) Further studies on the caddisflies (Trichoptera) of the water system of rivers Mura and Kerka, Southwest Hungary. Folia Historico-naturalia Musei Matraensis 28:187–194
Uherkovich Á (2005) Further faunistic results of the caddisfly (Trichoptera) examinations of the Dráva region, South Hungary. Folia Historico-naturalia Musei Matraensis 29:165–168
Uherkovich Á, Nógrádi S (1992) Some data to the Trichoptera fauna of Drava river, Hungary. Somogyi Múzeumok Közleményei 9:269–278
Uherkovich Á, Nógrádi S (1997) *Platyphylax frauenfeldi* Brauer, 1857 (Trichoptera, Limnephilidae) in Hungary. Braueria 24:13–14
Uherkovich Á, Nógrádi S (2005) Middle-term changes in caddisfly (Trichoptera) communities of the Hungarian part of Dráva river during the years 1992–2004. Natura Somogyiensis 7:49–62
Varga A (1995) A Dráva menti puhatestű (Mollusca) fauna kutatásának eddigi eredményei (Results of the malacological studies [Mollusca] of Dráva Region [Hungary] up to day). Dunántúli Dolgozatok, Természettudományi sorozat 8:9–19 (in Hungarian)
Varga A, Uherkovich Á (1998) A Dráva menti puhatestű (Mollusca) fauna kutatásának újabb eredményei (Recent studies on the malacofauna [Mollusca] of Dráva Region, South Hungary). Dunántúli Dolgozatok, Természettudományi sorozat 9:43–68

Chapter 17
Fishes of the Drava River

Péter Sály

Abstract The chapter presents the fish fauna of the Croatian-Hungarian section of the river by overview the fish faunistic literature of studies conducted between 1992 and 2016, provides an example for littoral fish assemblages, and evaluates the ecological status of a river reach on this basis. It seems that 66 fishes, most of them belong to the family Cyprinidae and one cyclostomata species occur in the studied river section. However, the number of species regularly inhabiting the Croatian-Hungarian section could be about 51, because some species require different habitat type than the main channel of the studied section, or they are not able to reach the studied section due to migration barriers. There are also some taxa with unclear taxonomic status. Twenty-two species are listed in one of the annexes of the European Union Habitats Directive. The ratio of native to non-native species is 52:15. Monkey goby (*Neogobius fluviatilis*) and western tubenose goby (*Proterorhinus semilunaris*), two non-native Ponto-Caspian gobies, appear to be among the most abundant fishes in the littoral zone. Recently, other goby species (*Ponticola kessleri*, and *N. melanostomus*) formerly not known from the Drava have been found at the lower end of the studied river section and results anticipate their potential future spreading upstream. Fish assemblages tend to mirror an overall good ecological status and the rich fish fauna is of considerable nature conservation value due to the minimum alteration of habitats and the relative geographical proximity of the Danube. Therefore, for an effective conservation of the fish fauna the actual seminatural status of the Drava riverscape is to be maintained.

Keywords Biodiversity · Fish faunistics · Non-native species · Ecological connectivity · Ecological status

---

P. Sály (✉)
Department of Hydrobiology, Institute of Biology, Faculty of Sciences, University of Pécs, Ifjúság útja 6, Pécs 7624, Hungary
e-mail: psaly@gamma.ttk.pte.hu

D. Lóczy (ed.), *The Drava River*, Springer Geography,
https://doi.org/10.1007/978-3-319-92816-6_17

17.1 Introduction

Fish play a great socio-economic role in our life. In addition to recreational and fishery interests, the structure of fish assemblages can be used as an indicator of the environmental status of rivers. To monitor fish in a river system, researchers usually make faunistic stock surveys in the field. However, it is very challenging to compile a 'scientifically valid and relevant' species list since the spatial distribution of species is dynamic, some species can appear in and disappear from a focal stream system due to natural spreading and/or anthropogenic activity. For example, the spectacular upstream spreading in the Danube catchment and colonization of the Rhine basin by Ponto-Caspian gobies experienced since the last decades of the 20th century is probably a result of an interaction of a quasi-natural range expansion and human-aided distributions (Roche et al. 2013; Manné et al. 2013). In contrast, the disappearance of native anadromous sturgeon species from the middle Danube catchment could be attributed primarily to the construction of the Iron Gate Barrages on the Danube.

Alongside the dynamic nature of the geographical range of species, evolution of scientific methodology could be another relevant issue in faunistics. The molecular techniques used more and more frequently in taxonomy tend to reveal diverse, hidden phylogenetic relationships within species that can be considered as a single taxonomic unit on the basis of morphology (e.g. Mendel et al. 2008; Takács 2012; Takács et al. 2014; Antal et al. 2016). Changes in taxonomy and nomenclature induced by molecular phylogeny generally originate from how biologists interpret the species concept (Agapow et al. 2004). Fish taxonomists operated mainly with morphological characters before the flourishing era of molecular biology, but nowadays molecular biotechnology has deeply penetrated into systematics and taxonomy too and make biologists reconsider what a species is and what relationships exist among taxa.

Nature conservation legislation generally lags behind the frontiers of taxonomy and use names no longer valid in the light of modern taxonomy. Meanwhile, the fishes living in the focal stream system themselves are the same, but the enumeration and nature conservation evaluation of the fauna members could rarely be complete or correct because of all these above mentioned facts.

Comprehensive information on the fauna of a stream system is not sufficient to assess the quantitative characteristics of the fish stocks in detail. Naturally, if a species is frequently caught in faunistic surveys, it suggests us that the species is probably common in terms of abundance. However, different faunistic studies usually use different or even combined methods to catch fish (e.g. electrofishing by wading, from the riverbank or a boat; seine netting) in order to being able to sample most of the habitats in the study area. Moreover, authors can also incorporate additional information, like anglers' reports to make their species list 'complete'. Despite multifaceted sampling, data on the number of the individuals caught are hardly numerically comparable, sometimes even within a single faunistic study, not to mention cross-study comparisons. A simply reason for the lack of comparability is that the probability of detection of a fish species also depends, among other factors, on sampling method. Therefore, corresponding to the type of the water body to be surveyed and the region (i.e. country), various standardized sampling protocols allow the quantitative investigation of fish assemblages.

The application of standardized fish sampling protocols combined with data on the abiotic environment of the habitats help elaborate environmental bioassessment methods (Karr 1981, 1991). Actually, fish are among the five groups of organisms which are suggested as biological quality elements for the classification of ecological status of rivers by the Water Framework Directive of the European Union (European Commission 2000).

As a major right-bank tributary of the Danube River, the Drava River has a rich fish fauna. Along the upper course of the river in Austria, Slovenia and Croatia, a series of hydropower plants alter natural hydromorphological connectivity (see Chap. 9 in this volume). Above the dams, reservoirs are highly unfavorable for rheophilic riverine fish species, often utilized by fishery, so game fish species are usually introduced. Dams prevent potamodromous and diadromous fishes from migrating between their upstream spawning and downstream feeding, overwintering habitats (Baxter 1977). Although fish passes are usually built in barrages to mitigate the unwanted effects of river impoundment, hydropower plants modify riverine fish fauna. The Croatian-Hungarian section of the Drava River is much less modified. In this chapter, the fish fauna of the Croatian-Hungarian section is presented.

17.2 Faunistic Literature

Harka (1992) surveyed the fish fauna between two Hungarian settlements, Őrtilos and Gordisa, at five locations, in July 1990. On the basis of his own samples and inspection of anglers' catches, he reports the presence of 48 species.

Majer (1995) collected data on the fish fauna between Őrtilos and Drávaszabolcs (Hungary) at seven locations. Unfortunately, he does not mention the date of the field surveys, and he does not distinguish the species recorded by him in the field from the species reported in the literature. Majer (1998) reports the occurrence of 47 fish species after surveying the main channel, side arms and oxbow lakes at 14 locations, inspecting anglers' catches between 1995 and 1997.

Croatia planned to build a new hydropower plant near Novi Virje, but it has not been realized to date. As a baseline environmental assessment conducted before the potential building started, Majer and Bordács (2001) had sampled the river upstream of the barrage site, between Őrtilos and Zákány villages (Hungary) and downstream, between Bélavár and Vízvár (Hungary) at altogether ten sites. They report 40 species but only 39 are listed in Table 3. Moreover, because the pumpkinseed (*Lepomis gibbosus*) is mentioned twice in the table, the actual number of the species they found must be 38.

Until now, the most detailed fish faunistic survey of the Croatian-Hungarian Drava section has been completed by Sallai (2002a, b). He sampled a wide variety of habitats, for example, gravel bars and rip-rap banks in the main channel, side-arms and oxbow lakes. Unfortunately, it is not easy to identify the exact number of the sampling locations from the report. Nonetheless, it seems that samples were taken at about 19 locations between Őrtilos and Matty, from 11 April 2001 to 19 October 2001. Synthesizing his own field data and anglers' reports verified by photo documentation, Sallai (2002b) proves the presence of 57 species in the studied section, although there are only 52 denoted in Table 5 as occurring in the Drava. Among the species with proved occurrence, eight had been unknown from the studied Drava section. Considering literature and his own data, Sallai (2002b) suggests a total of 64 fish species of occasional or regular occurrence.

The possible construction of the new, aforementioned hydropower plan near Novi Virje, Croatia, motivated Sallai (2004) to review the literature and his own data on fish fauna of the Drava River. Apart from the Eurasian minnow (*Phoxinus phoxinus*), the species listed in this study are actually the same as those enumerated by Sallai (2002b).

Soon after, Sallai and Kontos (2005) report the faunistic results from 11 monitoring sites on the Hungarian bank between 1999 and 2004. In fact, this study includes the report of Sallai (2002a, b), supplemented with new information. Contrary to Sallai (2002b), in this study 63 fish species are presumed to occur occasionally or regularly in the study region. The occurrence of the Eurasian minnow was not confirmed.

The fish monitoring of the Croatian Drava sections started in 2007 and the data were presented by Sallai and Kontos (2008). The monitoring sites were located between Őrtilos and Barcs, but their exact number is difficult to reconstruct, because the resolution of the site map is poor and no coordinates are indicated. Authors report field data obtained not only from the 2007 monitoring but from 2004 as well. All in all, they report the direct observation of 40 fish species, including the Eurasian minnow form the Croatian section.

Jelić et al. (2012) conducted a faunistic survey near Donji Miholjac (Croatia) at 18 sampling sites in 2006. They sampled the main types of habitats, such as the main channel, side-arms, backwaters and artificial channels, and also obtained information from local anglers. Authors give a species list in Table 1 with the species from their field research and literature since 1985. In this list, 46 species are denoted as a species recorded in their study, but, strangely, they report a total of only 44 species in the English summary of the article. The occurrence of six species was based on information from anglers. Compared to the previously mentioned faunistic studies, Jelić et al. (2012) mention the presence of the Kessler goby (*Ponticola kessleri*) for the first time.

In their short communication, Csipkés et al. (2012) report the first known occurrence of the western three-spined stickleback, *Gasterosteus gymnurus* Cuvier, 1829, in the Drava River. They caught one specimen near Matty on 7 September 2010. The taxonomic identification was based on the incomplete armoration of the fish. Formerly, the less armored form of the three-spined stickleback was considered as a subspecies, *G. aculeatus gymnurus*, however, this taxon was raised to species level rank (FishBase).

As a most recent faunistic result, Sallai (2016) reports the presence of the Kessler goby and round goby (*Neogobius melanostomus*) from Matty. The gobies were caught on 22 September 2015. To our knowledge, this is the first detection of round goby in the Drava.

17.3 Methods

We compiled a list on fish species known to occur in the Croatian-Hungarian section of the Drava River on the basis of the relevant faunistic literature published between 1992 and 2016 (Harka 1992; Majer 1998; Majer and Bordács 2001; Sallai 2002a, b; Sallai 2004; Sallai and Kontos 2005, 2008; Jelić et al. 2012; Csipkés et al. 2012; Sallai 2016). The literature written before 1992 on the fish fauna of the Drava is overviewed by Sallai and Kontos (2005). Fishes were classified according to Nelson et al. (2016), and the nomenclature of the species used in FishBase (www.fishbase.org), an ichthyological database, was followed. Only observations made directly by the authors of the faunistic studies or made by a third person and reported to the authors with photodocumentation, or data obtained from fishery database by the authors were considered as valid.

The biogeographical status of the species in the Drava River was characterized as follows. A species was considered native if the Drava belongs to its natural range; endemic if the Drava belongs to the natural range of the species, and the species is a biogeographic endemism of the Danube catchment, or in other words, the natural range of the species is restricted to the Danube River Basin; and non-native if the Drava does not belong to the original native range, so the presence of the species in the Drava River is highly supposed to be the result of some kind of human intervention. To highlight conservation importance, we indicated if a species is included in any of the annexes of the European Union Habitats Directive (European Commission 2000). Annex II lists species of community interest whose core habitats are included in the Natura 2000 network (Natura 2000 indicator species). Annex IV lists species of community interest in need of strict protection across their entire natural range within the EU. Annex V lists species of community interest whose taking in the wild and exploitation must be compatible with maintaining their favorable conservation status.

To illustrate the relative abundances and densities of the fishes, we present an example for a rank abundance distribution of fish assemblage of the Drava littoral zone using formerly unpublished original data from Sály et al. Data were collected by boat electrofishing of three 500 m reaches (subsamples) along the Hungarian bank of the river, near Barcs (between the endpoints of 45.950053°N, 17.432136°E and 45.942783°N, 17.484209°E), daytime, on 19 October 2016, in accordance with the suggested sampling protocol for the Hungarian lowland rivers (Sály and Erős 2016).

The width of the effective zone of the electrofishing gear (Hans Grassl EL 64II, SDC, 300 V, 10 A) was about 2 m, hence the sampled area can be roughly estimated at $1500 \times 2 = 3000$ m^2. The effectiveness of electrofishing, however, is generally influenced by several factors (e.g. conductivity, water level of the river, depth, current velocity, turbidity). Rank abundance distribution was constructed from the full sample, i.e. the pooled data of the three 500-m-long subsamples.

Finally, the ecological status of the sampled Drava section was assessed. We applied the Hungarian Multimetric Fish Index (Sály and Erős 2016), a recently developed biotic index family for evaluating the ecological status of surface running waters on the basis of fish assemblages in Hungary in accordance with the EU Water Framework Directive, to the above mentioned data of Sály et al. Assessment was made for the three subsamples separately, and, to obtain a more solid picture, also for the full sample data. The Hungarian Multimetric Fish Index provides two kinds of relevant information. The first one is an Ecological Quality Ratio (EQR), ranging from zero to one, a standardized quantitative indicator of ecological status; the closer to one, the better the ecological status is. The second information is the so-called Ecological Quality Class, which can be bad, poor, moderate, good or high, determined according to a conversion rule (not presented here) of the EQR.

17.4 Results and Discussion

17.4.1 Faunistics

The overview of the above-cited faunistic studies shows that representatives of 11 orders and 17 families have been known from the Croatian-Hungarian Drava section. The number of species occurring regularly or occasionally in the region seems to be about 67, including one cyclostomata (*Eudontomyzon mariae*) too. The fauna is dominated by the members of the Cyprinidae family (32 species, ca. 48%), which is a common characteristic for European lowland rivers. Altogether, there are 22 species of community interest. Eighteen species are listed in Annex II, one in Annex IV, and ten in Annex V. Seven species out of the 22 are listed in two annexes. A slightly more than three-fourths of the 67 species are native, including four Danubian endemisms. There are 15 non-native species (Table 17.1).

Here some remarks are due on changes in nomenclature and taxonomy to help understand the nature conservation interest of some species.

Leuciscus aspius is listed as its synonym, *Aspius aspius*, in the Annexes of the EU Habitats Directive.

Rhodeus amarus is listed as a subspecies, *R. sericeus amarus*, in Annex II, however, this taxon has been raised to species level rank (FishBase).

Romanogobio vladykovi was listed as *Gobio albipinnatus* in earlier studies (see e.g. Harka 1992; Sallai 2002b; Sallai and Kontos 2005), and in Annex II. However, according to the FishBase, the presence of *Romanogobio albipinnatus* (a synonym

Table 17.1 Fish species with verified occurrence in the Croatian-Hungarian Drava section, 1990–2016. Biogeographical status refers to the nativeness of the species. Habitats Directive denotes if the species is listed in any annexes of the European Habitats Directive. See the Methods section for details

No. of species	Taxon	Common name	Biogeographical status	Habitats directive
	ordo Petromyzontiformes	**lampreys**		
	familia Petromyzontidae	**northern lampreys**		
1	[a]*Eudontomyzon mariae* (Berg, 1931)	Ukrainian brook lamprey	Native	Annex II
	ordo Acipenseriformes	**paddlefishes and sturgeons**		
	familia Acipenseridae	**sturgeons**		
2	*Acipenser nudiventris* Lovetsky, 1828	fringebarbel sturgeon	Native	Annex V
3	*Acipenser ruthenus* Linnaeus, 1758	sterlet sturgeon	Native	Annex V
	ordo Anguilliformes	**eels**		
	familia Anguillidae	**freshwater eels**		
4	*Anguilla anguilla* (Linnaeus, 1758)	European eel	Native	
	ordo Cypriniformes	**carps, loaches, minnows**		
	familia Cyprinidae	**carps, loaches, minnows**		
5	*Abramis brama* (Linnaeus, 1758)	freshwater bream	Native	
6	*Alburnoides bipunctatus* (Bloch, 1782)	schneider	Native	
7	*Alburnus alburnus* (Linnaeus, 1758)	bleak	Native	
8	*Ballerus ballerus* (Linnaeus, 1758)	zope	Native	
9	*Ballerus sapa* (Pallas, 1814)	white-eyed bream	Native	
10	*Barbus barbus* (Linnaeus, 1758)	barbel	Native	Annex V
11	*Blicca bjoerkna* (Linnaeus, 1758)	white bream	Native	
12	*Carassius carassius* (Linnaeus, 1758)	Crucian carp	Native	
13	*Carassius gibelio* (Bloch, 1782)	Prussian carp	Non-native	
14	*Chondrostoma nasus* (Linnaeus, 1758)	common nase	Native	
15	*Ctenopharyngodon idella* (Valenciennes, 1844)	grass carp	Non-native	

(continued)

Table 17.1 (continued)

No. of species	Taxon	Common name	Biogeographical status	Habitats directive
16	*Cyprinus carpio* Linnaeus, 1758	common carp	Native	
17	[b]*Gobio obtusirostris* Valenciennes, 1842	gudgeon	Native	
18	*Hypophthalmichthys molitrix* (Valenciennes, 1844)	silver carp	Non-native	
19	*Hypophthalmichthys nobilis* (Richardson, 1845)	bighead carp	Non-native	
20	*Leucaspius delineatus* (Heckel, 1873)	belica	Native	
21	*Leuciscus aspius* (Linnaeus, 1758)	asp	Native	Annex II, V
22	*Leuciscus idus* (Linnaeus, 1758)	ide	Native	
23	*Leuciscus leuciscus* (Linnaeus, 1758)	common dace	Native	
24	*Pelecus cultratus* (Linnaeus, 1758)	sichel	Native	Annex II, V
25	*Phoxinus phoxinus* (Linnaeus, 1758)	Eurasian minnow	Native	
26	*Pseudorasbora parva* (Temminck & Schlegel, 1846)	stone moroko	Native	
27	*Rhodeus amarus* (Bloch, 1782)	European bitterling	Native	Annex II
28	*Romanogobio kesslerii* (Dybowski, 1862)	Kessler's gudgeon	Native	Annex II
29	*Romanogobio uranoscopus* (Agassiz, 1828)	Danubian longbarbel gudgeon	Endemic	Annex II
30	*Romanogobio vladykovi* (Fang, 1943)	white-finned gudgeon	Native	Annex II
31	*Rutilus rutilus* (Linnaeus, 1758)	roach	Native	
32	*Rutilus virgo* (Heckel, 1852)	Danubian roach	Endemic	Annex II, V
33	*Scardinius erythrophthalmus* (Linnaeus, 1758)	rudd	Native	
34	*Squalius cephalus* (Linnaeus, 1758)	chub	Native	
35	*Tinca tinca* (Linnaeus, 1758)	tench	Native	
36	*Vimba vimba* (Linnaeus, 1758)	vimba bream	Native	

(continued)

Table 17.1 (continued)

No. of species	Taxon	Common name	Biogeographical status	Habitats directive
	familia Cobitidae	**loaches**		
37	*Cobitis elongatoides* Băcescu & Mayer, 1969	spined loach	Native	Annex II
38	*Misgurnus fossilis* (Linnaeus, 1758)	weatherfish	Native	Annex II
39	[c]*Sabanejewia aurata* (De Filippi, 1863)	golden spined loach	Native	Annex II
	familia Nemacheilidae	**stone loaches**		
40	*Barbatula barbatula* (Linnaeus, 1758)	stone loach	Native	
	ordo SILURIFORMES	**CATFISHES**		
	familia Siluridae	**sheatfishes**		
41	*Silurus glanis* Linnaeus, 1758	wels catfish	Native	
	familia Ictaluridae	**North American catfishes**		
42	*Ameiurus melas* (Rafinesque, 1820)	black bullhead	Non-native	
43	*Ameiurus nebulosus* (Lesueur, 1819)	brown bullhead	Non-native	
	ordo SALMONIFORMES	**TROUTS, SALMONS, AND WHITEFISHES**		
	familia Salmonidae	**trouts, salmons, and whitefishes**		
44	*Hucho hucho* (Linnaeus, 1758)	huchen	Endemic	Annex II, V
45	*Oncorhynchus mykiss* (Walbaum, 1792)	rainbow trout	Non-native	
46	*Salmo trutta* Linnaeus, 1758	brown trout	Native	
47	*Salvelinus fontinalis* (Mitchill, 1814)	brook trout	Non-native	
48	*Thymallus thymallus* (Linnaeus, 1758)	grayling	Native	Annex V
	ordo ESOCIFORMES	**PIKES AND MUDMINNOWS**		
	familia Esocidae	**pikes**		
49	*Esox lucius* Linnaeus, 1758	northern pike	Native	
	familia Umbridae	**mudminnows**		
50	*Umbra krameri* Walbaum, 1792	European mudminnow	Native	Annex II
	ordo GADIFORMES	**CODS AND HAKES**		
	familia Gadidae	**cods**		

(continued)

Table 17.1 (continued)

No. of species	Taxon	Common name	Biogeographical status	Habitats directive
51	*Lota lota* (Linnaeus, 1758)	burbot	Native	
	ordo Gobiiformes	**gobies**		
	familia Gobiidae	**gobies**		
52	*Neogobius fluviatilis* (Pallas, 1814)	monkey goby	Non-native	
53	*Neogobius melanostomus* (Pallas, 1814)	round goby	Non-native	
54	*Ponticola kessleri* (Günther, 1861)	bighead goby	Non-native	
55	*Proterorhinus semilunaris* (Heckel, 1837)	western tubenose goby	Non-native	
	ordo Perciformes	**perches**		
	familia Centrarchidae	**sunfishes**		
56	*Lepomis gibbosus* (Linnaeus, 1758)	pumpkinseed	Non-native	
57	*Micropterus salmoides* (Lacepède, 1802)	largemouth bass	Non-native	
	familia Percidae	**perches**		
58	*Gymnocephalus baloni* Holčík & Hensel, 1974	Danube ruffe	Native	Annex II, IV
59	*Gymnocephalus cernua* (Linnaeus, 1758)	ruffe	Native	
60	*Gymnocephalus schraetser* (Linnaeus, 1758)	schraetzer	Endemic	Annex II, V
61	*Perca fluviatilis* Linnaeus, 1758	European perch	Native	
62	*Sander lucioperca* (Linnaeus, 1758)	pikeperch	Native	
63	*Sander volgensis* (Gmelin, 1789)	Volga pikeperch	Native	
64	*Zingel streber* (Siebold, 1863)	Danube streber	Native	Annex II
65	*Zingel zingel* (Linnaeus, 1766)	zingel	Native	Annex II, V
	ordo Scorpaeniformes	**mail-cheeked fishes**		
	familia Gasterosteidae	**sticklebacks**		

(continued)

Table 17.1 (continued)

No. of species	Taxon	Common name	Biogeographical status	Habitats directive
66	*Gasterosteus gymnurus* Cuvier, 1829	western three-spined stickleback	Non-native	
	Familia Cottidae	**sculpins**		
67	*Cottus gobio* Linnaeus, 1758	bullhead	Native	Annex II

[a]Some ichthyologists tend to consider the lampreys living in the Drava River as *E. vladykovi* (Oliva & Zanandrea, 1959), a distinct species from *E. mariae* (see e.g. Povz 2011). In fact, *E. vladykovi* has been considered as a subspecies of *E. mariae*. Hence, further studies are expected to clear the taxonomical status of the lampreys of the Drava. If results support *E. vladykovi* being a distinct species, it should be considered as an endemism of the Danube catchment

[b]Takács et al. (2014) argue that gudgeons in the Central and Southern Transdanubia region, including the Drava catchment too, in Hungary are genetically different from gudgeons identifiable as *G. obtusirostris* living in Northern Transdanubia. However, the differences are probably not enough to make a species-level distinction

[c]See Faunistics section in Results and Discussion

for *Gobio albipinnatus*) in Central Europe and Germany is questionable, and the species is denoted as native to Russia and Kazakhstan. Furthermore, FishBase denotes *R. vladykovi* as native to Austria, Hungary, and Romania. Consequently, *R. vladykovi* should also be considered a species of community interest. (The reference to gudgeons in the FishBase is Kottelat and Freyhof (2007), a not peer-reviewed handbook.)

Harka (1992) listed the spined loach as *Cobitis taenia*. Likewise, Annex II also contains the name *C. taenia*. Actually, the species *C. elongatoides* was formerly considered a subspecies of *C. taenia*. In accordance with this, *Cobitis taenia elongatoides* is denoted as a synonym for *C. elongatoides* in the FishBase. Thus, *C. elongatoides* is another species of community interest (see e.g. Sallai 2002b; Tóth et al. 2007).

Formerly, it was accepted that the golden spined loach, *Sabanejewia aurata,* has two subspecies, *S. a. balcanica* and *S. a. bulgarica*. However, it has become more and more widely accepted that these subspecies are actually two distinct species on the basis of colouration pattern. Correspondingly, Sallai and Kontos (2005, 2008) report *S. bulgarica* (Drensky 1928), and Jelić et al. (2012) report *S. balcanica* (Karaman 1922). But the phylogenetic relationships of *Sabanejewia* has not been fully clarified yet. It could be possible that *S. balcanica* and *S. bulgarica* are really distinct from *S. aurata*, but they are probably members of a species complex (Danubian Balkanian complex) rather than being two fully distinct species (see Perdices et al. 2003, 2016; The PLOS ONE Staff 2016). Because of this taxonomic uncertainty, we listed golden spined loaches as *S. aurata* in Table 17.1. Whatever the real taxonomic relationships between the golden spined loaches living in the Drava are, they should be considered as community interest fishes.

Although faunistic investigations suggest 67 fish species, the number of species actually living and reproducing in the Croatian-Hungarian Drava section can be supposed to be about 51. For the sake of simplicity, these regularly occurring species can be considered as the core of the fish fauna, and other occasionally appearing species function as colorizing satellite species of the fauna. The occurrence of species can be occasional because, for instance, they are not able to reach this section from downstream habitats (diadromous fishes like sturgeons and European eel [*Anguilla anguilla*]) due to migration barriers. Some fishes, like Salmonid species, natives and non-natives as well, and probably the Danubian longbarbel gudgeon (*Romanogobio uranoscopus*) too, find more suitable habitats on the upper reaches of the Drava, hence they have a naturally rare occurrence in the lower section. Although the regular downstream drift of some of their individuals is probably also negatively affected by barrages, and these species would likely be detected a little more frequently in the Croatian-Hungarian section as they are now if the upstream dams did not exist.

It is widely accepted that some non-native species, such as grass carp (*Ctenopharyngodon idella*), bighead carp (*Hypophthalmichthys nobilis*), silver carp (*H. molitrix*), do not have self-sustaining populations in the Drava or other rivers in the region. Therefore, their occurrences depend primarily on the human stocking activities directly into the main channel and/or fishponds from where they can escape into the main channel via inflow streams. The populations of another non-native species, the largemouth bass (*Micropterus salmoides*), although assumed to reproduce in the Drava (Povž and Šumer 2005; Sallai and Kontos 2008), typically have low density. A possible reason for this could be the extreme daily fluctuation of water level due to the operation of the upstream hydropower plants, which makes the main channel an unfavorable habitat for reproduction to the largemouth bass.

Other species require for their entire life-cycle habitats remarkably different from those in the main Drava channel. The Eurasian minnow can typically be found in fast flowing, cool, sub-mountainous brooks and clear, gravel, and stony-bottomed lakes. The crucian carp (*Carassius carassius*) and belica (*Leucaspius delineatus*) prefer slow-flowing and still water, so they are most likely to occur in densely vegetated oxbow lakes and backwaters. For them the main channel of high current velocity functions as a matrix habitat through which they can disperse among their real habitat patches (see Erős and Campbell Grant 2015). Hence, they are usually also regarded occasionally occurring species in the main channel.

The ongoing spreading of the Ponto-Caspian gobies in the Danube catchment could lead not only to the enrichment of the local fauna but the remarkable alteration of the fish abundances both in the offshore zone (Szalóky et al. 2015) and, especially, in the littoral zone of the rivers (Erős et al. 2005). The results of the faunistic surveys point to the forthcoming upstream expansion of the Ponto-Caspian gobies in the Drava as well. This prediction is supported by the following observations. The Kessler goby was first found at Donji Miholjac, Croatia, the most downstream location investigated in the studies reviewed here, and furthermore, this species or any other new goby species was not detected upstream of Donji

Miholjac until 2015. Subsequently, Sallai (2016) found the Kessler goby along with the round goby, a new species to the fish fauna of the Drava, upstream of Donji Miholjac at Matty, but he did not recover either of them more upstream, at Drávakeresztúr-Révfalu, Hungary. Along with the upstream spreading of the goby species already present in the Drava, other goby species, for example the racer goby (*Babka gymnotrachelus* [Kessler 1857]), also could colonize the Drava from the Danube.

17.4.2 Qualitative Aspects and Ecological Assessment

Turning to the quantitative aspect, the sample of Sály et al. contains 474 individuals of 22 fish species. The rank abundance distribution (Fig. 17.1) constructed from the data shows that the bleak (*Alburnus alburnus*), the most abundant species in the assemblage, represents two fifths of all individuals (41.6%), while the second most abundant chub (*Squalius cephalus*) only one fifth of the specimens (18.6%). Eight species show a relative abundance between ten and one percent. It is less fortunate, that among these species the three most abundant are non-native (Prussian carp [*Carassius gibelio*], monkey goby [*Neogobius fluviatilis*] and western tubenose goby [*Proterorhinus semilunaris*]), although there are three species with community interest (European bitterling [*Rhodeus amarus*], white-finned gudgeon [*Romanogobio vladykovi*], barbel [*Barbus barbus*]). The relative abundance of the remaining 12 species is less than 1%. Among them, there are four species with community interest (spined loach, zingel [*Zingel zingel*], Ukrainian brook lamprey [*Eudontomyzon mariae*], and Danube streber [*Zingel streber*]), and two non-native ones (stone moroko [*Pseudorasbora parva*] and pumpkinseed [*Lepomis gibbosus*]).

The ecological assessment of the three 500-m-long subsamples and that of the full sample containing the pooled data (i.e. 3 × 500 m) with the Hungarian Multimetric Fish Index (HMMFI) resulted in EQR values of 0.42, 0.47, 0.64, and 0.69. These figures correspond to the ecological classes of moderate, moderate, good, and good, respectively. Out of the three subsamples, the third (EQR = 0.64) was located the most downstream, the farthest from the built-up area of Barcs, whereas the other two subsamples were much closer to the inhabited region. This may, at least in part, explain the higher EQR value for the third subsample.

The HMMFI index applicable to lowland rivers like the Croatian-Hungarian Drava is primarily sensitive to the species number of the sample to be assessed. However, pooling the three subsamples into a single full sample can increase the robustness and information content of the data. Therefore, it appears that the ecological status of the assessed section near Barcs could fit in with the ecological status of the surface water bodies of the Hungarian Drava, which were estimated as good and high (see the coloured map in Sály and Erős 2016, Appendix 3).

Certainly, the analysis of this example data set provides only a snap-shot illustration on the abundances of the fish species and on the ecological status of the sampled habitat. Because many species are represented in this sample only by one

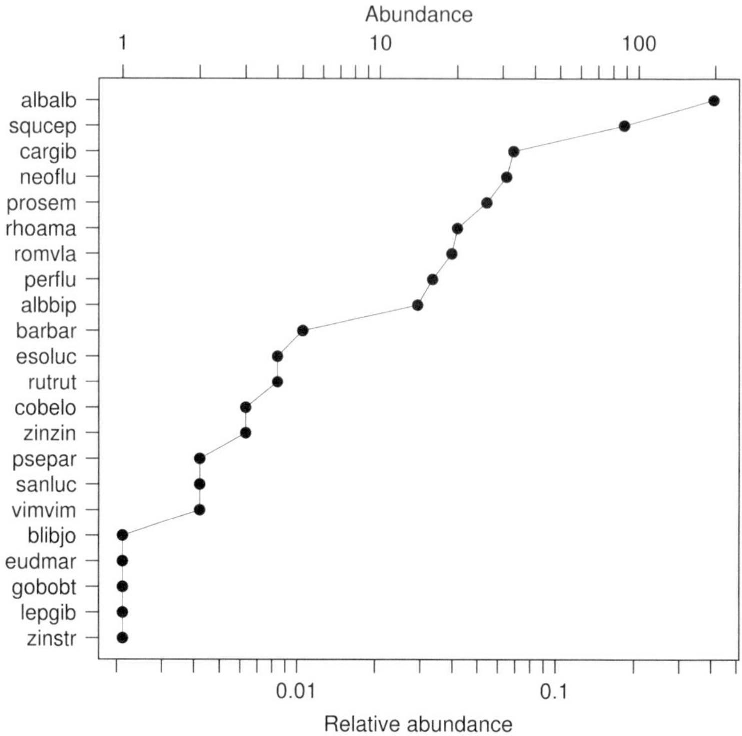

Fig. 17.1 Rank abundance distribution of the fish assemblage in the littoral zone of the Drava (original data of Sály et al.). The total number of the individuals caught was 474, which belonged to a total of 22 species. Upper x-axis represents the number of individuals (abundance), and the lower x-axis shows the relative abundance. Note that both the x-axes are logarithmic to the base of 10. Albalb, bleak (*Alburnus alburnus*); squcep, chub (*Squalius cephalus*); cargib, Prussian carp (*Carassius gibelio*); neoflu, monkey goby (*Neogobius fluviatilis*); prosem, western tubenose goby (*Proterorhinus semilunaris*); rhoama, European bitterling (*Rhodeus amarus*); romvla, white-finned gudgeon (*Romanogobio vladykovi*); perflu, European perch (*Perca fluviatilis*); albbip, schneider (*Alburnoides bipunctatus*); barbar, barbel (*Barbus barbus*); esoluc, northern pike (*Esox lucius*); rutrut, roach (*Rutilus rutilus*); cobelo, spined loach (*Cobitis elongatoides*); zinzin, zingel (*Zingel zingel*); psepar, stone moroko (*Pseudorasbora parva*); sanluc, pikeperch (*Sander lucioperca*); vimvim, vimba bream (*Vimba vimba*); blibjo, white bream (*Blicca bjoerkna*); eudmar, Ukrainian brook lamprey (*Eudontomyzon mariae*); gobobt, gudgeon (*Gobio obtusirostris*); lepgib, pumpkinseed (*Lepomis gibbosus*); zinstr, Danube streber (*Zingel streber*)

or two specimens, it can be assumed that additional species would have been caught if further effort had been made to sample. Yet, the survey is informative because sampling was made by using a single method. On the other hand, further effort probably would yield rare species, which could not profoundly modify the relative abundance of the more common species.

However, in the light of abundance the evaluation of commonness and rarity could be misleading when one uses a single sampling method only in case of large rivers. For instance, Szalóky et al. (2014) highlight the importance of offshore sampling to evaluate the abundance of benthic fish species in large rivers. Compared to shoreline boat electrofishing, a sampling method used commonly in lowland rivers, they detected the sterlet (*Acipenser ruthenus*) and caught much more individuals of the Danube streber by sampling with an electrified benthic frame trawl offshore in the Danube. Similarly, sampling with this recently developed method, Szalóky et al. (2015) point out the intense and somewhat species differentiated offshore habitat use of the Ponto-Caspian gobies. Therefore, the application of such formerly not used methodological approaches to sample the fishes of the Drava could refine our picture on the abundance structure and maybe even on the presence of some species in the Drava as well.

17.5 Conclusions

The Croatian-Hungarian section of the Drava River seems to have a reasonable good ecological status and provides home for a rich fish fanua, valuable for nature conservation. This richness is mainly due to that this river section has not been subjected to such large-scale human modifications as other Drava sections. Another control of its fish fauna and the density of the fish populations is the proximity to the Danube as a source region. Primarily potamodromous species can move between the two rivers. Consequently, the cornerstone of successful conservation of fish fauna could probably be the prevention of any further anthropogenic alteration of the Drava and its banks. In addition, improving longitudinal connectivity not only on the upper Drava but on the lower Danube too, a more frequent appearance of occasional satellite fauna members and the possible re-emergence of indigenous diadromous fishes could be achieved.

Acknowledgements Author thanks András Specizár and Tibor Erős for their essential help in field survey and Zoltán Sallai for his valuable remarks to improve the manuscript.

References

Agapow P-M, Bininda-Emonds ORP, Crandall KA, Gittleman JL, Mace GM, Marshall JC, Purvis A (2004) The impact of species concept on biodiversity studies. Q Rev Biol 79(2): 161–179

Antal L, László B, Kotlík P, Mozsár A, Czeglédi I, Oldal M, Kemenesi G, Jakab F, Nagy SA (2016) Phylogenetic evidence for a new species of in the Danube River basin. Mol Phylogenet Evol 96:187–194

Baxter RM (1977) Environmental effects of dams and impoundments. Annu Rev Ecol Syst 8(1): 255–283

Council Directive 92/43/EEC of 21 May 1992 on the conservation of natural habitats and of wild fauna and flora

Csipkés R, Szatmári L, Soós N (2012) Nyugati pikó (Gasterosteus gymnurus) a Drávában. Halászat 105(1):17–18 (in Hungarian)
Erős T, Campbell Grant EH (2015) Unifying research on the fragmentation of terrestrial and aquatic habitats: patches, connectivity and the matrix in riverscapes. Freshw Biol 60(8): 1487–1501
Erős T, Sevcsik A, Tóth B (2005) Abundance and night-time habitat use patterns of Ponto-Caspian gobiid species (Pisces, Gobiidae) in the littoral zone of the River Danube, Hungary. J Appl Ichthyol 21(4):350–357
European Commission (2000) Directive 2000/60/EEC. Establishing a framework for community action in the field of water policy. Official J European Communities, Luxemburg L327:1–71
Harka Á (1992) A Dráva halai. Halászat 85(1):9–12 (in Hungarian)
Jelić M, Jelić D, Žutinić P, Ćaleta M (2012) Značajke faune riba donjeg toka rijeke Drave i okolnih poplavih staništa kod donjeg miholjca (istočna Hrvatska) (Fish fauna of the lower reaches of the River Drava and surrounding marshland habitats near Donji Miholjac, Eastern Croatia). Croatian J Fisheries 70(4):153–167 (In Croatian with English abstract)
Karr JR (1981) Assessment of biotic integrity using fish communities. Fisheries 6(6):21–27
Karr JR (1991) Biological integrity: a long-neglected aspect of water resource management. Ecol Appl 1(1):66–84
Kottelat M, Freyhof J (2007) Handbook of European freshwater fishes. Publications Kottelat, Cornol and Freyhof, Berlin, 646 pp
Majer J (1995) Adatok a Dráva halfaunájához és egyes holtágak vízminőségéhez (Data to the fish fauna [Pisces] of Dráva river [Hungary] and to the quality of backwaters). Dunántúli Dolgozatok Természettudományi Sorozat 8:189–202 (in Hungarian with an English abstract)
Majer J (1998) Adatok a Dráva és a Dráva menti területek hal-, kétéltű- és hüllőfaunájához (Pisces, Amphibia, Reptilia) (Contributions to data on the fish, amhibian and reptile fauna [Pisces, Amphibia, Reptilia] of river Dráva and adjacent areas). Dunántúli Dolgozatok Természettudományi Sorozat 9:431–440 (in Hungarian with an English abstract)
Majer J, Bordács M (2001) A Dráva magyarországi felső szakaszának természetvédelmi értékelése halfauna alapján [Conservation status of fish fauna in the upper part of the Dráva river in Hungary]. Természetvédelmi Közlemények 9:251–263 (in Hungarian with an English abstract)
Manné S, Poulet N, Dembski S (2013) Colonisation of the Rhine basin by non-native gobiids: an update of the situation in France. Knowl Manag Aquatic Ecosyst 411:02. https://doi.org/10.1051/kmae/2013069
Mendel J, Lusk S, Vasil'eva ED, Vasil'ev VP, Lusková V, Ekmekci FG, Erk'akan F, Ruchin A, Koščo J, Vetešník L, Halačka K, Šanda R, Pashkov AN, Reshetnikov SI (2008) Molecular phylogeny of the genus *Gobio* Cuvier, 1816 (Teleostei: Cyprinidae) and its contribution to taxonomy. Molecular Phylogenetics and Evolution 47(3):1061–1075
Nelson JS, Grande TC, Wilson MVH (2016) Fishes of the world, 5th edn. John Wiley & Sons Hoboken, NJ
Perdices A, Bohlen J, Šlechtová V, Doadrio I (2016) Molecular evidence for multiple origins of the European spined loaches (Teleostei, Cobitidae). PLOS ONE 11(1):e0144628. https://doi.org/10.1371/journal.pone.0144628
Perdices A, Doadrio I, Economidis PS, Bohlen J, Bănărescu P (2003) Pleistocene effects on the European freshwater fish fauna: double origin of the cobitid genus *Sabanejewia* in the Danube basin (Osteichthyes: Cobitidae). Mol Phylogenet Evol 26(2):289–299
Povž M (2011) Distribution and conservation status of the lampreys in Slovenia. Bulletin Lampetra VII:84–91
Povž M, Šumer S (2005) A brief review of non-native freshwater fishes in Slovenia. J Appl Ichthyol 21:316–318
Roche KF, Janač M, Jurajda P (2013) A review of Gobiid expansion along the Danube-Rhine corridor – geopolitical change as a driver for invasion. Knowl Manag Aquatic Ecosyst 411:01. https://doi.org/10.1051/kmae/2013066

Sallai Z (2002a) A Dráva-Mura vízrendszer halfaunisztikai vizsgálata (Fish faunistical investigation of the Drava-Mura river system) I. Halászat 95(2):80–91. http://www.nimfea.hu/tagszervezeteink/halfaunisztika/drava1.htm (in Hungarian)
Sallai Z (2002b) A Dráva-Mura vízrendszer halfaunisztikai vizsgálata II. Fajlista, következtetések (Fish faunistical investigation of the Drava-Mura river system II. Species list and conclusions). Halászat 95(3)3:119–140. http://www.nimfea.hu/tagszervezeteink/halfaunisztika/drava2.htm (in Hungarian)
Sallai Z (2004) A Dráva folyó és hazai vízrendszerének halfaunája (Fish fauna in the Drava and its tributaries in Hungary). In: Sallai Z (ed) Mit veszíthetünk a Drávára tervezett horvát erőművel? (What can we lose with the planned hydropower plant on the Drava). Nimfea Természetvédelmi Egyesület, Túrkeve, pp 42–68 (in Hungarian with an English summary)
Sallai Z (2016) Inváziós gébfajok megjelenése a Drávában (Invasive goby species in the Drava). Halászat 109(2):14 (in Hungarian)
Sallai Z, Kontos T (2008) Data to the fish fauna of Croatian Drava sections. In: Purger JJ (ed) Biodiversity studies along the Drava river. University of Pécs, Pécs, Hungary, p 328
Sallai Z, Kontos Z (2005) Fishfaunistical monitoring of the Hungarian part of the River Drava (1999–2004). Natura Somogyiensis 7:75–104
Sály P, Erős T (2016) Vízfolyások ökológiai állapotminősítése halakkal: minősítési indexek kidolgozása (Ecological assessment of running waters in Hungary: compilation of biotic indices based on fish). Pisces Hungarici 10:15–45 (in Hungarian with an English abstract)
Szalóky Z, György ÁI, Tóth B, Sevcsik A, Specziár A, Csányi B, Szekeres J, Erős T (2014) Application of an electrified benthic frame trawl for sampling fish in a very large European river (the Danube River) – Is offshore monitoring necessary? Fish Res 151:12–19
Szalóky Z, Bammer V, György ÁI, Pehlivanov L, Schabuss M, Zornig H, Weiperth A, Erős T (2015) Offshore distribution of invasive gobies (Pisces: Gobiidae) along the longitudinal profile of the Danube River. Fund Appl Limnol 187(2):127–133
Takács P (2012) Morphometric differentiation of gudgeon species inhabiting the Carpathian Basin. Annales de Limnologie—Int J Limnol 48(1):53–61
Takács P, Bihari P, Erős T, Specziár A, Szivák I, Bíró P, Csoma E (2014) Genetic heterogeneity reveals on-going speciation and cryptic taxonomic diversity of stream-dwelling gudgeons (Teleostei, Cyprinidae) in the Middle Danubian hydrosystem (Hungary). PLOS ONE 9(5): e97278. https://doi.org/10.1371/journal.pone.0097278
The PLOS ONE Staff (2016) Correction: molecular evidence for multiple origins of the European spined loaches (Teleostei, Cobitidae). PLOS ONE 11(3):e0151228. https://doi.org/10.1371/journal.pone.0151228
Tóth B, Sevcsik A, Erős T (2007) Natura 2000-es halfajok elő fordulása a Duna hazai szakaszán (Occurrence of protected fishes of the Natura 2000 system in the Hungarian section of the Danube). Pisces Hungarici 2:83–94 (in Hungarian with an English abstract)

Chapter 18
Floodplain Forests

Balázs Kevey

Abstract In spite of centuries of forest exploitation and indirect human intervention into their physical environment, the forests along the Hungarian Drava section have preserved much of their natural character. Particularly along the northernmost reaches, where the river arrives from Croatian territory (the Zákány–Őrtilos area of the Somogy Hills), more than 100 protected herbaceous plant species are found in the forests. This chapter provides systematic descriptions of the forest communities found in the Drava Plain: two types of willow shrubs, three associations of riverine softwood groves, three hardwood forest communities and four classes of bog forests. Their successional evolution, significance in the landscape, threats imposed by human activities, proper and improper management and nature conservation aspects are also briefly treated. Coupled with microtopography, pedodiversity and microclimate, the distribution of forest communities creates a special landscape pattern.

Keywords Floodplain communities · Succession · Vegetation history
Nature conservation

18.1 Introduction

The forests in the Drava floodplain, both within and beyond the flood-control dykes, i.e. in the active and in the 'protected' floodplain, show great variety (Figs. 18.1 and 18.2). The succession of vegetation in the active floodplain was first studied by Klujber et al. (1963) and Vöröss (1964), followed by Kovács and Kárpáti (1973, 1974). They set up a schematic representation of the succession process, which—supplemented with recent research findings (Kevey 1998a, 2008) —can be outlined as follows. The first woody communities to appear along the Drava and its side-arms are represented by riparian willow shrubs

B. Kevey (✉)
Department of Ecology, Institute of Biology, Faculty of Sciences,
University of Pécs, Ifjúság útja 6, Pécs 7624, Hungary
e-mail: keveyb@gamma.pte.ttk.hu

D. Lóczy (ed.), *The Drava River*, Springer Geography,
https://doi.org/10.1007/978-3-319-92816-6_18

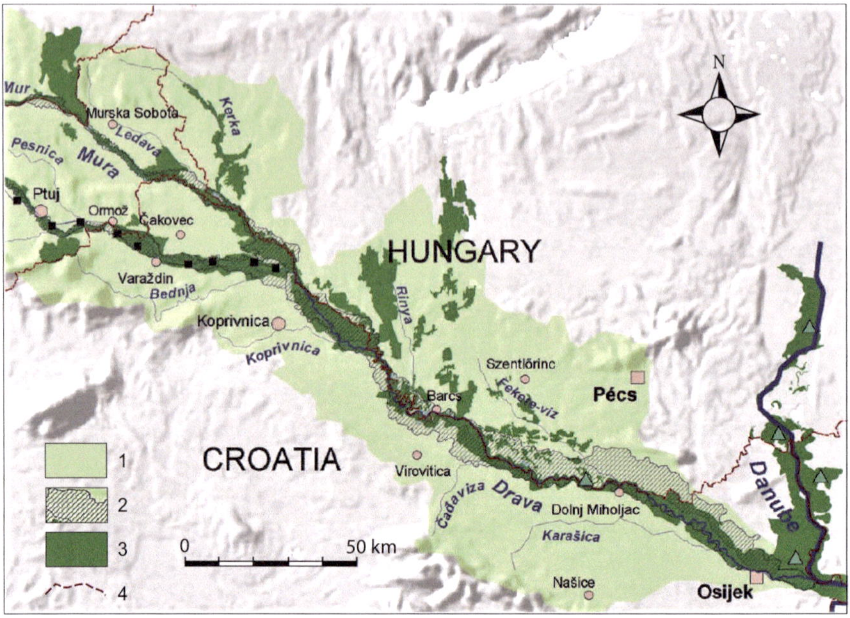

Fig. 18.1 Riparian forests along the Hungarian Drava section. *Source* Drava River Vision (2008). 1, Drava River catchment; 2, Drava River floodplain; 3, forests; 4, national border. Black squares mark hydroelectric plants and green triangles protected areas

(*Rumici crispi-Salicetum purpureae* and *Polygono hydropiperi-Salicetum triandrae*), and at waterlogged sites, willow swamps (*Calamagrostio canescenti-Salicetum cinereae*). Along the grove branch of the succession, the communities of *Rumici crispi-Salicetum purpureae* develop into black poplar gallery forests (*Carduo crispi-Populetum nigrae*) and from *Polygono hydropiperi-Salicetum triandrae* into continental willow gallery forests (*Leucojo aestivi-Salicetum albae*) and both softwood forests change to white poplar forests (*Senecioni sarracenici-Populetum albae*). In another type of succession, *Calamagrostio canescenti-Salicetum cinereae* communities turn into alder swamp forests (*Carici elongatae-Alnetum glutinosae*), and through further sedimentation, alder forests (*Paridi-Alnetum glutinosae*). On the top of the pyramidal succession, both branches meet in oak-ash-elm forests (*Carici brizoidis-Ulmetum*). Further on, the succession continues horizontally towards floodplain hornbeam-oak forests (*Veronico montanae-Carpinetum*) (Fig. 18.3).

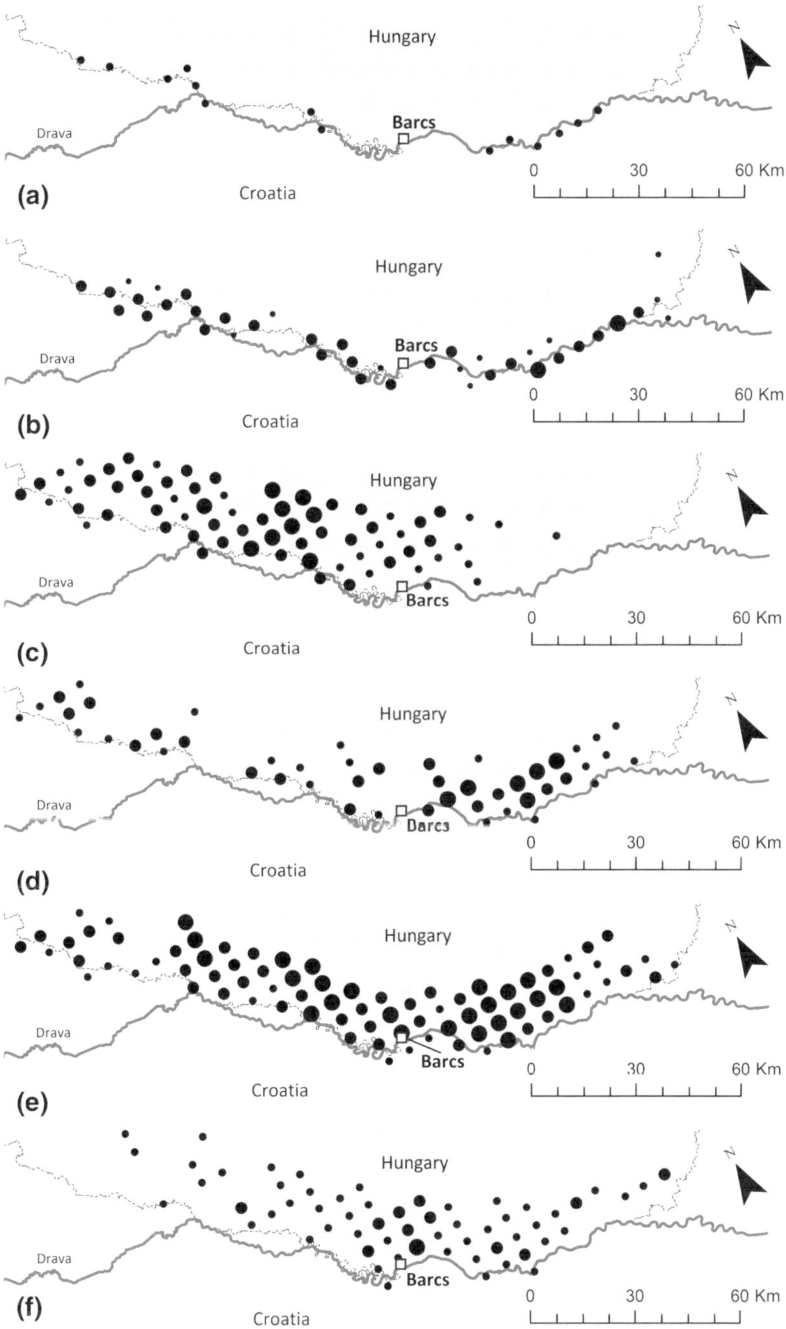

Fig. 18.2 Distribution of riparian forest communities. *Source* MÉTA database, Institute of Ecology, Hungarian Academy of Sciences. **a** Willow shrubs; **b** Softwood (willow and poplar) forests; **c** Alder forests; **d** Oak-ash-elm forests; **e** Hornbeam-oak forests; **f** Bog forests. The size of the circles is proportional to the areal extent of the given forest community in the 35-hectare hexagons

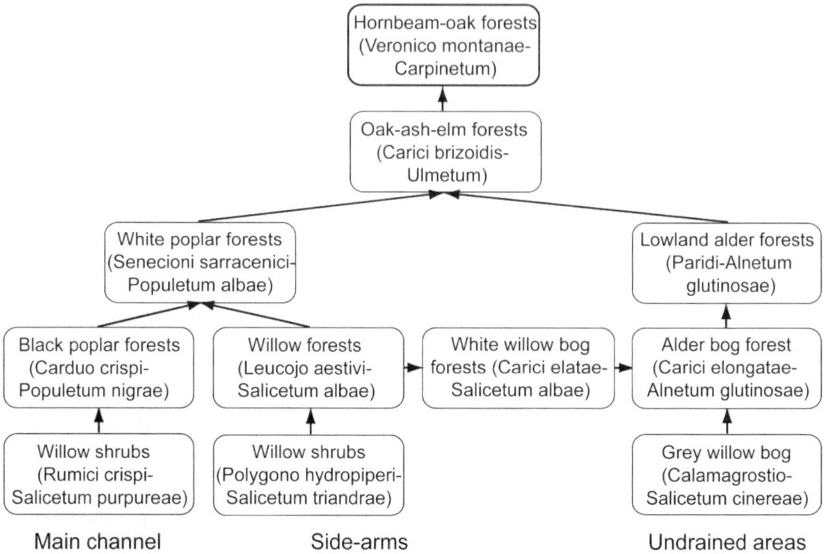

Fig. 18.3 Schematic general succession in the forests of the Drava Plain (by B. Kevey)

18.2 Willow Shrubs

On the banks and bars of the Drava channels and occasionally on the shores of oxbow lakes in the floodplain, pioneer herbaceous vegetation is gradually replaced by willow shrubs. They are usually inundated for 5–7 months in a year, but in drought periods, inundation can be much shorter or fall off completely. River flow keeps their water fresh and rich in oxygen. Dead organic matter fallen into the water is partly decomposed, and in a major part transported downstream by floods. Therefore, willow shrubs are poor in nutrients. Frequent inundation prevents soil formation (Pécsi 1959).

In Hungary (including the Drava floodplain), two willow shrub communities can be identified at present: the *Rumici crispi-Salicetum purpureae* and the *Polygono hydropiperi-Salicetum triandrae* types (Fig. 18.2a).

18.2.1 Rumici Crispi-Salicetum Purpureae *Kevey in Borhidi and Kevey (1996)*

Such communities are formed on bars and in bank zones built of gravels and coarse sands (Kárpáti 1957, 1982), where the current is rapid. In lack of gravel, coarse sand surfaces of high permeability, left dry during low-water stages, are also

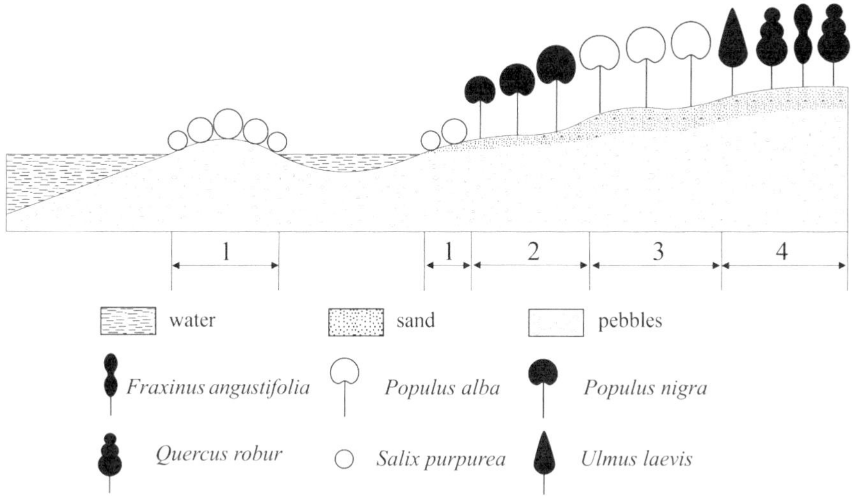

Fig. 18.4 Cross-section of vegetation from gravels bars to higher floodplain terrain (by B. Kevey). 1, *Rumici crispi-Salicetum purpureae*; 2, *Carduo crispi-Populetum nigrae*; 3, *Senecioni sarracenici-Populetum albae*; 4, *Carici brizoidis-Ulmetum*

suitable habitats for this community (Kárpáti 1985). Its development is controlled by periodical flooding and the rate of sedimentation. Floodplain ruderal herbaceous communities (*Bidention tripartiti, Chenopodion fluviatile*) and semiruderal herbaceous communities (*Agropyro-Rumicion crispi*) are the first to appear. With the spreading of shrubs 1.5–2 m high, woody vegetation develops on the bars in 2 years (Fig. 18.4). After some more years, the shrub may reach heights of 4–5 m. Further sedimentation leads to black poplar forests (*Carduo crispi-Populetum nigrae*—Kevey 1998a, 1999a), which often occur next to willow shrubs.

Willows in the community are 1.5–5 m high with 50–80% surface cover. Sporadically low trees—mostly, *Populus nigra* and *Salix alba*—also occur among them. The species composition of shrubs depends on river regime related to ripening. In addition to *Salix purpurea*, *S. alba,* and *Populus nigra*, and even the protected *S. elaeagnos* (along the uppermost Hungarian section of the Drava) occur. The percentage of grass cover (mainly of *Agrostis stolonifera*, *Phalaris arundinacea* and *Poa palustris*) is primarily controlled by river regime and shading by shrubs: in dry years with open shrub, it amounts to 90%, while with regularly recurring floods and dense undergrowth, it is non-existent. Further herbaceous plants, like *Matricaria maritima*, *Phragmites australis*, *Persicaria lapathifolia*, *P. dubia* and *Rorippa sylvestris* also occur. *Aster* × *salignus* indicates degraded stands. *Rumici crispi-Salicetum purpureae* communities are generally characterized by ruderal (*Bidention tripartiti, Chenopodion fluviatile*) and semiruderal (*Agropyro-Rumicion crispi*) elements.

Floods as a major selective factor control the species composition of willow shrubs. In the undergrowth predominant species, which tolerate extreme water regime, i.e. inundations alternating with desiccation. This partly explains the widespread occurrence of ruderal elements. The propagules of plants in the willow shrub community reach the bars by wind and flood transport, while their further fate is determined by the occurrence of flood waves. In wet years, they are regularly inundated and the community remains poor in species. In drought years, flooding is short-lived or does not happen at all and numerous plant species survive below the shrub layer. The dependence of species number on river regime underlines the rather unstable nature of such communities.

Typical *Rumici crispi-Salicetum purpureae* communities are only known from the Drava section in Somogy County. Plants of local occurrence are *Chlorocyperus glomeratus*, *Myricaria germanica*, and the locally widespread *Salix elaeagnos*.

18.2.2 Polygono Hydropiperi-Salicetum Triandrae *Kevey in Borhidi and Kevey (1996)*

This type of riverine willow shrub lines the side-arms and oxbows of the Drava. Since water flow is negligible or intermittent, in areas such as sandy silts or silts are deposited (Kárpáti 1985; Borhidi et al. 2012; Kevey 2008). With better water retention of fine sediments, water availability is more balanced here than on gravel and coarse sand surfaces. No desiccation occurs even during prolonged drought. Such habitats are populated with swamp and silt vegetation (*Nanocyperion flavescentis*) with additional paludal (*Phragmition, Magnocarition*), floodplain ruderal (*Bidention tripartiti, Chenopodion fluviatilis*) and semiruderal (*Agropyro-Rumicion crispi*) species. The spreading of shrubs leads to the development of *Polygono hydropiperi-Salicetum triandrae* community of ca 2 m height within 2 years (Fig. 18.5). In the forthcoming years, shrubs grow to 4–5 m height and then succession is directed towards willow forests (*Leucojo aestivi-Salicetum albae*—Kevey 1998a). Flooding here also lasts for 5–7 months annually and prevents soil formation.

This community also consists of shrub-size (1.5–5 m high) plants with 60–90% cover. In older stands, low *Salix alba* trees also grow. Species composition is controlled by the ripening of seeds and river regime. In addition to *Salix triandra*, other willow species *S. viminalis*, *S. alba*, and rarely *S. cinerea*, *S. fragilis* and *S. purpurea* also occur. Grass cover is primarily influenced by shading by shrubs and river regime: in drought years and with open shrub layer cover amounts to 90%, while regularly occurring flood waves and closed shrub layer reduce it to nil. Most common grasses are *Myosotis nemorosa*, *Phalaris arundinacea* and *Rorippa amphibia* with additional herbaceous plants as *Agrostis stolonifera*, *Carex acuta*,

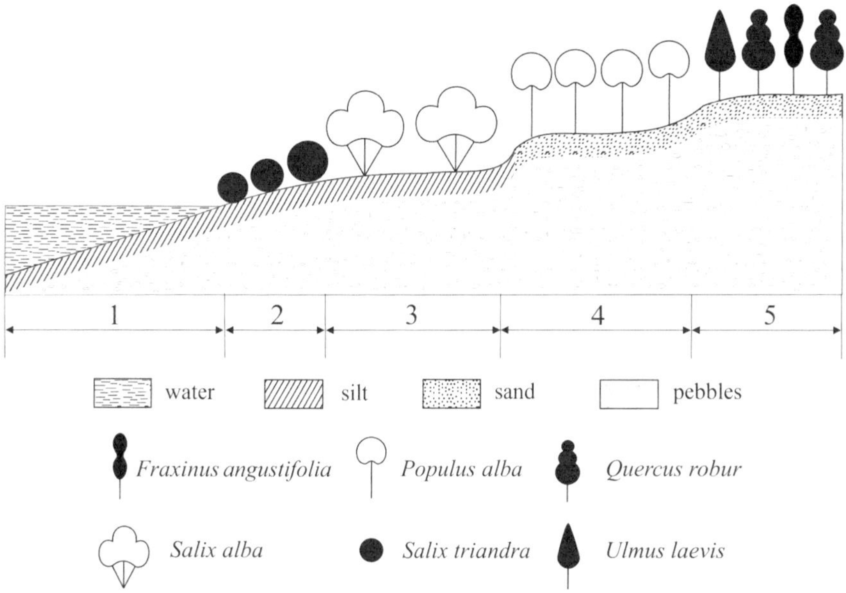

Fig. 18.5 Cross-section of vegetation from the silty bank zone to the higher floodplain terrain (by B. Kevey). 1, *Open water surface*; 2, *Polygono hydropiperi-Salicetum triandrae*; 3, *Leucojo aestivi-Salicetum albae*; 4, *Senecioni sarracenici-Populetum albae*; 5, *Carici brizoidis-Ulmetum*

Lythrum salicaria, *Phragmites australis*, *Poa trivialis*, *Persicaria hydropiper*, *Persicaria minor*, *Rorippa sylvestris*, *Typha angustifolia*, *T. latifolia*. In more degraded stands, *Aster* × *salignus* and *Urtica dioica* prevail.

Because of the higher water retention capacity of silty deposits, for *Polygono hydropiperi-Salicetum triandrae* communities, river regime is less influential in plant composition and make the community more stable with less ruderal elements. Paludal and silt-dwelling plants, however, are more widespread. Trees typical of softwood forests (*Salicion albae*) are also better represented, while the number of adventive species is very low (Kevey 2008).

18.3 Willow Shrubs in Landscape History

Fluvial accumulation continuously builds new bars in the channel and shapes river banks. Bars develop into islands within some decades. In parallel, their willow shrubs turn into softwood forests (*Leucojo aestivi-Salicetum albae, Carduo crispi-Populetum nigrae, Senecioni sarracenici-Populetum albae*). Since tree increments here are minimal, no commercial forestry is practised. The conditions of willow shrubs have been influenced by grazing, angling and gravel extraction, leading to degradation, expansion of weeds, fragmentation or even extermination of stands.

Willow shrubs are important landscape elements with locally protected species, including *Myricaria germanica* and *Salix elaeagnos* in the *Rumici cripi-Salicetum purpureae* community and *Lindernia procumbens*, *Leucojum aestivum* and the temporarily occurring *Selaginella helvetica* in the *Polygono hydropiperi-Salicetum triandrae* community. Most of these species are transported here from the mountains upstream by river flow. The significance of willow shrubs for nature conservation is underlined by their function in succession, i.e. their further development into black poplar forests (*Carduo crispi-Populetum nigrae*) willow forests (*Leucojo aestivi-Salicetum albae*).

18.3.1 Threats to Willow Shrubs

The deposition of gravel, and thus, the year-to-year generation of willow shrubs are hindered by river regulation, gravel extraction, and the operation of hydroelectric plants in Croatia. In spite of these circumstances, fine stands occur along the less regulated section between Őrtilos and Heresznye, Somogy County, Hungary. The *Polygono hydropiperi-Salicetum triandrae* community is also regularly created in inundated areas, side-arms, and oxbows. The implementation of the Djurdjevac barrage in Croatia is a threat to both communities as the drop in water levels could lead to degradation, similarly as it happened in the Szigetköz floodplain along the Danube (Kevey 2002d, 2003). In addition, water pollution (e.g. plastic bottles) drifted with flood waves and stuck during low-water stages also cause damage to habitats.

As it was mentioned above, willow shrubs represent no economic value for forestry. Their dynamics are controlled by fluvial processes and no human intervention is needed. Since the existence of willow shrubs is a necessary precondition to the formation of willow and black poplar forests, rehabilitation efforts are required. Such communities are infected by the invasion of species like *Acer negundo*, *Fraxinus pennsylvanica*, *Amorpha fruticosa*, and *Vitis riparia*, in the ground layer *Aster × salignus*, *Echinocystis lobata*, *Impatiens glandulifera*, *Solidago gigantea* and *Ambrosia artemisiifolia*. Unfortunately, there is no appropriate method to keep these plants under control.

18.4 Softwood Forests

With further deposition, bars are filled-up and willow shrubs gradually develop into softwood forests at lower floodplain level. Therefore, their understorey is of strongly hygrophile character. Since the humus produced is drifted away by the river or buried under flood deposits, the raw alluvial soil is short of appropriate time for soil formation. Only layered alluvia are accumulated. The river continuously carries propagules away and deposits them at low-water stages. Species composition is influenced by the vegetation of the drainage basin, current velocity, the nature of sediment load, and depth to average groundwater table.

Formerly 'willow-poplar forests' (*Salici-Populetum*) were regarded as a single plant community, but recent research (Tóth 1958; Kevey 1993; Borhidi and Kevey 1996; Kevey 2008) pointed out that there are three associations in the Great Hungarian Plain (including the Drava Plain): willow forests (*Leucojo aestivi-Salicetum albae*), black poplar forests (*Carduo crispi-Populetum nigrae*), and white poplar forests (*Senecioni sarracenici-Populetum albae*) (Fig. 18.3).

*18.4.1 Willow Forests (*Leucojo Aestivi-Salicetum Albae *Kevey in Borhidi and Kevey 1996)*

Willow forests occupy floodplain sections (backswamps) which are inundated for 3–4 months in a year (Figs. 18.5 and 18.6). The silty alluvial soils have a balanced water supply. In the bank zones of islands, willow forests grow on more permeable soils, but their understorey is less typical and degraded. The ground layer is affected by groundwater and floods.

Fig. 18.6 Willow forest (Leucojo aestivi-Salicetum albae) on an island of the Drava channel at Vízvár during the 2013 flood (photo by Z. Karancsi)

The top canopy layer of willow trees of up to 20 m height is medium closed (50–70%). The forests mostly consist of *Salix alba* with some *S. fragilis*. *Populus nigra* and *P. alba* are usually missing or occur sporadically. The lower canopy layer has a cover of 5–30% at 10–15 m height. With young *Salix alba* and *S. fragilis* trees, other arboreous species (*Alnus glutinosa, Ulmus laevis*) are sporadic. The shrub layer is insignificant or missing. The diverse ground layer covers 60–100% of the surface: primarily paludal facies (*Phragmition, Magnocaricion*) with *Carex acuta*, *C. riparia*, *C. vesicaria*, *Galium palustre*, *Myosotis nemorosa*, *Phragmites australis*, *Phalaris arundinacea*, *Poa palustris*, *Persicaria hydropiper, P. dubia*, *Rorippa amphibia,* and *Stachys palustris.* Occasionally, the protected *Leucojum aestivum* occurs densely in the ground layer of the forest.

Among the softwood forests, the willow forests show the most abundant paludal (*Phragmition, Magnocaricion*) and softwood (*Salicion albae*) elements. At the same time, the share of ruderal plants is the lowest in this community. This fact supports the conclusion that willow forests are closest to the natural conditions of softwood forests.

The evolution of willow forests from *Polygono hydropiperi-Salicetum triandrae* vegetation through the sedimentation of their habitats takes about 25 years to complete. *Salix alba* or more seldom *S. fragilis*, common elements of willow shrubs, overgrow and then shade the shrub-size willows (*S. triandra, S. viminalis*) of high light demand and forces them out from the habitat. In the meantime, the understorey is also transformed, a process driven by further sedimentation and shading (Kevey 1998a).

Over a longer successional period, willow forests (*Leucojo aestivi-Salicetum albae*) turn into white poplar forests (*Senecioni sarracenici-Populetum albae*) (for details, see white poplar forests—Kevey 1998a). The habitats of willow forests (e.g. oxbows) may become isolated from the river channel and spared from inundation. In this case, water demand is satisfied from groundwater (see Chap. 13). As a consequence of stagnant water, there is some peat accumulation and transformation of willow forests into white willow swamp forest (*Carici elatae-Salicetum albae*—Kevey 2008).

Willow forests in seminatural condition occur at Kisszentmárton along the Drava (Ortmann-Ajkai 1998a). Willow forests are poor in protected species but locally large masses of Summer Snowflake (*Leucojum aestivum*) and Swamp Ragwort (*Senecio palustris*) are found.

*18.4.2 Black Poplar Forests (*Carduo Crispi-Populetum Nigrae *Kevey in Borhidi and Kevey 1996)*

Black poplar forests are also associated with low-lying floodplain areas inundated by even minor floods. Their habitats are coarse sand surfaces overlying gravel beds and loose layered alluvial soils with raw humus (Fig. 18.4). The water retention capacity of such soils is poor. They desiccate to a larger extent than in the case of willow forests.

The top canopy layer of black poplar forests is somewhat more closed (60–75%) than that of willow forests and mature trees rise to 30 m height. In the forests, *Populus nigra* are predominant with some *Salix alba* stands. *P. alba* trees are missing or only sporadic. The lower canopy layer at 10–20 m height has a cover of 5–30%. In addition to younger specimens of trees of the upper canopy layer, *Ulmus laevis* and more seldom *U. minor* occur here. Among lianas, *Humulus lupulus* is common and *Clematis vitalba* is rare. The shrub layer is 1.5–3 m high, less developed (1–10%) in young stands, medium developed (20–40%) in older stands. In stands with a well-developed shrub layer, *Cornus sanguinea* and *Sambucus nigra* are often found. The cover of the lower shrub layer (regrowth) is 1–30% with masses of *Rubus caesius*. The cover of the developed ground layer is 80–100%. It has few natural facies (*Lamium maculatum* and *Phalaris arundinacea*). Some herbaceous plants represent facies in smaller patches (*Aegopodium podagraria*, *Elymus caninus*, *Calystegia sepium*, *Glechoma hederacea*, *Phragmites australis*, *Poa palustris*, *P. trivialis*, *Persicaria dubia*). *Galium aparine*, *Urtica dioica* and the invasive *Aster* × *salignus* and *Impatiens glandulifera* are common facies.

Formed on sand-covered gravel beds, the understorey of black poplar stands shows only moderately wet character. Raw alluvial soils intensively desiccate during drought periods. This explains why the understorey of black poplar forests can contain mesophile plants. Thus, species composition is transitional between white willow forests and white poplar forests, reflected by the shares of paludal (*Phragmition, Magnocaricion*), softwood (*Salicion albae*) and hardwood (*Alnion incanae, Ulmenion*) plants. Among the three softwood associations, ruderal plants are most common in the black poplar forests. This also proves their origin from *Rumici crispi-Salicetum purpureae* communities (Kevey 1998a), where *Populus nigra* and *Salix alba* are common and overgrowing, shading and forcing out the shrub-size *Salix purpurea* change willow forests into black poplar forests (*Carduo crispi-Populetum nigrae*) in about 25 years. If *Populus nigra* has been abundant in the willow forest, a typical black poplar forest comes about. There are, however, old stands with *Salix alba*. Black poplar forests develop into white poplar forests (*Senecioni sarracenici-Populetum albae*—Kevey 1998a).

Typical black poplar forests occur in the Somogy section of the Drava floodplain at Zákány, Vízvár and Babócsa. Protected plants include *Equisetum hyemale*, *Peucedanum verticillare* and the very rare *Myricaria germanica* and *Salix elaeagnos*.

*18.4.3 White Poplar Forests (***Senecioni Sarracenici-Populetum Albae** *Kevey in Borhidi and Kevey 1996)*

White poplar forests are found at relatively (1–1.5 m) higher elevations of the low floodplain (Figs. 18.3, 18.4, 18.5 and 18.7) than the previously mentioned communities and are only inundated during major floods. This lasts for 1–2 months, but in drought periods, flooding can fall out for years. The community occupies loose soils, transitional towards alluvial forest soils. The occurrence of white poplar forests was

Fig. 18.7 White poplar forest (*Senecioni sarracenici-Populetum albae*) (photo by B. Kevey)

first recorded along the Drava by Klujber et al. (1963) and Vöröss (1964), but they were surveyed only recently (Kevey and Tóth 2006). Similar forests are also found along the Mura River (Kevey 2014).

White poplar forests differ from both willow forests and oak-ash-elm forests (Kevey 2016a, b). The upper canopy of 20–30 m height covers 60–80% of the surface. *Populus alba* is predominant, subordinately with *P. nigra* and *Salix alba*. The cover of the lower canopy amounts to 5–40% and it rises to 10–18 m. Along with young *Populus alba* specimens, *Alnus incana, Ulmus laevis* and *U. minor* also occur. The shrub layer of 2–3 m height and 40–70% cover is primarily composed of *Cornus sanguinea* with *Crataegus monogyna* and *Euonymus europaeus* also common, and among the typical arboreous plants of hardwood forests, *Padus avium* and *Viburnum opulus*. In the lower shrub layer (regrowth), locally larger masses of *Rubus caesius* are found. The rather diverse ground layer may have a cover of 10–100% with *Glechoma hederacea*, *Impatiens noli-tangere*, *Lamium maculatum* and *Ranunculus ficaria* facies. The following plants occur in patches: *Aegopodium podagraria*, *Circaea lutetiana*, *Galanthus nivalis*, *Myosoton aquaticum*, *Phalaris arundinacea*, *Poa trivialis* and *Carex remota*. In the most degraded sites, *Galium aparine* and *Urtica dioica* are abundant and invasive elements are represented by *Impatiens glandulifera* and *Impatiens parviflora*.

In white poplar forests aspect changes, which are completed in mesophile deciduous forests (*Paridi-Alnetum glutinosae, Carici brizoidis-Ulmetum, Veronico montanae-Carpinetum*) are remarkable. Thus, *Ranunculus ficaria*, often occuring in masses, represents the early spring aspect with *Galanthus nivalis*, *Scilla vindobonensis* and *Gagea lutea*.

Compared to black poplar forests, in white poplar forests, the proportion of paludal elements (*Phragmition, Magnocaricion*) is reduced, but they are still important: *Carex gracilis*, *Iris pseudacorus*, *Lycopus europaeus*, *Phragmites australis*, *Solanum dulcamara*, *Stachys palustris*, etc. The plants typical of softwood forests (*Salicion albae*) are *Alnus incana*, *Cucubalus baccifer*, *Dipsacus pilosus*, *Humulus lupulus*, *Impatiens noli-tangere*, *Leucojum aestivum*. More abundant are elements of the hardwood and (*Alnion incanae*) and mesophile deciduous forests (*Fagetalia*): *Aegopodium podagraria*, *Anemone ranunculoides*, *Carex brizoides*, *C. remota*, *C. strigosa*, *Circaea lutetiana*, *Corydalis cava*, *Dryopteris filix-mas*, *Elymus caninus*, *Equisetum hyemale*, *Milium effusum*, *Moehringia trinervia*, *Paris quadrifolia*, etc. Their occurrence is evidence of the successional relationship with oak-ash-elm forests (*Carici brizoidis-Ulmetum*).

The transformation of black poplar forests and partly of willow forests into white poplar forests is a long process which extends over human lifespans. The knowledge on floodplain levels and the juxtaposition of associations allows conclusions on this succession process (Kevey 1998a, 2008). In older willow and black poplar forests incapable for reforestation, *Salix alba* or *Populus nigra* saplings are rare in the understorey. There are several possible explanations. On the one hand, the seeds of these species can only germinate on freshly deposited silt or sand. On the other hand, the ground layer of these softwood forests is dense and the competition of herbaceous plants can inhibit the germination of *Salix alba* and *Populus nigra* seeds and the growth of seedlings. If floods deposit fresh silt on the grass, shading by old trees can prevent the strengthening of seedlings. Finally, the surface of the habitats of these softwood forests is gradually elevated through sedimentation year by year. If no forestry interventions took place in such forests, long-term monitoring could reveal the whole transformation process, starting with the collapse of old *Salix alba* or *Populus nigra* trees. The higher floodplain level does not provide an opportunity for the germination of *Salix alba* and *Populus nigra* seeds. In contrast, *Populus alba* seeds germinate under the altered conditions. The process is also influenced by the frequency and duration of inundation. It has been observed that on higher surfaces *Salix alba* and *Populus nigra* seeds do not germinate so easily as *Populus alba* seeds. Competition may also play a part in the formation of the canopy layer of white poplar forest. It is possible that on 1–1.5 m higher terrain, *Populus alba* suppresses *Salix alba* and *Populus nigra* specimens forces out or attenuate the latter on the higher terrain. Since the higher terrain is flooded on much fewer occasions, the further growth of *Populus alba* saplings is ensured. The continuously disappearing old *Salix alba* and *Populus nigra* trees are replaced by young *Populus alba* specimens and the willow and black poplar forests turn into white poplar forests. This hypothesis is supported by the sporadic occurrence of *Populus alba* shrubs in older willow and black poplar forests on elevated terrain and some plants in the herbaceous layer which are characteristic of white poplar forests (e.g. *Aegopodium podagraria*, *Arum maculatum*, *Paris quadrifolia*, *Stachys sylvatica*, etc.).

The succession of white poplar forests is directed towards oak-ash-elm forests (*Carici brizoidis-Ulmetum*—Kevey 1998a). In the Drava active floodplain, both associations are in contact locally. White poplar forests along the Drava include numerous protected species: *Carex strigosa*, *Carpesium abrotanoides*, *Equisetum hyemale*, *Fritillaria meleagris*, *Galanthus nivalis*, *Leucojum aestivum*, *Listera ovata*, *Lonicera caprifolium*, *Ophioglossum vulgatum*, *Petasites albus*, *Peucedanum verticillare*, *Scrophularia scopolii,* and *Tamus communis*.

18.4.4 Softwood Forests in Landscape History

Since ancient times, the low Drava floodplain has been occupied by various types of softwood forests (*Leucojo aestivi-Salicetum albae, Carduo crispi-Populetum nigrae, Senecioni sarracenici-Populetum albae*). The willows and poplars were used for fuel, tool making and building. Animals grazed on floodplains often incurred into softwood forests, inflicted gnawing and trampling damage in the understorey and caused degradation of various extents. In recent decades, numerous alien tree species were introduced to Hungary to grow in forest habitats. The species include *Acer negundo* and *Fraxinus pennsylvanica*, and in the twentieth century, *Populus* × *euramericana* varieties were planted in huge areas. In the meantime, other invasive plants also appeared in the Hungarian flora, like *Aster* × *salignus*, *Echinocystis lobata*, *Impatiens glandulifera*, *I. parviflora*, *Solidago gigantea*,

Amorpha fruticosa and *Vitis riparia*. These rapidly reproducing species spread out from gardens and have become pests.

River regulation and land drainage since the 1800s reduced wetlands along the Drava and divided the floodplain into the active floodplain between the river channel and the flood-control dykes and the protected floodplain beyond the dykes (see Chap. 14). The frequently inundated narrow active floodplain is vegetated by softwood forests, while in the drier habitats of the protected floodplain succession is directed from softwood forests towards hardwood forests.

Softwood gallery forests are also valuable habitats for a limited number of protected plant species, like *Leucojum aestivum* and *Senecio paludosus* in willow forests, *Equisetum hyemale*, *Myricaria germanica* (at Zákány), *Petasites albus* (at Drávasztára), *Peucedanum verticillare* (at Zákány, Barcs and Tótújfalu) and *Salix elaeagnos* (at Zákány). White poplar gallery forests (*Senecioni sarracenici-Populetum albae*) are home to species like *Equisetum hyemale*, *Leucojum aestivum* and *Peucedanum verticillare* (at Szentborbás). Sporadically and rarely *Carex strigosa*, *Carpesium abrotanoides*, *Cephalanthera longifolia*, *Dryopteris carthusiana*, *Fritillaria meleagris* (at Matty), *Galanthus nivalis*, *Listera ovata*, *Lonicera caprifolium*, *Neottia nidus-avis*, *Omphalodes scorpioides* (at Babócsa), *Ophioglossum vulgatum* (at Babócsa), *Platanthera bifolia*, *Polystichum aculeatum*, *Tamus communis* (at Matty and Zákány) also attest to the significance of the community for botany and conservation.

18.4.5 Threats to Softwood Forests

Among softwood forests, willow groves react most sensitively to water-level fluctuations, black poplar groves are less sensitive in this respect, while white poplar groves are least effected by water deficit (Kevey 1999b, 2002b, c). Invasions endanger the naturalness of softwood forests: among trees *Acer negundo* and *Fraxinus pennsylvanica*, whose seeds are spread by both wind and floods. *Acer negundo* often appears in masses in the shrub layer of older willow groves, threatening the survival of the community. With old willows disappearing, such communities easily turn into green maple forests alien to the landscape. The influence of *Morus alba* specimens is limited. *Populus* × *euramericana* agg. does not reproduce spontaneously, but forestry plantations represent a threat to the survival of softwood groves. Another problem is the hybridization of *Populus* × *euramericana* and *Populus nigra* which is a hazard to pure *Populus nigra* stands. *Amorpha fruticosa* is a dangerous invasive species, which spreads similarly to *Acer negundo* and *Fraxinus pennsylvanica*. Masses of *Vitis riparia* locally veil trees and hybridizes with the protected *Vitis sylvestris*. Among invasive herbs, *Aster* × *salignus*, *Echinocystis lobata*, *Impatiens glandulifera* and *Solidago gigantea* appear in masses. *Solidago canadensis* is somehwat less aggressive. *Impatiens parviflora* is a disturbance in white poplar forests. The sporadic occurrence of *Stenactis annua* does not present a problem.

Another danger is grazing on floodplains since domesticated animals intrude into softwood forests and cause gnawing and trampling damage. Plastic bottles can also be deposited here by floods.

18.4.6 Forestry and Nature Conservation

Softwood forests have been cut and replaced by quickly growing tree species over large areas. Their old stands are gradually felled, roots are removed, they are deep ploughed, and the forest is renewed with *Populus* × *auramericana* agg. or *Salix alba* cultivars with straight trunks. This kind of tree replacement has taken place along the Drava too. In the place of white poplar forests, *Fraxinus pennsylvanica* monocultures were established.

Since the extension of softwood groves in the floodplain has diminished considerably, a new approach to forestry and nature conservation is needed to preserve the rest of the stands or to reconstruct degraded forests.

There are several problems associated with the reforestation and preservation of willow forests. The old timber willow forests grow on grounds higher than young willow groves, elevated by the deposition during repeated floods. On the one hand, willow seeds can only germinate on fresh alluvial silt or sand, and in old forest, this is precluded by the competition with dense herbaceous vegetation. On the other hand, it is often observed that under the foliage, a shrub layer first of all consisting of *Acer negundo* takes shape. If such willow groves are not logged for some nature conservation consideration, old stands are turned into green maple forests. It is advisable to plant *Populus alba* trees in the stand since the *Salix alba* habitats have changed into *Populus alba* habitats and succession points towards white poplar communities. Therefore, willow groves should be reproduced, which is possible through the continuous development of willow shrubs. To this end, new silty surfaces have to be created in active floodplains and oxbow systems (Kevey 2008, 2016c), which is a restoration task.

There are sporadic surviving black poplar groves along the Őrtilos–Babócsa section of the Drava. In the frame of the forestry programme to save the Hungarian gene stock of *Populus nigra* (Bach 1998), no logging should be allowed there. After clearfelling and in sites of overaged black poplar stands, reforestation with *Populus alba* should be favoured. The reproduction of black poplar forests is ensured by the continuous emergence of willow shrubs (*Rumici crispi-Salicetum purpureae*—Kevey 1998a, 2008, 2016c). It is possible as such shrubs in almost natural conditions exist along the relatively unregulated Hungarian Drava section. Some of them lie on the Croatian bank so isolated that no forest management takes place there.

The logging of old white poplar stands is usually followed by the spreading of *Populus* × *euramericana* agg. cultivars or *Juglans nigra*. Since their ground layer is natural to some extent, the former plantations can be easily retransformed into softwood gallery forests through reforestation by *Populus alba*. In heavily filled-up habitats, the species of oak-ash-elm forests of the next community in succession can also be planted (Kevey 1998a, 2008). The spreading of invasive alien trees (*Acer negundo, Fraxinus pennsylvanica*) as well as of *Amorpha fruticosa* is a problem. These trees rapidly grow and ripen many seeds. Although there are numerous attempts to repress them, the appropriate method has not been invented yet. In areas which have been cleared of these species, their propagules reappear with the next flood and colonize areas where they have not been planted at all. To avoid this situation, seed-bearing trees have to be cut immediately. The liana *Vitis riparia* of North American origin appears in masses, but can be repressed by cutting at ground level. The intervention should be timed in autumn, when its yellow leaves make its distinction from *Vitis sylvestris* with red leaves. Among invasive herbs *Aster salignus*, *Echinocystis lobata*, *Impatiens glandulifera*, *I. parviflora* and *Solidago gigantea* present great problems. Their repression is not yet solved, but conservationists are searching solutions very actively (Kevey 2016c). Grazing animals should be excluded from white poplar groves bordering on floodplain pastures using wire fences.

18.5 Forests on Higher Floodplain Level

On higher terrains of the floodplain, forests are only inundated by extremely high floods. Their communities include lowland alder forests (*Paridi-Alnetum glutinosae*), oak-ash-elm groves (*Carici brizoidis-Ulmetum*) and lowland hornbeam-oak forests (*Veronico montanae-Carpinetum*) (Fig. 18.3).

18.5.1 *Lowland Alder Forests (*Paridi Quadrifoliae-Alnetum Glutinosae *Kevey in Borhidi and Kevey 1996)*

Lowland alder forests had long been regarded as a special variety (Soó 1940, 1943, 1958) or sub-association (Jurko 1958) of oak-ash-elm forests. Recent findings (Kevey 1993, 1995, 1997a, b, 1999c, 2008), however, show that they are an association, which occurs along both the Baranya (Ortmann-Ajkai 1998b) and Somogy sections of the Drava Plain (e.g. at Vajszló, Bürüs, Gyékényes, Somogyudvarhely, Bélavár, Babócsa, Komlósd, Barcs, Kisdobsza), mostly next to alder bogs (Fig. 18.2c) or oak-ash-elm forests. Because of relatively infrequent inundation, on the ground layer, mesophile elements are more common. Since their understorey is strongly influenced by the groundwater table, they grow on heavier alluvial forest soils with higher humidity. In the vicinity of alder bogs, soils even include some decomposing peat. Most of the stands are found on terrain protected from floods.

The upper canopy layer, developed to various degrees, usually has a cover of 60–80%, and a height of 18–25 m. Trunk diameters range from 35 to 60 cm. Most common trees are *Alnus glutinosa* and even more often *Fraxinus angustifolia*. A similar diversity characterizes the lower canopy (cover: 5–30%, height: 10–15 m), where among young *Alnus glutinosa Acer campestre* occurs in large numbers, and locally *Padus avium*, *Ulmus laevis* and *Ulmus minor* are also present. In the shrub layer (cover: 15–70%, height: 2–4 m), masses of *Cornus sanguinea, Corylus avellana, Crataegus monogyna, Padus avium, Sambucus nigra* and *Viburnum opulus* are found. In the lower shrub layer (regrowth), *Hedera helix* and *Rubus caesius* are locally abundant. The ground layer of 30–95% is composed of *Aegopodium podagraria, Allium ursinum, Cardamine amara, Carex remota, Chrysosplenium alternifolium, Corydalis cava, Leucojum vernum, Galeobdolon luteum, Mercurialis perennis, Oxalis acetosella, Ranunculus ficaria,* and others. Typically for mesophile deciduous forests, the spring aspect is well developed with *Adoxa moschatellina, Allium ursinum, Anemone nemorosa, A. ranunculoides, Corydalis cava, C. solida, Erythronium dens-canis, Gagea lutea, Galanthus nivalis, Isopyrum thalictroides, Lathraea squamaria, Leucojum vernum, Ranunculus ficaria, Scilla drunensis*.

In alder forests plants of mesophile deciduous forests (*Fagetalia*) and hardwood elements (*Alnion incanae*) are prevalent, but paludal (*Phragmition, Magnocarition*) and moor elements (*Molinio-Juncetea*) also appear. The most typical stands develop from alder bogs (*Carici elongatae-Alnetum*) (Figs. 18.3, 18.8, 18.9). When these habitats are supplied with less and less water, the aeration of soils intensifies, oxygen induces peat decomposition and the formation of dark muck, transitional towards alluvial forest soils. The transformation of soils involves alterations in the ground layer: herbs from neighbouring higher ground with oak-ash-elm forests colonize the upfilled alder bog and turn it into alder forest. As along abandoned channels water supply becomes less favourable for *Alnus glutinosa* trees, which are gradually replaced by *Fraxinus angustifolia*, *Ulmus laevis,* and *Quercus robur*, alder forests further develop towards oak-ash-elm groves (*Carici brizoidis-Ulmetum*). The process is evidenced by the common juxtaposition of the *Carici elongatae-Alnetum, Paridi quadrifoliae-Alnetum* and *Carici brizoidis-Ulmetum* associations (e.g. the Lankóci Forest near Gyékényes). In addition, there are ageing alder forests, where regeneration is mainly through *Fraxinus angustifolia*.

*18.5.2 Oak-Ash-Elm Forests (***Carici Brizoidis-Ulmetum** *Kevey 2008)*

Oak-ash-elm forests occur on the mostly flood-free higher terrain in the Drava floodplain. Before flood-control measures, they had been only inundated during extremely high flood waves (Figs. 18.2d, 18.3, 18.9 and 18.10). The moisture content of their humic alluvial soil (transitional to brown forest soils) is lower than

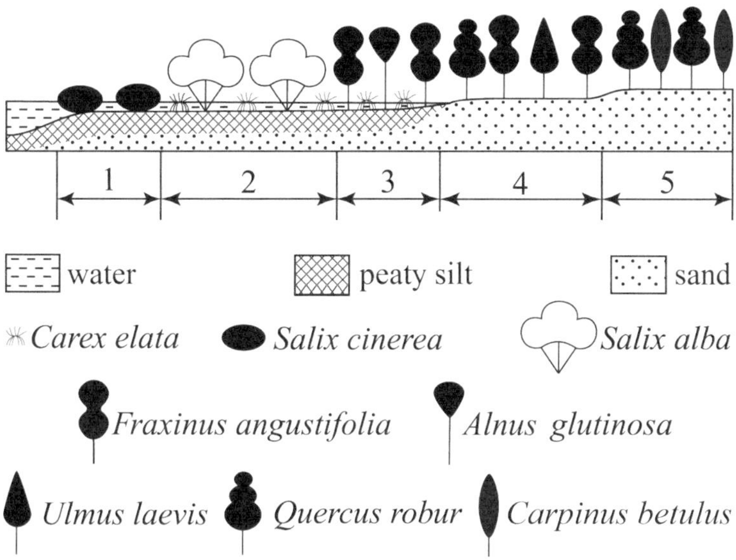

Fig. 18.8 Cross-section of vegetation from waterlogged habitats to the higher floodplain terrain I. (by B. Kevey). 1, *Calamagrostio canescentis-Salicetum cinereae*; 2, *Carici elatae-Salicetum albae*; 3. *Fraxino pannonicae-Alnetum glutinosae*; 4, *Carici brizoidis-Ulmetum*; 5, *Veronico montanae-Carpinetum*

in alder forests. The community is transitional between hygrophile and mesophile deciduous forests. Water budget is controlled by the water regime of the Drava and its tributaries, groundwater table and floodplain deposits (gravel, sand, silt or loess).

Canopy layer cover is 60–80%, tree height (depending on age) is 25–30 m and trunk diameter ranges from 40 to 60 cm. Predominant trees are *Quercus robur*, *Fraxinus angustifolia* and *Populus alba*, previously with masses of *Ulmus laevis* and *U. minor*, but the Dutch elm disease decimated them. The lower canopy layer is developed to various degrees with a cover of 15–50% and heights from 10 to 20 m. In addition to *Acer campestre*, *Ulmus laevis* and *U. minor*, *Padus avium* and *Tilia cordata* are occasionally major constituents. The shrub layer is of variable importance. Its cover amounts to 20–70% and shrub heights to 2–5 m. With generally spread species (e.g. *Cornus sanguinea*, *Corylus avellana*, *Crataegus monogyna*), the shrubs of hardwood forests (*Frangula alnus, Padus avium, Ribes rubrum, Viburnum opulus*) also occur. The ground layer reaches 70–100% cover and includes species of mesophile deciduous forests (*Asarum europaeum, Cardamine bulbifera, Galanthus nivalis, Galeobdolon luteum, Galium odoratum, Isopyrum thalictroides, Lathraea squamaria, Maianthemum bifolium, Mercurialis perennis, Oxalis acetosella, Pulmonaria officinalis, Ranunculus lanuginosus, Salvia glutinosa, Sanicula europaea, Veronica montana*, etc.). They are partly so-called demountainous-adventive elements with fluvial transport and partly relics from the Beech I stage with cool, humid and equable climate (2500 BC–800 BC)

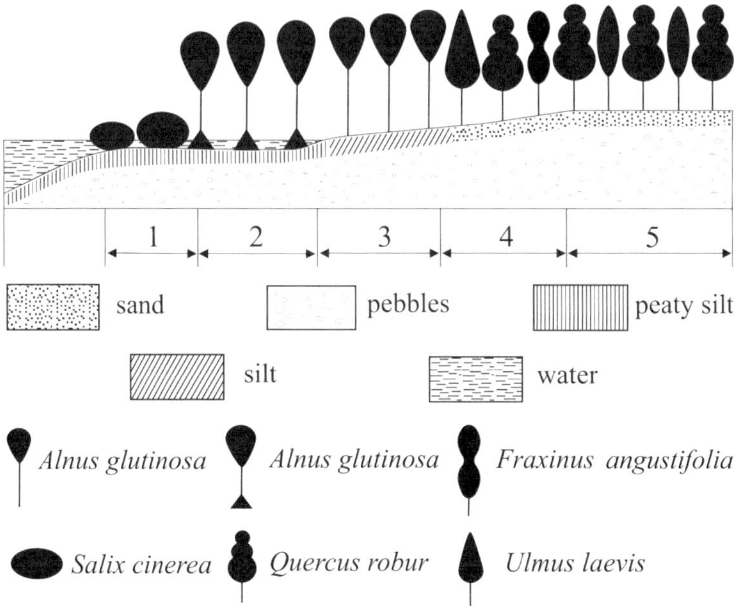

Fig. 18.9 Cross-section of vegetation from waterlogged habitats to the higher floodplain terrain II. (by B. Kevey). 1, *Calamagrostio canescentis-Salicetum cinereae*; 2, *Carici elongatae-Alnetum glutinosae*; 3, *Paridi quadrifoliae-Alnetum glutinosae*; 4, *Carici brizoidis-Ulmetum*; 5, *Veronico montanae-Carpinetum*

(Zólyomi 1936, 1952). In addition, characteristic species of hardwood forests are also present (*Carex brizoides, C. remota, Cerastium sylvaticum, Impatiens noli-tangere*, etc.). Bulbous plants present a typical early spring aspect: *Adoxa moschatellina, Allium ursinum, Anemone nemorosa, Anemone ranunculoides, Corydalis cava, Erythronium dens-canis, Fritillaria meleagris, Gagea lutea, Galanthus nivalis, Isopyrum thalictroides, Leucojum vernum, Ranunculus ficaria, Scilla drunensis*, etc. Mesophile elements include *Aegopodium podagraria, Allium ursinum, Corydalis cava, Galeobdolon luteum, Galium odoratum, Hedera helix, Ranunculus ficaria.*

Oak-ash-elm forests develop either from white poplar groves or alder groves over long periods (Fig. 18.3). Parallel to sedimentation of habitat ageing, *Populus alba* trees are replaced by *Quercus robur, Fraxinus angustifolia, Ulmus laevis* and *Ulmus minor*. In the meantime, the shrub and ground layers are also transformed, paludal (*Magnocaricion*) and softwood (*Salicion albae*) elements disappear and species of mesophile deciduous forests (*Fagetalia*) and hardwood species (*Alnion incanae*) appear. Oak-ash-elm forests turn into hornbeam-oak forests (*Veronico montanae-Carpinetum*) probably over centuries. The stages of succession can be reconstructed from the floodplain layers and the parallel occurrence of associations (Figs. 18.3 and 18.2e).

Fig. 18.10 Oak-ash-elm forest (*Carici brizoidis-Ulmetum*) near Somogyudvarhely (photo by B. Kevey)

Before river regulations, the oak-ash-elm groves had been inundated only in case of exceptionally high floods. Therefore, they grow on more mature alluvial forest soils. After the human interventions, the differences in the properties of such forests inside and beyond flood-control dykes became more marked. The stands in the active floodplain are approaching the white poplar forests, while those in the protected floodplain begin to assimilate to hornbeam-oak forests. Transitional types are common and it is difficult to distinguish them (Kevey 2007a, b, c, d).

In addition, submontane influence from west and southwest, the flora of the Drava Plain also shows some submediterranean and West Balkan character. The main submediterranean elements are: *Helleborus dumetorum, Knautia drymeia, Primula vulgaris, Ruscus aculeatus, Tamus communis, Tilia tomentosa* (Kevey 2008). These hardwood forests have been poorly studied (Kárpáti in Horvát 1972). Detailed surveys began two decades ago (Ortmann-Ajkai 1998a, b; Kevey 2007a, b; Kevey and Kovács 2011). On the Somogy section, most of the oak-ash-elm forests are located on the Croatian bank, but some stands are on Hungarian territory. On the Baranya section, most of these forests are closer to hornbeam-oak forests. The most typical oak-ash-elm forest is the Ataki Forest near Kisszentmárton (Ortmann-Ajkai 1998a) and other examples are also known from the Ormánság region (Ortmann-Ajkai 1998b; Kevey 2008).

18.5.3 *Hornbeam-Oak Forests* (Veronico Montanae-Carpinetum *Kevey 2008)*

Hornbeam-oak forests are related to the very similar oak-ash-elm groves by succession. They are exclusively found on flood-free terrain, 1–1.5 m higher than oak-ash-elm groves (Figs. 18.3, 18.11 and 18.2e). In the Drava Plain, gradually sloping from northwest to southeast, hornbeam-oak forests lie at 125 m elevation at

Fig. 18.11 Hornbeam-oak forest (*Veronico montanae-Carpinetum Allium ursinum facies*) near Bürüs (photo by B. Kevey)

Gyékényes and at 95 m at Beremend. The streams crossing the forests and the relatively high groundwater table ensure a cool and moist microclimate. The stands mostly occupy semi-humid brown forest soil.

The closed (75–85%) upper canopy layer of hornbeam-oak forests is up to 30–35 m high and predominantly consists of *Quercus robur* and *Carpinus betulus* locally with numerous *Acer campestre, Fraxinus angustifolia* and *Tilia cordata.* The lower canopy layer is 15–20 m high with 10–50% cover (mostly *Carpinus betulus*, but *Acer campestre*, *A. tataricum*, *Tilia cordata* and *Ulmus minor* are also important). Also the variably developed shrub layer (1–5 m high with 5–50% cover) is controlled by forestry. *Corylus avellana*, *Cornus sanguinea* and *Ligustrum vulgare* occur in masses with young specimens of *Acer campestre, Carpinus betulus* and *Tilia cordata.* The lower shrub layer with *Hedera helix* has a cover of 1–70%. Extreme cover percentages (10–100%) characterize the ground layer of *Aegopodium podagraria, Allium ursinum, Anemone nemorosa, A. ranunculoides, Carex brizoides, Corydalis cava, Ficaria verna, Galeobdolon luteum, Galium odoratum, Lamium maculatum, Mercurialis perennis, Oxalis acetosella.*

Similarly to oak-ash-elm forests, this community also includes a range of submotane elements of *Fagetalia* character: *Acer platanoides, A. pseudo-platanus, Adoxa moschatellina, Aegopodium podagraria, Allium ursinum, Anemone nemorosa, A. ranunculoides, Arum maculatum, Asarum europaeum, Athyrium filix-femina, Cardamine impatiens, Carex pilosa, C. sylvatica, Carpinus betulus, Cerastium sylvaticum, Cerasus avium, Circaea lutetiana, Corydalis cava, Daphne mezereum, Dentaria bulbifera, Dryopteris filix-mas, Epilobium montanum, Euphorbia amygdaloides, E. dulcis, Fagus sylvatica, Gagea lutea, Galanthus nivalis, Galeobdolon luteum, Galeopsis speciosa, Galium odoratum, Geranium phaeum, Glechoma hirsuta, Hedera helix, Hordelymus europaeus, Isopyrum thalictroides, Lathraea squamaria, Lathyrus vernus, Listera ovata, Luzula pilosa, Majanthemum bifolium, Mercurialis perennis, Milium effusum, Moehringia trinervia, Oxalis acetosella, Paris quadrifolia, Pulmonaria officinalis, Ranunculus lanuginosus, Rubus hirtus, Salvia glutinosa, Sanicula europaea, Scilla vindobonensis. Senecio nemorensis* ssp. *nemorensis, Stachys sylvatica, Tilia platyphyllos, Ulmus glabra, Veronica montana, Vinca minor, Viola sylvestris.*

Kinship with oak-ash-elm forests is indicated by some species of *Alnion incanae* character: *Carex brizoides, C. pendula, C. remota, C. strigosa, Frangula alnus, Fraxinus angustifolia, Impatiens noli-tangere, Padus avium, Populus* × *canescens, Ribes rubrum, Ulmus laevis, Viburnum opulus, Vitis sylvestris*, etc.

In the southernmost plain of Hungary, located next to the South Transdanubian *Praeillyricum* floral province, hornbeam-oak forests also show submediterranean character, indicated by species like *Carex strigosa, Helleborus dumetorum, H. odorus, Knautia drymeia, Lonicera caprifolium, Polystichum setiferum, Primula vulgaris, Ruscus aculeatus, R. hypoglossum, Tamus communis* and *Tilia tomentosa*, which are not present in other lowland hornbeam-oak associations (Horvát and Kevey 1983; Ortmann-Ajkai 1998b; Kevey 2007c, d; Kevey 2008; Kevey and Kovács 2010).

In Baranya county, hornbeam-oak forests are common in the Ormánság region. Between Dencsháza and Bürüs, they occur with *Fagus silvatica*, sometimes regarded an association of its own (*Carici strigosae-Fagetum*—Kevey 1998b, 2008).

18.5.4 Forests of the Higher Floodplain in Landscape History

Once floodplain forests covered much larger areas and human settlements occupied higher terrain of the floodplain. Traditional floodplain economy spread at the expense of forests over the centuries. As a consequence of expanding built-up areas, grazing and arable lands, the extension of floodplain forests became fragmented and isolated, they impoverished in species and floral exchange between forest patches became limited.

Initially, forests were only utilized to satisfy immediate demands (fuel, building material, tools, industrial timber). After the 1500s, however, the expanding villages and towns required more and more wood. Under the Ottoman occupation (1542–1668), deforestation continued in the Drava Plain, but even on the maps of the First Military Survey (1783) showed much more forests than we find today. Later, random tree felling was replaced by clearfelling and without reforestation, the regeneration of forests only took place through natural processes. Planned forest management was only introduced after the Compromise with Austria (1867) and reforestation became an important guideline. Coppicing or artificial reforestation were practised. Around 1900, the agricultural use of floodplain forests was widespread. Clear-cut areas were sown with grain crops (wheat, barley and rye) and then reforested (cf. Molnár 1996). Where similar interventions happened, the understorey became impoverished in species. Fortunately, the Drava Plain was less affected. In the decades, after World War II, profit-oriented forest management enterprises applied large-scale clearfelling and then new technologies were introduced in reforestation (use of chemicals, stump removal by dozer blade, deep ploughing and planting alien tree species—Fidlóczky 1995; Gyöngyössy 2003; Oroszi 2010).

With river regulation and land drainage, the groundwater table in floodplain forests sank and the species composition of forests on higher terrain adjusted to this change. Hygrophile elements (e.g. *Carex remota, Cerastium sylvaticum, Impatiens noli-tangere*) lost ground. Lowland alder groves are of lesser importance in the landscape. The largest are the Lankóci Forest of Gyékényes and the Vecsenye Forest of Somogyudvarhely. They are rich in protected plants (*Carex strigosa, Dryopteris carthusiana, D. dilatata, Epipactis helleborine, Erythronium dens-canis, Leucojum aestivum, L. vernum, Listera ovata, Omphalodes scorpioides, Platanthera bifolia, Primula vulgaris, Ruscus aculeatus, Scilla drunensis,*

S. vindobonensis, Scrophularia scopolii, Tamus communis, Vitis sylvestris, etc.). The diverse alder groves may develop into oak-ash-elm forests.

Oak-ash-elm forests are of decisive importance in the landscape of the Hungarian Drava Plain. They abound in protected species: *Carex strigosa, Carpesium abrotanoides, Dryopteris carthusiana, Epipactis helleborine, Erythronium dens-canis, Galanthus nivalis, Leucojum aestivum, L. vernum, Listera ovata, Neottia nidus-avis, Omphalodes scorpioides, Platanthera bifolia, Primula vulgaris, Ruscus aculeatus, Scilla vindobonensis, Scrophularia scopolii, Tamus communis, Vitis sylvestris*, etc. Their natural succession is towards hornbeam-oak forests.

On the higher terrain of the Drava, floodplain hornbeam-oak forests are the most widespread. Their protected species are: *Carex strigosa, Carpesium abrotanoides, Cephalanthera damasonium, Cephalanthera longifolia, Daphne mezereum, Digitalis ferruginea, Dryopteris carthusiana, Epipactis helleborine, Erythronium dens-canis, Galanthus nivalis, Helleborus dumetorum, H. odorus, Listera ovata, Lonicera caprifolium, Muscari botryoides, Neottia nidus-avis, Omphalodes scorpioides, Ornithogalum sphaerocarpum, Platanthera bifolia, Polystichum aculeatum, Polystichum setiferum, Primula vulgaris, Ruscus aculeatus, Ruscus hypoglossum, Scilla vindobonensis, Scrophularia scopolii, Tamus communis, Vitis sylvestris*, etc. Hornbeam-oak forests are regarded as the climax association of the floodplain succession.

18.5.5 Threats to Forests on Higher Terrain

In the past, forests on higher ground (*Paridi quadrifoliae-Alnetum, Carici brizoidis-Ulmetum, Veronico montanae-Carpinetum*) were primarily endangered by fragmentation and isolation due to deforestation. Damage was made to species richness and floral exchange between isolated patches. With lowered groundwater table (Kiss 2004), in alder groves and oak-ash-elm forests, hygrophile elements (e.g. *Carex remota, C. strigosa, Cerastium sylvaticum, Impatiens noli-tangere*) became sparse, some stands lost their character. A more severe threat would be the construction of the barrage at Djurdjevac or Novo Virje, Croatia. (Although experience from the Szigetköz floodplain of the Danube indicates no major deterioration in the conditions of oak-ash-elm forests there—Kevey 2002a.) In degraded stands, *Galium aparine* and *Urtica dioica* may occur in masses. Invasive elements are less common here than in softwood forests, but the spreading of *Impatiens parviflora, Solidago gigantea* and *Vitis riparia* presents problems.

In recent decades, forest management has also involved numerous degradation processes (Kevey and Buzássy 2003; Timár 2016), including stump removal by dozer blade, deep ploughing, planting non-native tree species, chemical weed control and harrowing for the protection of seedlings (Bartha 2012). Such interventions degrade or even exterminate the ground layer of native species. In forests, many timber loading facilities were established and timber transport considerably damaged the soil and the ground layer (Bartha 2001a).

Alien species formerly introduced to forests on higher floodplain terrain have fundamentally changed the landscape and ecological conditions. Species like *Acer negundo* and *Fraxinus pennsylvanica* are not planted anymore but *Juglans nigra*, *Populus × euramericana* and *Robinia pseudo-acacia* monocultures still exist and locally *Quercus rubra* is planted. Mention has to be made of the overpopulous game stock, which has increased manifold in recent decades. Large game (roe deer, red deer and wild boar) cause gnawing and trampling damage. At wallowing sites, the understorey is degraded, the consumption of acorn and feeding on saplings hinders natural reforestation (Baranyai-Nagy 2012; Szmorad 2014).

18.5.6 Forestry and Nature Conservation

A good part of forests along the Drava are incorporated into the Danube-Drava National Park. Unfortunately, the most valuable forests of the Ormánság do not belong to the Park. In the protected areas, forest management has to be improved to a great deal since nature conservation legislation is far from being fully implemented. A major deficiency of present forestry is that forest use ensuring shelterwood tree cover is not yet universally practised (Forest Commission 2011; Duryea and Dougherty 2012). The naturalness of forests is rather low as few dead tree trunks are preserved, and thus, few habitats are provided for cavity-dwelling birds, bats and numerous insects (Bartha 2001a; Varga 2009). It is a misconception by foresters that dead trees should be removed from the forest because they spread fungal diseases. In reality, fungi living on dead wood are saprophitic and do not cause infection. Moreover, fungi which spread infection only attack weakened or diseased trees. Finally, the spore of such fungi are present overall in the air and through the removal of dead wood, the amount of spores in the forest cannot be reduced.

Alder groves have never been decisive in the landscape, but oak-ash-elm forests have shrunk in area as well as softwood forests. Oak-ash-elm forests are almost exclusively managed applying clearfelling (Bartha 2001b) with stump removal, deep ploughing, harrowing between rows of planted trees and application of chemicals (Kevey and Buzássy 2003; Kevey 2016c). Even today reforestation takes place with *Populus × euramericana* agg., *Juglans nigra, Quercus rubra* or *Robinia pseudo-acacia* species. Forest management and conservation should allow the preservation and reconstruction of the remaining oak-ash-elm stands, avoiding any forestry intervention which could lead to further degradation. No plantation of non-native *Juglans nigra, Populus × euramericana* agg., *Quercus rubra* or *Robinia pseudo-acacia* should happen or their proportion should be reduced. In the cases of *Juglans nigra, Populus × euramericana* agg. and *Quercus rubra*, this is not difficult as a simple exchange of species is feasible. *Robinia pseudo-acacia*, however, sprouts extremely well. The same applies to *Acer negundo* and

Fraxinus pennsylvanica. Their spreading can only be hindered by the total extermination of trees with ripe seeds and the cutting of young specimen in the shrub layer using much manual labour, which is expensive. In some oak-ash-elm forests, the spreading of *Amorpha fruticosa* causes problems in the shrub layer, similar to *Acer negundo* and *Fraxinus pennsylvanica*. Locally, masses of *Vitis riparia* climb onto trees. To prevent this, the liana should be cut in autumn, when its yellow leaves are easily distinguished from the red leaves of *Vitis sylvestris*. Similar problems can occur through the presence of some invasive herbs (*Echinocystis lobata, Impatiens noli-tangere, Solidago gigantea*). Unfortunately, in spite of ongoing experiments, there is no way to reduce them as yet. In practice, in undisturbed hardwood forests, these herbaceous plants cannot become predominant. As a consequence, if the naturalness of the forest is restored, some of these species are forced out in the future.

For oak-ash-elm and hornbeam-oak forests, cutting age is of special significance. According to Papp (1975), *Quercus robur* ripens seeds at 60–80 years of age, while *Quercus robur* ssp. *slavonica* at 90–110 years. Large amounts of ripe acorns can only be expected after this age. Formerly, the cutting age of *Quercus robur* trees had been planned to 120–150 years, but in the 1970s, it was reduced to 80–100 years. The trunk diameter of oaks cut at such a young age is only 30–40 cm, and they are of low profile in the timber industry. *Quercus robur* can be kept until 120–140 years without any damage and even grows best at this age. Since an 80-year-old oak stand has hardly any chance to produce considerable amounts of acorn, natural reforestation is regarded subordinate in the practice of forest management (Papp 1975). Another factor is the intolerably high number of large game.

In the future, the protection of valuable forests on higher floodplain terrain should be ensured (first of all in the Ormánság). The protected forest stands, core areas and the network of forest reserves should be expanded. The target is to create oak ash-elm and hornbeam-oak stands of diverse composition and age with standing or fallen dead wood of various diameter (Szmorad and Király 2014; Frank and Szmorad 2014). Clearfelling should be stopped and cutting ensuring natural restoration preserving understorey, but selection cutting would be optimal. Sufficient light is needed for good germination. However, far-fetched tree removal may involve too much light and the spreading of weeds. Tree cutting should be executed in winter with snow cover and dragging of trees should be avoided (Bartha 2001a).

Since for most of the forests on higher floodplain water balance also presents a problem, an important task of nature conservation should be the restoration of optimal groundwater depth within the rehabilitation programme of the side-arm system.

18.6 Bog Forests

Bog forests are found in oxbows disconnected from the Drava and in other undrained depressions (Fig. 18.2f). Their water is rich in dissolved oxygen. Dead plant material fallen into the water does not decompose but began to rot. Anaerobic processes lead to peat accumulation in the water, which is discoloured by humin substances and saturated with various gases (CH_4, NH_4, H_2S, SO_2). The soil contains more nutrients than the alluvial soils of riverine groves. The seasonal oscillation of the groundwater table often generates gleyification and iron precipitation. In the understorey, true bog forest elements also occur.

*18.6.1 Grey Willow Bog (*Calamagrostio Canescenti-Salicetum Cinereae *Soó and Zólyomi in Soó 1955)*

The first tree association of the bog succession is represented by willow bogs (Figs. 18.3, 18.8, 18.9). They develop from high sedge-beds, reed-beds or tussocks on the banks of bog lakes through shrub growth. Their habitat is constituted of bog or gley soils developed on sedge and moss peat or peaty mud. There is virtually no canopy layer, although sporadically lower trees (such as *Alnus glutinosa, Betula pubescens*) may rise above the 2–5 m high shrub layer. The most abundant shrub is *Salix cinerea* along with *Frangula alnus, Viburnum opulus* or the protected *Salix aurita* and *Spiraea salicifolia* trees. These shrubs disperse vegetatively. Since the shrub layer is fairly closed, strong shading and frequent waterlogging make the ground layer sparse. The understorey comprises floating (*Lemnetea*) and rooting pondweeds (*Potametea*) and some paludal elements (*Phragmition, Magnocaricion*). In addition, bog species are of nature conservation importance: *Calamagrostis canescens, Carex appropinquata, C. elongata, C. lasiocarpa, C. nigra, C. pseudo-cyperus, Dryopteris carthusiana, D. dilatata, D. expansa, Galium uliginosum, Hottonia palustris, Hydrocotyle vulgaris, Ludwigia palustris, Menyanthes trifoliata, Peucedanum palustre, Ranunculus lingua, Salix aurita, Spiraea salicifolia, Thelypteris palustris, Urtica kioviensis*. The moss layer is usually sparse (Kevey 1995, 1997b; Borhidi et al. 2012). They are found at numerous sites along the Hungarian Drava (Ortmann-Ajkai 1998a; Kevey 2008).

*18.6.2 Alder Bog Forest (*Carici Elongatae-Alnetum *W. Koch 1926)*

The first forest association in the organogenic succession is alder bog forest (*Carici elongatae-Alnetum*—Figs. 18.3, 18.9 and 18.12). Their gleyic, acidic habitats poor in carbonate and bases is often waterlogged. Summer water recharge mostly

Fig. 18.12 Alder bog (*Carici elongatae-Alnetum*) at Lake Baláta (photo by B. Kevey)

depends on water levels in the Drava channel and oxbow lakes. Alder bogs consist of high trees, characteristically the so-called 'legged trees' with supporting roots and often multiple trunks. In the canopy layer, *Alnus glutinosa* mostly occurs exclusively. Their shrub layer is less developed and contains—in addition to *Frangula alnus* and *Salix cinerea*—the protected *Ribes nigrum* and *Spiraea salicifolia*. Alder bogs are often composed of a double habitat: one is the top of roots with saprophitic and bark-living species, mosses and horsetails (e.g. *Dryopteris carthusiana, D. dilatata, Thelypteris palustris*) and also flowering plants (e.g. *Carex elongata, Galium palustre, Stachys palustris*). Between the trees, waterlogged surfaces are speckled with tussocks of grasses and sedges (e.g. *Carex elata, C. paniculata, Deschampsia caespitosa*). Also, important are aquatic plants (*Lemnetea, Potametea*), while after the withdrawal of water annual ruderal floodplain plants (*Bidentetea*) shoot. Characteristics are the paludal (*Phragmition, Magnocaricion*), bog (*Caricion davallianae*) and bog forest species (*Alnetea, Alnion glutinosae*). The most important bog or bog forest plants are *Carex appropinquata, C. elata, C. elongata, C. paniculata, C. pseudocyperus, Deschampsia caespitosa, Dryopteris carthusiana, D. cristata, D. dilatata, D. expansa, Glyceria plicata, Hottonia palustris, Hydrocotyle vulgaris, Menyanthes trifoliata, Montia fontana, Osmunda regalis* (Royal Fern, the most valuable species with spectacular tussocks, only found at Darány), *Peucedanum palustre, Ranunculus lingua, Spiraea salicifolia, Thelypteris palustris, Urtica kioviensis* and others. Alder bog forests are quite rare in the Drava Plain today (Ortmann-Ajkai 1998a), only found in the Lankóci Forest of Gyékényes and the Ogreda of Felsőszentmárton.

*18.6.3 Hungarian Ash-Alder Bogs (***Fraxino Pannonicae-Alnetum Glutinosae** *Soó—Járai-Komlódi in Járai-Komlódi 1958)*

Hungarian ash-alder bogs are less rich in species that the Central European bog forests as with the exception of Hungarian ash (*Fraxinus angustifolia* ssp. *danubialis*), they do not have character species. They mainly occur in the subcontinental-submediterranean central part of the Great Hungarian Plain (Járai-Komlódi 1958) but are sporadically found along the Drava too (in the Ataki Forest of Kisszentmárton—Ortmann-Ajkai 1998a) and the Kobari Forest of Drávafok (Fig. 18.4). Their habitat lies on somewhat higher terrain than true bog forests and inundation is more restricted in time. Therefore, peat accumulation is more limited. In summer, they receive groundwater from the oxbow lakes. Their habitats mostly result from the sedimentation of abandoned riverbeds. In the canopy, *Alnus glutinosa* is mingled with *Fraxinus angustifolia* (Ortmann-Ajkai 1998a). The shrub layer is very similar to that of true alder bogs. In the ground layer, characteristic bog forest species (*Alnion glutinosae*) are less common, sporadically, however, *Carex elata, Dryopteris carthusiana, D. dilatata, Hottonia palustris, Peucedanum palustre, Thelypteris palustris,* and *Urtica kioviensis* do occur.

*18.6.4 White Willow Bog Forests (***Carici Elatae-Salicetum Albae** *Kevey 2008)*

This community is also typical of low-lying terrain with stagnant water, immediately line the reed-beds of bog lakes and often also bind grey willow bogs with Hungarian ash-alder bogs (Fig. 18.9). If their habitat is bordered on higher terrain where no waterlogging happens, they gradually turn into white willow groves. Their soil is moist throughout the year. Peat accumulation is more limited than in true bog forests. Therefore, the soil is less dark and less springy. As stagnant water is decisive in their formation, along with other bog forests, they belong to azonal associations.

The upper canopy of *Salix alba* is 15–20 m high and developed to variable degrees (40–75%). Other tree species (*Alnus glutinosa, Fraxinus angustifolia, Populus alba*) are mingled with white willow. There are many succumbed trees. The lower canopy layer is governed by the shading effect of high trees. Its cover is 10–40% and height 10–15 m, and primarily consists of young willows (*Salix alba, S. fragilis*). The shrub layer (cover: 1–10%, height: 1–2.5 m) is poorly developed or missing entirely. Regeneration is absent or minimal (1–3%). The well-developed ground layer with 70–95% cover and the following facies-forming herbs: *Carex elata, C. riparia, C. vesicaria, Glyceria maxima, Phalaris arundinacea, Phragmites australis.* The species *Galium palustre, Hottonia palustris, Iris pseudacorus,*

Mentha aquatica, Ranunculus repens, Rorippa amphibia, Symphytum officinale and *Urtica kioviensis* occur in larger patches.

Characteristic species of white willow bog forests are aquatic (*Lemnetea, Potametea*), paludal (*Phragmition, Magnocaricion*) and bog meadow elements (*Molinio-Juncetea*). Plants of softwood (*Salicion albae*) and hardwood (*Alnion incanae*) groves also appear in large numbers, but the character of the community is defined by bog forest species (*Alnetea glutinosae, Alnion glutinosae*). Common herbaceous plants are: *Alisma lanceolatum, A. plantago-aquatica, Alopecurus aequalis, Angelica sylvestris, Batrachium trichophyllum, Callitriche cophocarpa, Caltha palustris, Carex elata, C. pseudocyperus, C. remota, C. vesicaria, C. vulpina, Dryopteris carthusiana, Equisetum palustre, Euphorbia palustris, Frangula alnus, Galium palustre, Glyceria maxima, G. plicata, Hottonia palustris, Hydrocharis morsus-ranae, Hypericum tetrapterum, Iris pseudacorus, Lemna minor, L. trisulca, Mentha aquatica, Oenanthe aquatica, Phragmites australis, Ranunculus sceleratus, Rorippa amphibia, Rumex hydrolapathum, Salix alba, S. cinerea, Scrophularia umbrosa, Scutellaria galericulata, Senecio paludosus, Sium latifolium, Sparganium erectum, Spirodela polyrhiza, Stachys palustris, Thelypteris palustris, Typha latifolia, Urtica kioviensis, Utricularia vulgaris*, etc.

White willow bog forests are derived from grey willow bogs with moderate peat accumulation. Since *Salix alba* or *S. fragilis* saplings are present, during sedimentation they overgrow and shade *S. cinerea* stands, which thin out and are driven out from the habitat. Thus, the bog shrub transforms into willow bog forest (*Carici elatae-Salicetum albae*) and subsequently into Hungarian ash-alder bog. The old and succumbed willows are unable to regenerate and replaced by young *Alnus glutinosa* and *Fraxinus angustifolia* trees. Transitional types of these associations are observed at several sites. White willow bog forests may also develop from deeper-lying willow groves (*Leucojo aestivi-Salicetum albae*), which have lost their connectivity—naturally or through human intervention (e.g. dyke construction)—with the active floodplain of the Drava. During floods, groundwater rises to the surface and this waterlogging allows some peat accumulation. Then bog forest species may occur in the understorey and the white willow grove turns into white willow bog forests, which are thus transitional between softwood groves (*Salicion albae* Soó 1930) and bog forests (*Alnion glutinosae* Malcuit 1929).

Although such communities may have been widespread on seasonally waterlogged and undrained ground before river regulations, in the Drava Plain, they have only been found in the Ataki Forest of Kisszentmárton and at Lake Majláthpuszta. They have not been surveyed botanically in detail. They are not managed by forestry, and consequently, show wilderness locally. Their valuable vegetation includes protected species like *Dryopteris carthusiana, Hottonia palustris, Leucojum aestivum, Senecio paludosus, Thelypteris palustris* and *Urtica kioviensis*.

18.6.5 Bog Forests in Landscape History

In lowlands, human settlement was generally bound to the vicinity of rivers or other water surfaces. Over the centuries, land utilization has fundamentally transformed the neighbouring woodlands. A large part of riverine forests has been cut. Water management works in the 1700s, as well as river regulations in the 1800s, led to dropping groundwater table and the desiccation of vast bogs (e.g. Hanság, Ecsed Bog). As a consequence, the area of bog forests has also shrunk, the remaining stands began to be fragmented, isolated and deteriorated. Floral exchange between isolated patches became limited or even stopped. With lower groundwater table, alder and ash bogs became dewatered, the understory was transformed and bog plants were replaced by invasive elements (e.g. *Acer negundo, Fraxinus pennsylvanica, Impatiens glandulifera, Solidago gigantea*). Timber from the alder and ash bogs was used as fuel, for building, making tools or industrial products, while that from willow bogs was burned or used in households. In the remaining bog forests clearfelling is still practised with coppicing (Baranyai-Nagy 2012).

Due to soil properties and the cool microclimate, in the understorey of bog forests and bog shrubs, relict species from glacial and postglacial periods survived: *Hydrocotyle vulgaris, Ludwigia palustris, Menyanthes trifoliata, Montia fontana, Osmunda regalis, Ranunculus lingua, Salix aurita, Spiraea salicifolia, Urtica kioviensis* stb. In addition, other protected species also occur in these habitats: *Dryopteris carthusiana, Dryopteris dilatata, Dryopteris expansa, Hottonia palustris, Leucojum aestivum, Osmunda regalis, Senecio paludosus, Thelypteris palustris*, etc. Bog forests are valuable habitats for lichens, mosses, horsetails and animals. Standing and fallen deadwood is particularly appreciated by hollow-dwelling birds and insects. A good part of bog forests are integrated into the Natura 2000 network or belong to the Danube-Drava National Park.

18.6.6 Threats to Bog Forests

In the past fragmentation and isolation presented the greatest danger to bog forests and shrubs. Deforestation has virtually stopped and now water management, i.e. land drainage and sinking groundwater table are the main threats. In lowlands, drainage has reduced the groundwater table by several metres compared to the situation some 100 years ago (Kiss 2004). As a consequence, a considerable portion of bog forests have lost their valuable species and weeds (*Galium aparine, Rubus caesius* and *Urtica dioica*) have begun to spread in the understorey. They are invaded by *Acer negundo, Echinocystis lobata, Fallopia japonica, Fraxinus pennsylvanica, Helianthus tuberosus, Impatiens glandulifera, Parthenocissus inserta* and *Solidago gigantea*. Clearfelling changes the microclimate (light and moisture conditions). The pesticides applied to neighbouring agricultural areas and washed in into the soil of the bog forests of small extension are also damaging to the undergrowth. Finally, overpopulous large game (roe deer, red deer, wild boar) also contribute to degradation (grazing, trampling, wallowing) (Baranyai-Nagy 2012).

18.6.7 Forestry and Nature Conservation

The bog forests along the Drava belong to the Danube-Drava National Park. In spite of this high level of protection, unfortunately, clearfelling still damages the undergrowth. Since these forests are usually waterlogged, artificial reforestation through planting saplings is not feasible. Therefore, shootings in clear-cut forests are mostly preserved. It would be advisable to avoid commercial forestry in alder bogs. With their small extension, they have minor significance for timber production. After a certain period, the bogs abandoned by forestry could develop into the wilderness where standing and fallen dead trees provide diverse habitats for wildlife. This kind of bog forest management is an important task of nature conservation in Hungary (Baranyai-Nagy 2012).

18.7 Conclusions

The vegetation types outlined above in the botanist's view reflect the distribution of topographical, pedological and microclimatic properties. Their interaction results in very characteristic landscape patterns in the floodplain, best visible in some larger protected forests of the Drava Plain. As restricted border zones before 1990, these forests had not been intensively managed, and thus, preserved some spatial features predating human settlement. In the Lankóci Forest of Gyékényes, for instance, a range of associations from waterlogged habitats with alder bogs and alder forests to oak-ash-elm and hornbeam-oak forests can be studied in an area of 677 ha. Similarly to plant communities, animal life is of high species diversity closely associated with water. It includes 10 amphibian species, 7 reptiles (including lizards like *Lacerta viridis, Anguis fragilis;* snakes: *Coronella austriaca* and *Elaphe longissima*) and 69 birds (primarily woodpeckers, pipits, warblers). Riparian willow forests are valuable habitats for Night Heron (*Nycticorax nycticorax*) and Little Egret (*Egretta garzetta*) and poplar forests for Black Stork (*Ciconia nigra*).

In few instances, landscape ecological studies have been directed at describing the pattern and dynamics of riverine environments. In the Lankóci Forest, the investigations of small mammals (Horváth et al. 2004) revealed a rich network of ecotones between mosaics of forest types, high γ-diversity (landscape-level diversity), and consequently, high β-diversity (degree of community differentiation), particularly in the hornbeam-oak forests with drier microclimate, where the species *Clethrionomys glareolus* was found dominant. Large mammals also use the forests along the Drava as ecological corridors—freely crossing international borders (e.g. otters—Lanszki 2005).

There are new perspectives in ecologically oriented forestry to enhance biodiversity at all levels. One problem is presented by plantations of poplar cultivars which have replaced native softwood and hardwood forests in many sections of the floodplain. Forest management should aim at the reforestation of one-time hardwood habitats with oak species commercially and ecologically equally highly valuable. A precondition to successful reforestation is the regulation of large game stocks.

Woodlands promote adaptation to the impacts of climate change (Forestry Commission 2011) through regulating water balance. For the human population, they create environments with shelter and shade, cool and moist air, which are attractive for tourists arriving from urban areas.

References

Bach I (1998) A fekete nyár (*Populus nigra* L.) génmegőrzése, genetikai azonosításának elméleti alapjai és gyakorlati módszere (Theory and practice of gene preservation and genetic identification of black poplar). Erdészeti Lapok 133(9):287 (in Hungarian)

Baranyai-Nagy A (2012) A nyugat-dunántúli patakmenti égerligetek tájtörténeti, növényföldrajzi és társulástani vizsgálata (Landscape history, *phytogeography* and association study of alder groves along streams in Western Transdanubia). Manuscript PhD Thesis. Faculty of Forest Engineering, University of West Hungary, Sopron. 162 pp (in Hungarian)

Bartha D (2001a) A természetszerű erdők kezelése (Management of seminatural forests). TermészetBÚVÁR Alapítvány Kiadó, Budapest. 263–275 (A KöM Természetvédelmi hivatalának tanulmánykötetei 7) (in Hungarian)

Bartha D (2001b) Veszélyeztetett erdőtársulások Magyarországon (Endangered forest communities in Hungary). WWF Hungary, Budapest. 35 pp (WWF füzetek 18) (in Hungarian)

Bartha D (2012) A magyarországi erdők fenntartása (Maintenence of foredsts in Hungary). In: Kozák L (ed) Természetvédelmi élőhelykezelés (Habitat management for nature conservation). Mezőgazda Kiadó, Budapest, pp 195–198 (in Hungarian)

Borhidi A, Kevey B (1996) An annotated checklist of the Hungarian plant communities II. The forest communities. In: Borhidi A (ed) Critical revision of the Hungarian plant communities. Janus Pannonius University, Pécs, pp 95–138

Borhidi A, Kevey B, Lendvai G (2012) Plant communities of Hungary. Akadémiai Kiadó, Budapest. 544 pp

Drava River Vision (2008) International Symposium "Drava River Vision" 23–25 September 2008, Maribor, Slovenia. Final report. http://ec.europa.eu/environment/life/project/Projects/index.cfm

Duryea ML, Dougherty PM (eds) (2012) Forest regeneration manual. Springer Science & Business Media, Dordrecht, p 433

Fidlóczky J (1995) Az erdőgazdálkodás helyzete és annak természetvédelmi vonatkozásai (Forest management and its nature conservation aspects). WWF Hungary, Budapest. 18 pp (WWF füzetek 7) (in Hungarian)

Forestry Commission (2011) The UK Forestry Standard: The governments' approach to sustainable forest management Forestry Commission, Edinburgh. 105 pp

Frank T, Szmorad F (2014) Az optimális és elvárható élőhelyi állapot jellemzése (Optimal and expectable habitat conditions). In: Frank T, Szmorad F (eds), Védett erdők természetességi állapotának fenntartása és fejlesztése (Maintenance and development of naturalness in protected forests). Duna–Ipoly Nemzeti Park Igazgatóság, Budapest. 25–30 (Rosalia Kézikönyvek 2) (in Hungarian)

Gyöngyössy P (2003) Település és táj kapcsolata az Őrségben (Settlement and landscape in the Őrség region). Kerekerdő Alapítvány, Szombathely. 81 pp (in Hungarian)

Horvát AO (1972) Die Vegetation des Mecsekgebirges und seiner Umgebung. Akadémiai Kiadó, Budapest, p 376

Horvát AO, Kevey B (1983) Hornbeam-oak-forests in Ormánság. Macedonian Acad Sci Arts, Contrib 4(1–2):203–210

Horváth Gy, Sárkány H, Molnár D (2004) Kisemlősök közösségi szintű monitorozása két erdei élőhelyen (Bükkháti-erdő – Baranya megye, Lankóci-erdő – Somogy megye) (Community-level monitoring of small mammal in two forest habitats (Bükkhát Forest, Baranya County; Lankóci Forest, Somogy County). Somogyi Múzeumok Közleményei 16:421–430 (in Hungarian with English summary)

Járai-Komlódi M (1958) Die Pflanzengesellschaften in dem Turjángebiet von Ócsa-Dabas. Acta Botanica Academiae Scientiarum Hungaricae 4:63–92

Jurko A (1958) Podne ekologické pomery a lesné spolocenstvá Podunajskej níziny (Soil ecological conditions of forest communities in the Danubian Plain). Slovenská Akadémia Vied, Bratislava. 225 pp (in Slovakian)

Kárpáti I (1957) A hazai Duna-ártér erdei (Forests in the Danube floodplain of Hungary). Manuscript CSc Thesis, Budapest (in Hungarian)

Kárpáti I (1982) Die Vegetation der Auen-Ökosysteme in Ungarn. Veröffentlichung der Internationalen Arbeitsgemeinschaft für Clusius-Forschung, Güssing 4:1–24

Kárpáti I (1985) Az ártéri szintek geomorfológiai- és vegetáció-szukcessziójának kapcsolata (Geomorphological nad vegetation succession of floodplain surfaces). In: Fekete G (ed) A cönológiai szukcesszió kérdései (Issues in cenological succession). Biológiai Tanulmányok 12:73–81 (in Hungarian)

Kárpáti I, Kárpáti V (1969) Die zönologischen Verhältnisse der Donauauenwälder Ungarns. Verhandlungen der Zoologisch-Botanischen Gesellschaft in Wien 108–109(1968–1969): 165–179

Kevey B (1993) A Szigetköz ligeterdeinek összehasonlító-cönológiai vizsgálata (Comparative coenological study of grove forests in the Szigetköz). Manuscript CSc Thesis. Department of Botany, Janus Pannonius University, Pécs (in Hungarian)

Kevey B (1995) Síkvidéki ligeterdők (Lowland grove forests). Tilia 1:62–70 (in Hungarian)

Kevey B (1997a) Égerligetek (Alder groves). In: Fekete G, Molnár Zs, Horváth F (eds) Nemzeti biodiverzitás-monitorozó rendszer II. A magyarországi élőhelyek leírása, határozója és a Nemzeti Élőhely-osztályozási Rendszer (National biodiversity monitoring system II: Description and identification of habitats in Hungary and the National Habitat Classification System). Hungarian Natural History Museum, Budapest. 125–127 (in Hungarian)

Kevey B (1997b) Fűz- és nyírlápok (Willow and birch bogs). In: Fekete G, Molnár Zs, Horváth F (eds) Nemzeti biodiverzitás-monitorozó rendszer II. A magyarországi élőhelyek leírása, határozója és a Nemzeti Élőhely-osztályozási Rendszer (National biodiversity monitoring system II: Description and identification of habitats in Hungary and the National Habitat Classification System). Hungarian Natural History Museum, Budapest. 118–119 (in Hungarian)

Kevey B (1998a) A Szigetköz erdeinek szukcessziós viszonyai (Succession in the Szigetköz forests). Kitaibelia 3:47–63 (in Hungarian)

Kevey B (1998b) A szentegáti bükkállomány társulási viszonyai (Coenological features of the beech wood at Szentegát, County Baranya, South Hungary). Janus Pannonius Múzeum Évkönyve 41–42(1996–1997):13–26 (in Hungarian with English summary)

Kevey B (1999a) A Szigetköz erdei I. Ligeterdők (Forests of the Szigetköz I: Grove forests). Moson Megyei Műhely 2(1):59–82 (in Hungarian)

Kevey B (1999b) A Duna szlovákiai elterelésének hatása a Szigetköz növényvilágára (Impact of the new Danube channel in Slovakia on the flora of the Szigetköz). Moson Megyei Műhely 2(2):75–95 (in Hungarian)

Kevey B (1999c) Dombvidéki égerligetek (Alder groves in hills): *Aegopodio-Alnetum* Kárpáti V, Kárpáti I, Jurko 1961), *Carici pendulae-Alnetum* Borhidi – Kevey 1996), *Carici brizoidis-Alnetum* Horv. I 1938 em. Oberd. 1953), *Carici remotae-Fraxinetum* Koch ex Faber 1936, *Paridi quadrifoliae-Alnetum* Kevey in Borhidi – Kevey 1996. In: Borhidi A, Sánta A (eds), Vörös könyv Magyarország növénytársulásairól 2 (Red book on the plant associations of Hungary 2). TermészetBÚVÁR Alapítvány Kiadó, Budapest. 143–150 (in Hungarian)

Kevey B (2002a) A Duna szlovákiai elterelésének hatása a Felső-Szigetköz tölgy-kőris-szil ligeteire (*Pimpinello majoris-Ulmetum* Kevey in Borhidi, Kevey 1996) (Impact of the new Danube channel in Slovakia on the oak-ash-elm groves of the Upper Szigetköz). Kanitzia 9:227–249 (in Hungarian)

Kevey B (2002b) A Duna szlovákiai elterelésének hatása a Szigetköz ligeterdeire (Impact of the new Danube channel in Slovakia on the grove forests of the Szigetköz). In: Aktuális flóra- és vegetációkutatások a Kárpát-medencében V (Abstracts of the Conference "Actual flora and vegetation studies in the Carpathian Basin"). 8–10 March 2002, Pécs. p 63. (in Hungarian)

Kevey B (2002c) A Duna szlovákiai elterelésének hatása a Felső-Szigetköz fekete nyárligeteire (*Carduo crispi-Populetum nigrae* Kevey in Borhidi, Kevey 1996) (Impact of the new Danube channel in Slovakia on the black poplar groves of the Upper Szigetköz). In: Stirling J (ed) Hittel és alázattal. Köszöntő kötet Horvát Adolf Olivér ocist 95. születésnapjára (With faith and humbleness: Tribute to Adolf Olivér Horváth on his 95th birthday). Szent István Akadémia, Magyar Tudományos Akadémia Pécsi Területi Bizottsága, Pécsi Tudományegyetem Növénytani Tanszék és Botanikus Kert, Pécs. 181–200 (in Hungarian)

Kevey B (2002d) A Duna szlovákiai elterelésének hatása a Felső-Szigetköz csigolya bokorfüzeseire (*Rumici crispi-Salicetum purpureae* Kevey in Borhidi, Kevey 1996) (Impact of the new Danube channel in Slovakia on the willow shrubs of the Upper Szigetköz). In: Salamon-Albert É (ed) Magyar botanikai kutatások az ezredfordulón. Tanulmányok Borhidi Attila 70. születésnapja tiszteletére (Botanical studies in Hungary at the turn of the millenium: Tribute to Attila Borhidi on his 70th birthday). Department of Botany, University of Pécs, Pécs. 415–429 (in Hungarian)

Kevey B (2003) A Duna szlovákiai elterelésének hatása a Felső-Szigetköz mandulalevelű bokorfüzeseire (*Polygono hydopiperi-Salicetum triandrae* Kevey in Borhidi, Kevey 1996) (Impact of the new Danube channel in Slovakia on the willow shrubs of the Upper Szigetköz). Acta Óváriensis 45:137–154 (in Hungarian)

Kevey B (2007a) A Somogyi-Dráva-ártér tölgy-kőris-szil ligetei (*Fraxino pannonicae-Ulmetum* Soó in Aszód 1935 corr. Soó 1963) (Oak-ash-elm groves of the Somogy Drava Plain). Somogyi Múzeumok Közleményei 17(2006):103–122 (in Hungarian)

Kevey B (2007b) A baranyai Dráva-sík tölgy-kőris-szil ligetei (*Fraxino pannonicae-Ulmetum* Soó in Aszód 1935 corr. Soó 1963) (Oak-ash-elm groves of the Baranya Drava Plain). Natura Somogyiensis 10:11–39 (in Hungarian)

Kevey B (2007c) A Somogyi-Dráva-ártér gyertyános-tölgyesei (*Circaeo-Carpinetum* Borhidi 2003 em. Kevey hoc loco) (Hornbeam-oak floodplain forests of the Somogy Drava Plain). Somogyi Múzeumok Közleményei 17(2006):83–102 (in Hungarian)

Kevey B (2007d) A baranyai Dráva-sík gyertyános-tölgyesei (*Circaeo-Carpinetum* Borhidi 2003 em. Kevey 2006b) (Hornbeam-oak floodplain forests of the Somogy Drava Plain). Natura Somogyiensis 10:41–71 (in Hungarian)

Kevey B (2008) Magyarország erdőtársulásai (Forest associations of Hungary). Tilia 14:1–488 + CD database (230 tables + 244 figures) (in Hungarian)

Kevey B (2014) A hazai Mura-ártér fehérnyár-ligetei (*Senecioni sarracenici-Populetum albae* Kevey in Borhidi, Kevey 1996) (White poplar groves of the Mura floodplain in Hungary). Kaposvári Rippl-Rónai Múzeum Közleményei 3:29–56 (in Hungarian)

Kevey B (2016a) Puha-és keményfás ligeterdők kapcsolata a Szigetközben (Relationship between softwood and hardwood groves in the Szigetköz). Botanikai Közlemények 103(1):45–115 (in Hungarian)

Kevey B (2016b) A fekete-és fehérnyáras ligeterdők kapcsolata a Szigetközben (Relationship between black and white poplar forests in the Szigetköz). Botanikai Közlemények 103(2):195–212 (in Hungarian)

Kevey B (2016c) Ártéri erdők és cserjések (Riparian forests). In: Korda M (ed) Az erdőgazdálkodás hatása az erdők biológiai sokféleségére (Impact of forestry on the biodiversity of forests). Collection of papers. Danube-Ipoly National Park Directorate, Budapest. 459–484 (in Hungarian)

Kevey B, Buzássy B (2003) A Körös menti keményfás erdők természetvédelmi problémái (Nature conservation problems of the hardwood forests along the Körös rivers). Folia Comloensis 12:93–102 (in Hungarian)

Kevey B, Kovács JA (2010) A Mura-vidék gyertyános-tölgyesei (*Veronico montanae-Carpinetum* Kevey 2008) (Hornbeam-oak forests in the Mura region). Kanitzia 17:195–221 (in Hungarian)

Kevey B, Kovács JA (2011) A Mura-vidék tölgy-kőris-szil ligetei (*Carici brizoidis-Ulmetum* Kevey 2008) (Oak-ash-elm forests of the Mura region). Kanitzia 18:195–238 (in Hungarian)

Kevey B, Tóth V (2006) A Baranyai-Dráva-sík fehér nyárligetei (*Senecioni sarracenici-Populetum albae* Kevey in Borhidi, Kevey 1996) (White poplar groves in the Baranya Drava Plain). Natura Somogyiensis 9:47–62 (in Hungarian)

Kiss J (2004) Javaslatok a természetszerű erdőtelepítések fellendítésére (Proposals to promote the plantation of seminatural forests). In: Exner T, Máthé L (eds) A természetközeli erdőtelepítés lehetőségei Magyarországon (Opportunities for planting seminatural forests in Hungary). WWF Hungary, Budapest. 18–25 (WWF füzetek 23) (in Hungarian)

Klujber L, Tihanyi J, Vöröss LZs (1963) Adatok drávamenti holtágak cönológiai és florisztikai ismeretéhez (Data on the coenological and floristic knowledge of oxbows along the Drava). Pécsi Tanárképző Főiskola Tudományos Közleményei 1963:271–303 (in Hungarian)

Kovács M, Kárpáti I (1973) Untersuchung über die Zonations- und Produktionsverhältnisse im Überschwemmungsgebiet der Drau I. Verlandung der toten Arme und die Zonationen des Bodens und der Vegetation im Inundationsgebiet der Drau. Acta Botanica Academiae Scientiarum Hungaricae 18:323–353

Kovács M, Kárpáti I (1974) A Mura- és a Dráva-ártér vegetációja (Vegetation of the Mura and Drava floodplains). Földrajzi Értesítő 22:21–32 (in Hungarian with English summary)

Lanszki J (2005) Otter monitoring between 2000 and 2004 in the Drava Region (Hungary). Natura Somogyiensis 7:169–178

Molnár Zs (1996) Ártéri vegetáció Tiszadob és Kesznyéten környékén II. A keményfaliget-erdők (*Fraxino pannonicae-Ulmetum*) története és mai állapota (Floodplain vegetation near Tiszadob and Kesznyéten II: History and present condition of hardwood grove forests). Botanikai Közlemények 83(1–2):51–69 (in Hungarian)

Oroszi S (2010) Erdőgazdálkodás 1935–1990 (Forest management, 1935–1991). Erdészettörténeti Közlemények 82:127–135 (in Hungarian)

Ortmann-Ajkai A (1998a) Vegetation mapping as a base of botanical GIS applications I. Vegetation map of the Atak forest (Southwest Hungary). Acta Botanica Academiae Scientiarum Hungaricae 41:171–192

Ortmann-Ajkai A (1998b) Vegetation mapping as a base of botanical GIS applications II. Vegetation map of the Vajszló forest (Southwest Hungary). Acta Botanica Academiae Scientiarum Hungaricae 41:193–227

Papp T (1975) A természetes felújítás lehetőségei és módszerei a Drávamenti tölgyesekben (Oportunities and methods of natural reforestation in the oak forests along the Drava). Manuscript university doctorate thesis. University of Forestry and Timber Industry, Sopron (in Hungarian)

Pécsi M (1959) A magyarországi Duna-völgy kialakulása és felszínalaktana (Development and morphology of the Danube valley in Hungary). Akadémiai Kiadó, Budapest. 342 p (Földrajzi Monográfiák III) (in Hungarian)

Soó R (1940) Vergangenheit und Gegenwart der pannonischen Flora und Vegetation. Nova Acta Leopoldina, Halle. N. F. 9(56):1–49

Soó R (1943) A nyírségi erdők a növényszövetkezetek rendszerében (The Nyírség forests in the system of plant communities). Acta Geobotanica Hungarica 5:315–352 (in Hungarian)

Soó R (1958) Die Wälder des Alföld. Acta Botanica Academiae Scientiarum Hungaricae 4:351–381

Szmorad F (2014) Enyves éger (*Alnus glutinosa*) és magas kőris (*Fraxinus excelsior*) alkotta ligeterdők (*Alno-Padion, Alnion incanae, Salicion albae*) (Alder and ash forest groves). In: Haraszthy L (ed) Natura 2000 fajok és élőhelyek Magyarországon (Natura 2000 species and habitats in Hungary). Pro Vértes Közalapítvány, Csákvár, pp 881–887 (in Hungarian)

Szmorad F, Király G (2014) Az erdő természetvédelmi szempontú kezelése (Forest management for nature conservation). In: Haraszthy L (ed) Natura 2000 fajok és élőhelyek Magyarországon (Natura 2000 species and habitats in Hungary). Pro Vértes Közalapítvány, Csákvár, pp 758–760 (in Hungarian)

Timár G (2016) A jelenlegi erdőgazdálkodási módok áttekintése (Overview of present silvicultural methods). In: Korda M (ed) Az erdőgazdálkodás hatása az erdők biológiai sokféleségére (Impact of forestry on the biodiversity of forests). Collection of papers. Danube-Ipoly National Park Directorate, Budapest. 11–30 (in Hungarian)

Tóth I (1958) Az Alsó-Dunaártér erdőgazdálkodása, a termőhely- és az erdőtípusok összefüggése (Forestry in the Lower Danube floodplain: relationships between habitat and forest type). Erdészeti Kutatások 1958(1–2):77–160 (in Hungarian)

Varga G (2009) A folyamatos erdőborítás fenntartása melletti erdőgazdálkodás alapjai (Foundations of shelterwood management). Ministry of Agriculture and Rural Development, Budapest. 185 pp (in Hungarian)

Vöröss LZs (1964) Újabb adatok a szaporcai holtágak cönológiai és florisztikai ismeretéhez (New data on the coenology and floristics of the Szaporca oxbows). Szegedi Tanárképző Főiskola Tudományos Közleményei 1964:75–95 (in Hungarian)

Zólyomi B (1936) Tízezer év története virágporszemekben (The history of 10,000 years in pollen grains). Természettudományi Közlöny 68:504–516 (in Hungarian)

Zólyomi B (1952) Magyarország növénytakarójának fejlődéstörténete az utolsó jégkorszaktól (The evolution of vegetation in Hungary since the last glaciation). MTA Biológiai Osztályának Közleményei 1:491–530 (in Hungarian)

Chapter 19
Nature Conservation

Dénes Lóczy and Rok Ciglič

Abstract In the Drava catchment, nature conservation is practiced at various levels of institutionalization. In Austria, the uppermost valleys of the tributaries of the Upper Drava belong to the High Tauern National Park. In Carinthia further tributaries issue in the recently established Nockberge National Park and the Upper Drava European Reserve takes care of the preservation of natural values on the Drava floodplain. In Slovenia three Nature Parks (Lake Maribor, Drava and Šturnovec), two Nature Reserves (Zlatoličje and Lake Ormož), and in Hungary the Danube-Drava National Park ensure protection for the abiotic and biotic components of the riverine environment. One of the most valuable protected areas is the Kopački rit Nature Park in the confluence area. On the same Drava reaches river and floodplain restoration projects have taken place recently or are planned in the future. The new Transboundary Biosphere Reserve Mura-Drava-Danube (TBR MDD) unites all protected areas from Carinthia down to the Croatian-Serbian border and provides a chance for the better coordination of rehabilitation efforts.

Keywords Protected areas · Natural values · Wetland habitats
Biodiversity · Geoheritage · River regulation · Restoration programmes

D. Lóczy (✉)
Department of Physical and Environmental Geography, Institute of Geography and Earth Sciences, Faculty of Sciences, University of Pécs, Ifjúság útja 6, Pécs 7624, Hungary
e-mail: loczyd@gamma.ttk.pte.hu

R. Ciglič
Department of Geographic Information System, Anton Melik Geographical Institute, Research Centre of the Slovenian Academy of Sciences and Arts,
Novi trg 2, Ljubljana 1000, Slovenia
e-mail: rok.ciglic@zrc-sazu.si

D. Lóczy (ed.), *The Drava River*, Springer Geography,
https://doi.org/10.1007/978-3-319-92816-6_19

19.1 Introduction

Given their vulnerability to human interventions, rivers and the adjacent wetlands (floodplain lakes, backswamps, etc.) are important objects of nature conservation worldwide. Through their ecosystem services, including, among others, contribution to flood prevention, water purification for drinking water supply, maintenance of biodiversity and reduction of the harmful impacts of climate change (Rey Benayas et al. 2009) wetlands are also key sites for water management, regional planning, agriculture, fishery, forestry, recreation and tourism. Therefore, river restoration or rehabilitation has become a common response to declining river health and its importance to water resources management is increasing (Speed et al. 2016).

Similar to most European rivers, river regulation has affected virtually all the watercourses of the Drava-Mura drainage basin. Valuable habitats such as gravel banks, side-arms, riparian forests, waterlogged meadows, were lost or shrank considerably. To achieve the ecological integrity of impaired riverine landscapes ('good ecological status' according to WFD—European Commission 2000), the rehabilitation of the Drava River has to address the following problems (Muhar et al. 2008):

- bedload transport;
- longitudinal discontinuity caused by hydropower plants;
- intensive land use in the floodplain;
- loss of wetlands and reduced lateral connectivity between river and its floodplain and
- human-induced disturbances through hydro-peaking.

19.2 Nature Conservation and River Restoration in Austria

The first major river stabilization measures on the Drava date back to the mid-19th century. In the 20th century typical fluvial landforms were destroyed, braided sections transformed into channels of straight alignment and uniform width. Adjacent floodplains were used by intensive agriculture and forestry (Habersack and Nachtnebel 1995). Together with all other gravel-bed rivers with braided pattern, the Drava channel and its floodplain have also been heavily altered by channelization and construction of barrages (Muhar et al. 2008).

Since 2011 along a 68-km-long section of the trunk river downstream Oberdrauburg the *Upper Drava European Reserve* is under protection (RIS 2011; Michor 2013). It is also a Ramsar area (no 2208). The Reserve has an exemplary function in flood-control strategy for all countries on the Drava. For 20 years actions have been based on a novel stream development concept which aims at creating a close-to-natural water regime and incorporating synergies with the demands of nature conservation and recreation—instead of relying on'hard engineering' solutions (Michor 2013). The EU LIFE-Nature programmes implemented on this 68-km-long Drava section (*Auenverbund Obere Drau*, 1999–2003; *Lebensader Obere Drau*, 2006–2011—EC 2011) at first focussed on flood control and floodplain forest restoration and, in the next stage, on improving the ecological status of the river at local to reach scale according to the specifications of the EU Water Framework Directive (WFD—European Commission 2000; RIS 2011). Restoration efforts were concentrated on five short (400–2,000-m-long) reaches and involved channel widening on one bank or both banks. The revitalization of side-arms used the Military Survey between 1832 and 1834 and the historical description of the Drava River by Schmidt (1880) as references (Muhar et al. 2008) (Fig. 19.1). The projects affected a total of ca 1,000 ha area (ca 350 ha open water surface, ca 450 ha riparian forests and ca 200 ha pioneer vegetation, agricultural land, and others). The re-introduction of a typical floodplain plant, German tamarisk (*Myricaria germanica*) was an important goal.

The assessment of success was founded on comparing the post-rehabilitation quality of key habitat types to the type-specific reference conditions applying a scoring system (1–5) (Muhar et al. 2008). Channel widening was found to have an equally positive impact on aquatic biodiversity, flood control, fishery and recreation.

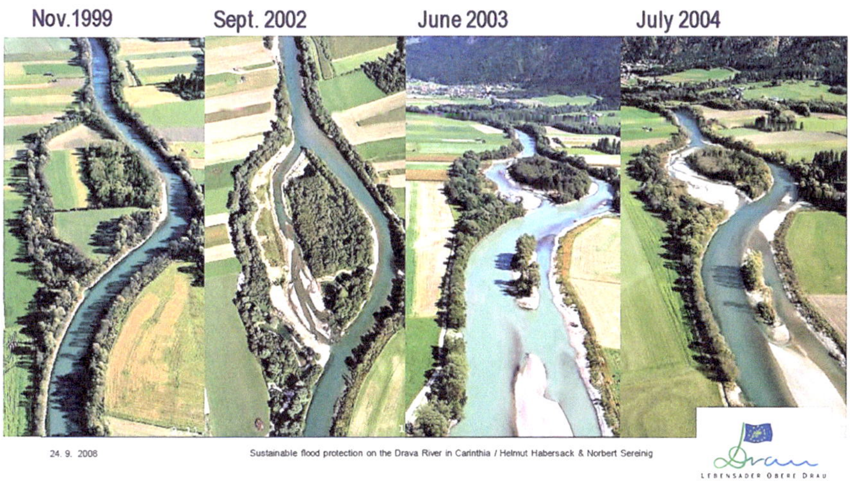

Fig. 19.1 The revitalization of the Drava side channel at Kleblach, Carinthia, 1999–2004 (after Habersack and Sereinig 2008)

19.3 Nature Conservation and Habitat Restoration in Slovenia and Croatia

The Slovenian section of the Drava has also been affected by river training and hydroelectric plant constructions. Along some sections, however, valuable semi-natural habitats are worth protecting (Fig. 19.2). In the environs of *Lake Maribor* (the reservoir of the Maribor Island hydropower plant, built in 1948) the riparian landscape is preserved in the frame of a *Nature Park.* Partly for nature conservation purposes the *Drava Nature Park* of 2,175 ha area was established in 1992 (Mencinger 2004). The Drava River section between Maribor and Zavrče is also included in the List of International Important Bird Areas (IBA). In the area 266 bird species have been recorded, of which 89 are regular or occasional breeders. The protection of native birds calls for some river restoration measures. High banks are regularly cleared to provide suitable sites for the breeding holes of the 9,000 sand martins and kingfishers which live here. Gravel banks are ripped of vegetation to allow the nesting of little ringed plovers. Through the restoration of three side channels, new spawning areas were created for fishes and feeding grounds for kingfishers and sandpipers.

Where canals were built to serve hydropower plants, floodplain forests are preserved and partially restored along the old Drava (e.g. *Zlatoličje Nature Reserve*, 122 ha) (Schneider-Jacoby 1995). The forests also have a botanical interest,

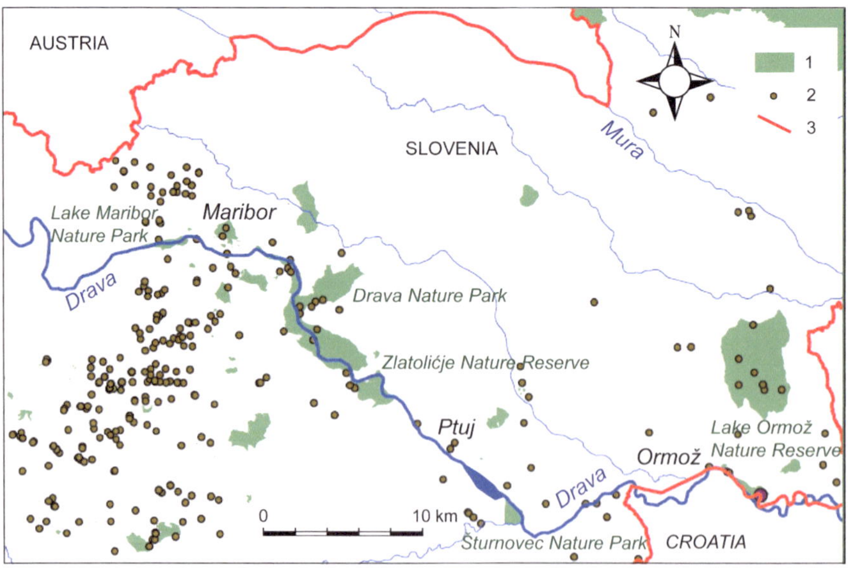

Fig. 19.2 Protected areas along the Slovenian Drava section (by Rok Ciglić). 1, Protected areas, 2, nature monuments, 3, national border

Common Buckthorn (*Hippophae rhamnoides*). The Natura 2000 sites along the 'Stara Drava' make up 9,525 ha area downstream Maribor to Ormož.

The protected Natura 2000 areas are heavily impacted by the reservoirs of the hydropower plants. Hydromorphological restoration works began in the pilot area Malečnik (downstream of Melje dam, near Maribor, 26.45 km reach) for flood protection, since has been regularly flooded in the last 15 years, particularly in 2012 (IzVRS 2017). With the introduction of the Natura 2000 system, restoration works were adjusted to nature conservation requirements. The interventions completed between 2002 and 2012 covered lowering or raising, narrowing or merging gravel bars; selective removal of vegetation (bushes, shrubs, tree cutting) and soil; digging lead trenches for and sediment reinjection into meanders). A physical monitoring system is designed to study river dynamics (sediment transport, gravel bar accumulation), the maintenance of environmental flow in the old Drava channel and habitat change on bars and in the riparian zone.

On the largest reservoir, Lake Ptuj (Firbas 2001), established behind the dam of the Formin hydropower plant in 1978, navigation has been restricted to avoid disturbance to the 120 bird species visiting the area. In 2004 an artificial breeding island was built by the hydropower company for common terns and later two man-made gravel islands for black-headed gulls. On the premises of the closed sugar factory at Ormož (former wastewater basins), constructed wetlands with water recharge from the Drava serve nature conservation purposes (*Lake Ormož Nature Reserve*). Grasslands along the Drava were restored with importing buffaloes which graze the land and maintain herbaceous vegetation.

Increasing attention is paid to the vegetation along flood-control dykes. Following the devastating floods in October 1998, flood-control measures began in the area of the two heavily hit settlements, Duplek and Dogoš (Juvan and Horvat 2015). The needed earth material was obtained from the excavations for the expansion of the head race of the Zlatoličje hydroelectric plant power canal. Although construction activities required the removal of forests and other natural vegetation, subsequently native trees and bushes were restored along the 5.1-km-long dyke.

Between 2000 and 2006, a Croatian-Hungarian Interreg project was performed with the purpose of maintenance of the Drava River channel for safer navigation along the international border from 70.2 to 198.6 rkm (https://www.keep.eu/keep/project-ext). It covered the removal of tree branches, stumps and regulation structures that are no longer in use and obstruct the safety of water traffic, including the remains of a stone dam on the left bank of the Drava River at 74.0 rkm. The project had a positive impact on the riverine environment.

In Croatia, however, plans for constructing new hydropower plants regularly re-occur in the news. It is claimed that the plants Molve 1 and 2 would help regulate the operation of the Dubrava plant, do not need to construct new dykes, but protect against floods, create new jobs, generate revenues and better conditions for irrigated agriculture (Pavlić 2017). Environmentalists strongly oppose such plans.

19.4 The Danube-Drava National Park

The National Park was founded with 49,752 ha area (ca 18,000 ha Ramsar area) in 1996 to ensure protection for riverine landscapes (Závoczky and Fenyősi 1999). Where it enters Hungary, the Drava is a relatively fast-flowing river with a steep slope and high sediment discharge. Reduced stream power results in the deposition of an ever-rearranging pattern of mid-channel sand bars, areas with pioneer vegetation. Larvae of caddisflies and mayflies inhabit the river channel. (For the description of macrovertebrate fauna see Chap. 16 and for fish fauna see Chap. 17). The mammals most frequently encountered are otters, wild cats and red deer. The National Park abounds in bat species which have become rare in other parts of Europe. The riparian white willow, white poplar, and black poplar forests (see Chap. 18) accommodate valuable herbs.

The partial areas of the National Park along the Drava include the Barcs Juniper Forest of acidophilous vegetation; the high banks of Zákány-Őrtilos, where sand martins and bee-eaters nest and the Cún-Szaporca oxbow (see Chaps. 12–14). Within birdlife, the corncrake (*Crex crex*) is a rare species of high conservation value. The protected sections of the active floodplain preserve traces of traditional occupations (cattle and sheep grazing, wooden trough making, basketwork) and floodplain economy.

The management plans of the Drava section of the National Park focus on actions aiming at the preservation of aquatic habitats, precluding the disturbance of nesting sites, regulating the populations of big game, stopping or at least slowing down the spreading of intrusive plants, encouraging traditional crafts and providing the necessary infrastructure for ecotourism. River restoration efforts concentrated on the revitalization of the side-arms of the Drava (see Chap. 20), while floodplain rehabilitation, within the frame of the Old Drava Programme, is underway in the Ormánság region, downstream of Barcs (see Chap. 21).

19.5 The Transboundary Biosphere Reserve Mura-Drava-Danube

The UNESCO Transboundary Biosphere Reserve Mura-Drava-Danube (TBR MDD) extends to more than 800,000 ha of highly valuable natural and cultural landscapes, uniting 13 areas of various protection level, in all five countries. The Biosphere Reserve concept defines ca 300,000 ha of core and buffer zones (existing protected area network) and ca 700,000 ha of transition zones. Natural habitats such as floodplain forests, river islands, gravel and sand banks, side branches and oxbows are protected. Habitat diversity maintains high biodiversity. For instance, in the Reserve white-tailed eagle-breeding pairs are found in the largest number in continental Europe.

The Biosphere Reserve opens new perspectives in the management of protected areas in the Drava region. Since river channelization and bed material extraction involve detrimental environmental impacts (channel incision, desiccation of wetlands and floodplain forests, decline of natural river habitats threatening rare species), these outdated practices are being replaced by landscape rehabilitation in order to achieve sustainable water management, also boosting economic growth and development in the region.

19.6 Kopački Rit Nature Park

One of the most valuable geoheritage areas within the TBR MDD, the Kopački rit Nature Park (23,894 ha, including 7,700 ha special zoological reserve in the southern portion with the highest biodiversity) was established in 1967 in the regularly inundated floodplains of the Danube and the Drava at the confluence of the rivers (Springer et al. 2003). From 1699 to 1918 this area belonged to the Bellye (now: Belje) landed property, originally owned by Eugene of Savoy and the Habsburg family as a hunting ground. Since 1989 it is Important Bird Area (IBA), since 1993 a Ramsar area, in 1999 on the waiting list for UNESCO World Heritage site and since 2008 part of the Croatian Natura 2000 network.

The Nature Park hosts the most extended softwood (willow) forests in the Danube Basin and ca 40 forest associations (such as *Genisto elatae-Quercetum roboris, Carpino betuli-Quercetum roboris, Fraxino-Ulmetum laevis* and others). In the Park 293 bird species occur, 141 occasionally or regularly nest there, including 23 breeding pairs of white-tailed eagle, 10 pairs of black stork, 40 pairs of little egrets and ca 100 pairs of wild geese. After the Danube Delta it is the second most important fish spawning area in Europe for 44 fish species. Five to seven thousand red deer live in the Kopački rit, including the best prize deer (See also Sect. 9.2).

19.7 Conclusions

River and floodplain rehabilitation is becoming an integral part of nature concervation. Where human interventions resulted in the degradation of the riverine environment, the restoration of geomorphological processes in the river channel and in the floodplains creates favourable conditions for the development of habitats which are hopefully self-sustaining over the long term. Among other benefits, rehabilitation efforts, such as the re-establishment of naturally disturbed and re-shaped gravel and sand bars, are capable of improving habitat conditions within the aquatic area, for instance, for gravel-spawning fish species or the development of pioneer vegetation. The projects along the Drava provide good examples for the regeneration of riparian pioneer vegetation and floodplain woodlands.

References

EC (2011) Life Project 'Lebensader Obere Drau'. Final Report. European Commission, Brussel. 120 pp. (LIFE06NAT/A/000127). http://ec.europa.eu/environment/life/project/Projects/index.cfm?fuseaction=home.showFile&rep=file&fil=LIFE06_NAT_A_000127_FTR_DE1.pdf

Firbas P (2001) Vsa slovenska jezera: leksikon slovenskih stoječih voda (All lakes in Slovenia: lexicon of Slovenian standing waters). Anton Melik Geographical Institute, Slovenian Academy of Sciences and Arts, Ljubljana. 368 pp (in Slovenian)

Habersack H, Nachtnebel HP (1995) Short term effects of local river restoration on morphology, flow field, substrate and biota. Regulated Rivers: Res Manag 10:291–301

Habersack H, Sereinig N (2008) Sustainable flood protection on the Drava River in Carinthia. Department of Water–Atmosphere–Environment, Institute of Water Management, Hydrology and Hydraulic Engineering, University of Natural Resources and Applied Life Sciences (BOKU), Vienna. 18 pp. http://www.crorivers.com/docs/20_Habersack_Sereinig_Sustainable_Flood_Protection_Carinthia.pdf

IzVRS (2017) Technical note about the monitoring of hydromorphological restoration/management of the Drava River in Slovenia. Interreg Alpine Space. Inštitut za vode Republike Slovenije/Institute of Water of the Republic of Slovenia, Ljubljana. 13 pp. http://www.alpine-space.eu/projects/hymocares/case-studies/drava_technical-reports.pdf

Juvan S, Horvat M (2015) Izgradnja visokovodnih nasipov ob reki Dravi v Dogoše in Dupleku (Construction of flood embankments along the Drava at Dogoš and Duplek). In: Aktualni projekti s področja upravljanja in urejanja voda (Current projects in the field of water management and regulation). 26th Mišičev Water Day, Maribor, 5 November 2015. 264–274 (in Slovenian)

Mencinger B (2004) Krajinski park Drava (Nature park Drava). In: Naravni parki Slovenije (Nature parks in Slovenia). Mladinska knjiga, Ljubljana. 218 pp (in Slovenian)

Michor K (2013) Die Obere Drau in Kärnten – ein Beispiel für Hochwasserschutz mit Mehrwert. Wasserwirtschaft 103(7–8):69–73. https://doi.org/10.1365/s35147-013-0654-1

Muhar S, Jungwirth M, Unfer G, Wiesner C, Poppe M, Schmutz S, Hohensinner S, Habersack H (2008) Restoring riverine landscapes at the Drau River: successes and deficits in the context of ecological integrity. In: Habersack H, Piégay H, Rinaldi M (eds) Gravel-Bed Rivers VI: from process understanding to river restoration, Chap 30. Elsevier, Amsterdam, pp 779–803

Pavlić V (2017) Two new hydroelectric power stations on Drava River. Total Croatia News, Jelsa, Hvar, Croatia. https://www.total-croatia-news.com/business/23890-two-new-hydroelectric-power-stations-on-drava-river

Rey Benayas JM, Newton AC, Diaz A, Bullock JM (2009) Enhancement of biodiversity and ecosystem services by ecological restoration: a meta-analysis. Science 325:1121–1124

RIS (2011) Europaschutzgebiet Obere Drau. Rechtsinformationssystem des Bundes (RIS), Bundeskanzleramt Österreich. https://www.ris.bka.gv.at/Dokumente/Lgbl/LGBL_KA_20110615_49/LGBL_KA_20110615_49.html

Schmidt F (1880) Die Drauregulierung in Kärnten. Seperatdruck aus Österreichisch-Ungarische Revue, Wien

Schneider-Jacoby M (1995) Razvojni koncept rečnega ekosistema Drava–Mura (Development concepts for the Drava–Mura river ecosystem). DOPPS, Ljubljana. 52 pp (in Slovenian)

Speed R, Li Y, Tickner D, Huang H, Naiman R, Cao J, Lei G, Yu L, Sayers P, Zhao Z, Yu W (2016) River restoration: a strategic approach to planning and management. Paris, UNESCO. 204 pp. http://unesdoc.unesco.org/images/0024/002456/245644e.pdf

Springer OP, Mihaljević M, Getz D, Božičević S, Bognar A, Topić J, Merdić E, Krčmar S (2003) Kopački rit park prirode – ekološki turistički vodič (Kopački rit Nature Park: ecotouristic guide). Barbat, Zagreb. 155 pp (in Croatian)

Závoczky Sz, Fenyősi L (1999) The messenger wild-water land: the Duna-Dráva national park. In: Tardy J (ed) Conserving Hungary's heritage. TermészetBÚVÁR Publishing, Budapest, pp 107–118

Chapter 20
Rehabilitation of Drava Side Channels

Tibor Parrag and Dénes Lóczy

Abstract Side-arms are water bodies which maintain active connection with the main river channel, have entrance (inflow) and exit (outflow) channels and receive recharge of water for most of the year. They are valuable habitats with multiple functions in the fluvial ecosystem. Flow regulation measures, the construction of weir closures, reduced their connectivity with the main channel and resulted in their rapid sedimentation. Today projects financed by the European Union encourage side channel revitalization in all countries along the Drava River. The chapter presents different approaches to revitalization, the environmental impacts of interventions (weir closure lowering, dredging, landscaping) and case studies to illustrate how successful such projects are. Hydromorphological and biological monitoring are necessary for the reliable evaluation of project efficiency. It is emphasized that properly designed revitalization extends the lifespan of these water bodies in the highly dynamic fluvial system, but it is only a medium-term solution.

Keywords Aquatic ecosystem · Side channel revitalization · Nature conservation

20.1 Introduction

Side channels differ from abandoned river channels not only in their hydromorphology and ecology (Riquier et al. 2017) but also from a nature conservation aspect (Nestler et al. 2016). Oxbows (abandoned channels) and side channels are distinguished since they need different management in nature conservation.

T. Parrag
Danube-Drava National Park Directorate, Tettye tér 9, Pécs 7625, Hungary
e-mail: tibor.parrag@ddnp.kvvm.hu

D. Lóczy (✉)
Department of Physical and Environmental Geography, Institute of Geography and Earth Sciences, Faculty of Sciences, University of Pécs, Ifjúság útja 6, Pécs 7624, Hungary
e-mail: loczyd@gamma.ttk.pte.hu

D. Lóczy (ed.), *The Drava River*, Springer Geography,
https://doi.org/10.1007/978-3-319-92816-6_20

In unconfined natural alluvial river systems, lateral migration constantly both creates and obliterates new side channel habitats.

Side channels are water bodies which maintain active connection with the main river channel, have entrance (inflow) and exit (outflow) channels and receive recharge of water over a considerable interval each year (Roni et al. 2008). Located between the river bank and the flood-control dykes, they often derive a major portion of their flow from either groundwater or seepage from the river.

As far as vegetation is concerned, in the side channels of the Drava River, particularly on the section upstream Barcs, relative high current flow does not allow the appearance of floating pondweeds, while on gravel and sand bars valuable pioneer species (such as *Salix purpurea*) and communities emerge. The fish fauna resembles to that of the main channel and includes numerous rheophile species (*Zingel streber, Z. zingel, Gobio uranoscopus, Gymnocephalus schraetzer*) (Sallai 2004).

Over decades or centuries side channels may evolve into isolated oxbows intermittently connected to the main channel during floods or exclusively through groundwater flow (Roni et al. 2002). Ecologically, the oxbow channels in the Drava floodplain belong to the plesiopotamon or parapotamon category. Mostly found beyond the flood-control dykes, but also occurring in the active floodplain (e.g. the Erzsébet Island oxbow near Babócsa), they are often infilled to a large extent (WWF 2001). Because of insufficient freshwater recharge, they are in various stages of bog formation. Lake vegetation is typical, aquatic plants are abundant, in the riparian zone willow bushes, reed and bulrush beds grow, while the open water surfaces are often fully covered by rooting pondweed (*Trapa natans, Salvinia natans, Stratiotes aloides*) during the summer period. Among fish species, *Rhodeus sericeus amarus* is common and *Umbra kameri* is rare (Sallai 2004).

The channel types, however, are not distinct in every case. There are gradual transitions, for instance, within larger water bodies habitats of rather potamic and rather lacustrine type equally occur (e.g. in the case of the side channel system of Vízvár). On the other hand, advancing succession also modifies the character of the individual water bodies.

Before river regulation, dynamic floodplain evolution created a constantly rearranging mosaic of wetland habitats. Along the Hungarian section, the meandering Drava was permanently reshaping the face of the floodplain, cutting off and developing meanders at different sites. Channel bars were wandering downstream and high bluffs were regularly undercut. Morphological changes involved alterations in the character of riverine habitats (see Chap. 5 in this volume).

For large rivers, general graphic models have been suggested for the succession from lower to higher terrain as well as for the stages of sedimentation for cut-off meanders from open water surfaces through the swamp stage to forests (see Figs. 18.4–18.9 in Chap. 18). Compared to the rate of geomorphic changes, succession takes place rapidly and easily observed by locals. River regulation and hydropower plant constructions upstream have upset dynamic equilibrium. Although succession was not stopped, it was restricted to an unidirectional process: increased current velocity and deficit of bedload leading to overall main channel entrenchment of locally variable extent. Accumulation prevails in the side channels, where bedload is also deposited during floods, and in the oxbows, where suspended and organic loads are predominant. With channel incision side channels acquired a ‘hanging’ position and the probability of their recharge became reduced and the spreading of aquatic vegetation enhanced.

Most of the side channels are associated with channel bars and islands. For nature conservation the gravel bars upstream Barcs with protected species like *Myricaria germanica* and *Sterna hirundo* are the most valuable (Purger 2013). Although willow bushes and willow-poplar forests on stabilizing gravel bars are highly appreciated, the reduction of open water surfaces is assessed as a negative tendency.

However, it is not only degradation that is observed on side channels and oxbows. The situation is more differentiated. According to the classification of the EU Habitat Protection Directive, four aquatic habitat types of community level protection are associated with side channels and oxbow lakes (Haraszthy 2014, pp. 770, 773, 780, 783) and they accommodate numerous protected species (Ábrahám 2005).

Side channels are exploited for spawning by rheophile fish species which lay eggs on hard bottom (*Phoxinus phoxinus, Gobio uranoscopus, Barbatula barbatula* and others) or on the temporarily inundated plants (like *Rutilus pigus virgo*). Lentic water bodies are also important for the reproduction of amphibians like *Triturus cristatus dobrogicus, Bombina bombina* and *Rana* sp. The specific fish fauna of oxbow lakes overgrown by vegetation includes *Rhodeus sericeus amarus, Umbra kameri, Misgurnus fossilis).* With regard to biota, there are transitions between the two types of water body. For instance, while from the interior of the Old Drava of Barcs *Rhodeus sericeus amarus* and *Cobitis taenia,* from the canal connecting the oxbow lake with the main channel *Gobio albipinnatus* have been recovered (Harka and Sallai 2004).

The same dichotomy is apparent in the occurrence of *Odonata* with the note that mobile imagos also hunt over water bodies which are not suitable for breeding their larvae. This circumstance makes the Odonata fauna even more diverse than expected.

20.2 Revitalization of Side Channels

River restoration projects have been launched for both the upland and lowland sections of the Drava River. It is estimated (WWF 2013) that along the Drava a total of 120 major side-channels with a length of 519 km could be reconnected with the rivers.

In Austria, within the frame of LIFE projects (‘*Connecting floodplains of the Upper Drava*’, 1999–2003; ‘*Life artery Upper Drava*’, 2006–2011) the revitalization of a 68-km long section of the river in Carinthia was completed and the monitoring of environmental impacts began (Lazowski and Schwarz 2011). The project included the rebuilding of a weir closure on the Feistritz Stream near Spittal to allow 5,000 $m^3\ y^{-1}$ of bedload input. At the same time, ca 100 ha new land area was created for gallery forests. The spreading of tamarisk (*Tamarix* sp.) was achieved at five sites, dwarf bulrush (*Typha minima*) at four sites and black poplar (*Populus nigra*) at three sites. Instead of the intensive use of floodplains for grazing, animal foraging in gallery forests is encouraged to prevent the expansion of invasive plants (Lazowski and Schwarz 2011).

On the Slovenian section, downstream Maribor, restoration affected the natural Drava channel, into which more water is conducted from the canal of the hydropower plant. Another intervention took place on the Croatian border in the framework of the LIFE project '*Riparian Ecosystem Restoration of the Lower Drava River in Slovenia*' (acronym: LIVEDRAVA). Its objectives included the conversion of 61 ha of the Ormož basins into a semi-natural wetland as a stopover site for migrating birds; improving the conditions of ca 15 ha of alluvial forests; creating new artificial breeding island; transforming river banks to allow breeding by the kingfisher (*Alcedo atthis*) and sand martin (*Riparia riparia*). Three side arms of 1.5 km total length were also opened up to enhance connectivity with the main channel (Life Slovenija 2017).

In Croatia, the restoration of side channels also promotes novel-approach flood control through diverting the water away from settlements, bridges, roads, and dykes (WWF 2016). The actions consist of the widening of side-arms, dismantling and modifying existing regulation structures.

The project will also enhance river water infiltration and raise the groundwater table. Seven locations are involved: the Virje island at Ormož (312–314.3 rkm), the old Drava at Varaždin (289.3–292 rkm), between the Donja Dubrava hydropower plant and Legrad (240–241.45 rkm), the Botovo bridge (226.6–227.9 rkm), at Novačka (214–217 rkm), at Miholjački Martinci (104–106 rkm) and at Podravska Moslavina near Gola (96–98 rkm).

The Lower Drava on the Croatian-Hungarian border is a relatively slightly modified river—but cannot be regarded natural. The drainage basin management plan (VKKI 2010), compiled by the guidelines formulated in the EU Water Framework Directive (European Commission 2000), referred the upper section of the Hungarian Drava into heavily modified, while the lower section into the natural category. Poorer naturalness on the upper section was explained by the peaking of the Dubrava hydropower plant (and the plants further upstream). Experience shows that, on the one hand, the natural values of the Drava are not yet appreciated so high that deserve revitalization and, on the other, it is modified to an extent that the restoration of habitats is necessary.

Targets and methods are both crucial aspects of revitalization schemes. In conservation biology (Sodhi and Ehrlich 2010) revitalization efforts are directed at endangered species or degraded habitats, restoration of previous conditions and creation of brand new habitats. In practice, a combination of the above is implemented. In the following, all these measures are called revitalization.

In designing revitalization, conservation biological requirements are fundamental: which species and habitats need improvement in their conditions, what the causes of degradation are, and how they can be remedied. Natural succession, e.g. infilling of an oxbow, is not considered harmful. However, we have to remember that no new oxbow will come about and occasionally nature conservation should attempt to decelerate or even block the process of natural succession. It is vital to maintain biodiversity through ensuring habitat diversity. The revitalization projects of the Danube-Drava National Park Directorate (DDNPI) (DDKÖVÍZIG 2004; DDNPI 2008, 2014, 2015) are not aimed at the preservation of individual habitats or species, but at the improvement of abiotic conditions (morphology, water availability) which provide the basis for predictable changes in biota (Purger 2013). The principle is also emphasized that the type of the habitat (side channel or abandoned channel) should not be changed.

Since degradation is mostly due to some human impact, it seems reasonable that the elimination of this impact will result in a more favorable environmental status. This is true in certain cases, e.g. demolishing a stone closure highly improves flow in the side channel, and rheophile fishes are able to return. In most of the cases, however, the situation is not so simple. Revitalization is restricted to short reaches and longer stretches cannot be reverted to more natural conditions. If the mouth reach of an abandoned channel is opened for water recharge, the incised main channel can only supply water during high water stages. Such an intervention may even bring negative consequences: the main channel could drain water from the environs of the abandoned channel through the restored watercourse during low water stages.

Another challenge is altered sediment transport. Through changing flow, revitalization affects sediment load. In a simple case the opened side channel receives large amounts of sediment with water during floods and, with dropping current velocity, deposition occurs. A famous example is the opened meander entrance channels on the Morava River in Slovakia (Holubova et al. 1999; Steiner et al. 2014). On the Drava it was also observed that the removal of weir closures modifies flow and result in new bars. It is evaluated negatively for nature conservation if the free flow of water is blocked.

Finally, it has to be mentioned that although the active floodplain on the Hungarian side is part of a national park, constitutes Natura 2000 areas, and since 2012 belongs to the Mura-Drava-Danube Transboundary Biosphere Reserve (UNESCO 2012), nature conservation is not the sole aspect in the utilization of the river. A fundamental consideration that habitat revitalization should not reduce protection against floods. A navigation route has been marked on the Drava to 198.6 river km from the confluence with the Danube. Although there is virtually no boat traffic on the river, the requirements of this navigation route have to be regarded in designing revitalization.

20.3 Alternative Approaches to Revitalization

Several alternatives for channel revitalization have been proposed worldwide (FISRWG 2001). The choice among them is based on partly ecological, partly economic considerations (Thomas et al. 2016). The simplest option is the partial or total removal of stone closures to ensure more favorable flow conditions. Although the dredging of the Drava channel had been deemed rather negative (Rákóczi and Szekeres 2008; WWF 2009), for highly infilled side channels the dredging of fine deposits is an accepted technique of revitalization. Sluices or weirs can also be applied to retain water in abandoned channels or water replenishment from other watercourses is also feasible.

The above-mentioned techniques are suitable for the revitalization of minor side channels, but unable to remedy the impact of large-scale degradation induced by main channel incision or sediment imbalance. Consequently, even successful revitalization projects are not more than symptomatic treatment which solve environmental problems for some decades at most. As an example, the opportunities for the revitalization of the Vízvár-Bélavár side-arm system (Figs. 20.1 and 20.2), cut off from the Drava in the 1970s and continuously infilling ever since, were investigated in detail (DDNPI 2014). The study revealed that, with the same rate of riverbed entrenchment, if the dredging to a depth which ensures water inflow throughout the year were implemented, the present-day situation would return in 30 years (Fig. 20.3).

The authors of the study propose that current velocity in the main channel should be reduced. Thus, erosion rates can be kept moderate and the side channels can be regularly flushed. A bottom weir could decrease stream power and divert higher discharge into the side channel. An intervention of this scale would allow the restoration of the dynamics of the former floodplain, but its implementation (after precise modeling) would require caution (DDNPI 2014).

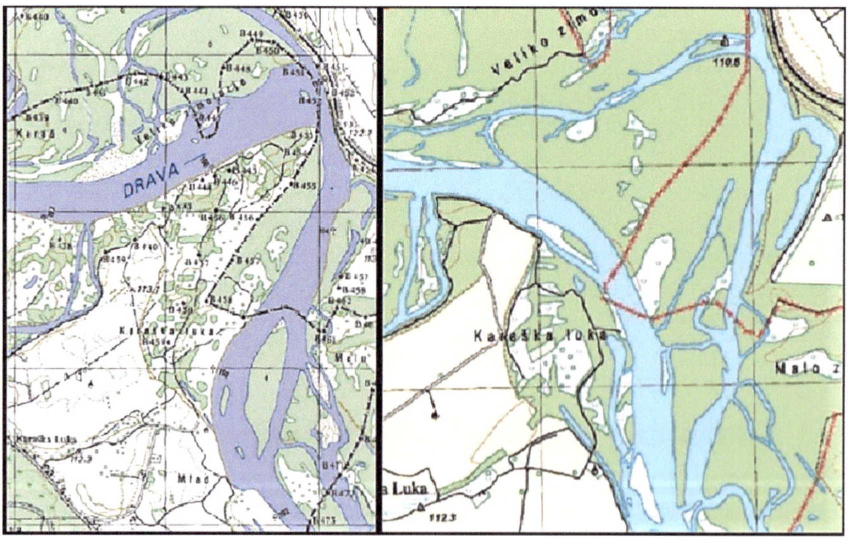

Fig. 20.1 The Vízvár meander of the Drava in 1979 (left) and 2003 (right). *Source* Gábor Andrási

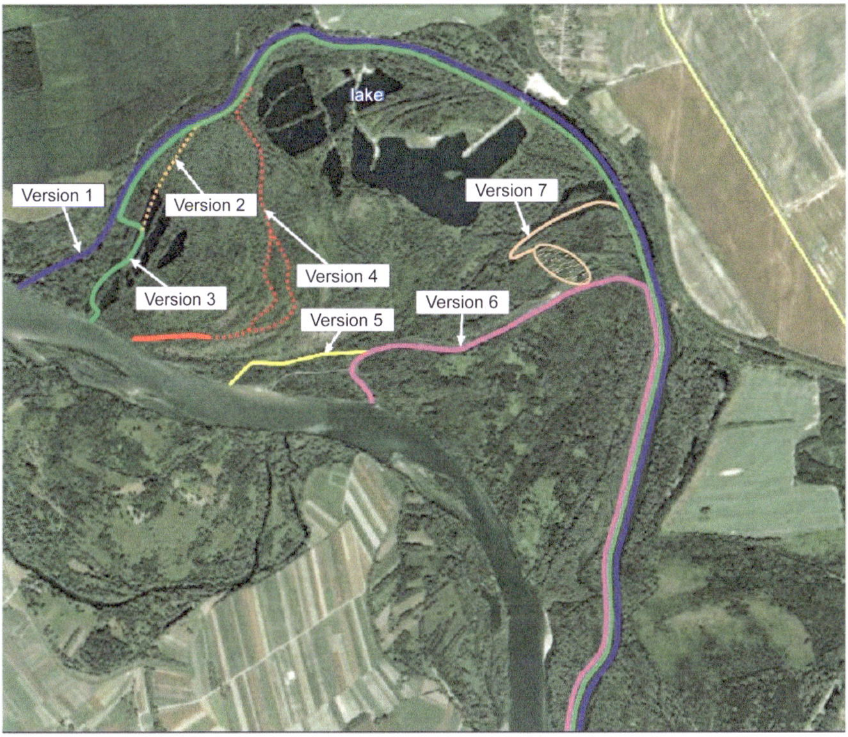

Fig. 20.2 Channel alignment options for the revitalization of the Vízvár-Bélavár side-arm system. *Source* DDNPI (2014)

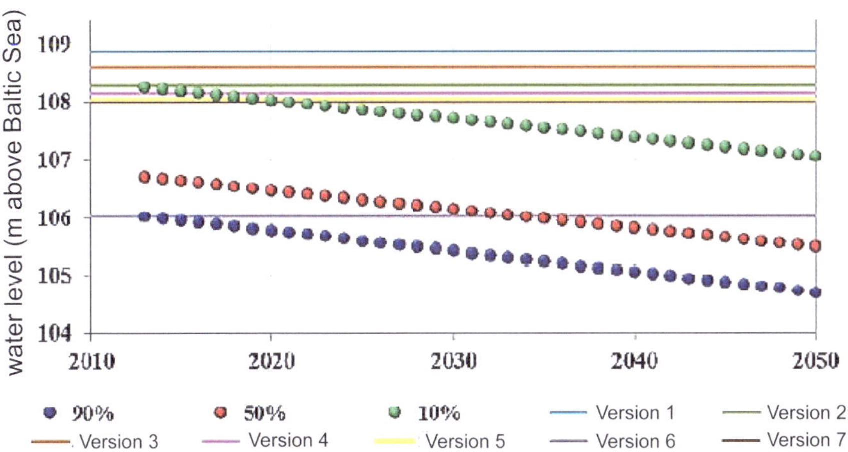

Fig. 20.3 Water levels predicted for the entrance points of the Vízvár side channels (2012–2050) plotted against the actual water levels of side channels (DDNPI 2014)

20.4 Case Studies in Hungary

Before Hungary's accession to the European Union (2004), within a Croatian-Hungarian Instrument for Pre-Accession Assistance (IPA) project, four side channels on the Hungarian bank were selected for revitalization directed by the DDNPI (DDKÖVÍZIG 2004). All four channels are side-arms in the classical sense: located in the main channel between an island and the left bank, closed with a stone structure. The weir closures had different crown heights and only allowed free flow above given water stages. Reducing current velocity, they fostered deposition in the side channels. The restoration of side channels ensures water inflow and outflow and maintains a lotic environment throughout the year. The revitalization intervention consisted of the reduction of weir crowns to allow flow for 200 days in a year or even for longer duration if culverts are built in. Since deposition had accumulated above the crown level, minor dredging was also needed. The dredged material was disposed in the main channel of the Drava. Among the four selected side channels, the Drávatamási Upper and Lower and the Tótújfalu arms are found next to reaches with gravel transport, while at Drávapalkonya river flow is slower and the sediment load is mostly composed of sand.

20.5 The Side-Arm at Drávatamási

Among the side channels, the Drávatamási Upper (between 147.8 and 146.1 rkm) had best preserved its eupotamon character (Fig. 20.4). Deposition here did not call for dredging. Channel width is ca 70 m–relatively broad compared to the 150 m

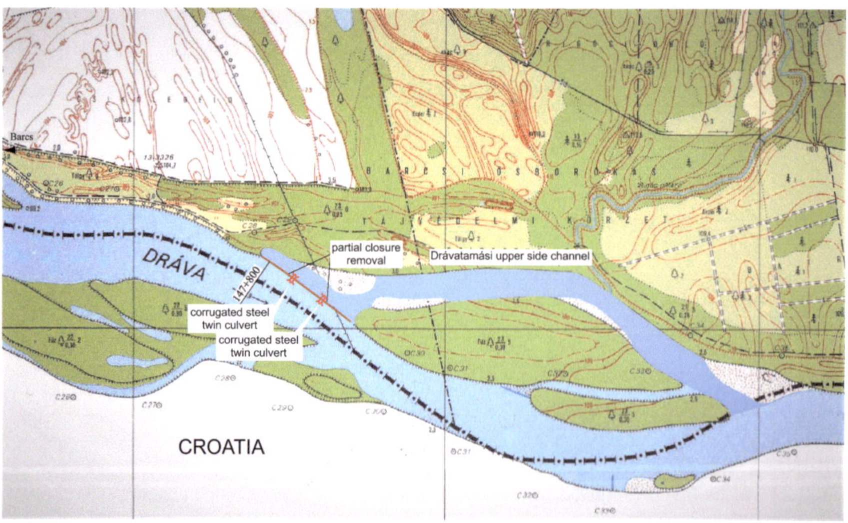

Fig. 20.4 Location of the Drávatamási Upper side channel with structures serving revitalization targets (DDNPI 2010)

wide main channel. The bar, which separates the two channels, is overgrown by a softwood gallery forest. At the mouth, the side channel was closed by a 350-m long stone structure.

In the course of revitalization, the crown of the closure was reduced from 98–99.75 to 97 m above the Baltic Sea level in 120 m width (Figs. 20.5, 20.6 and 20.7). This allows free flow in the side channel for 200 days annually. At 96.5 and 95.5 m levels, culverts were also installed. Since gravel surfaces were exposed,

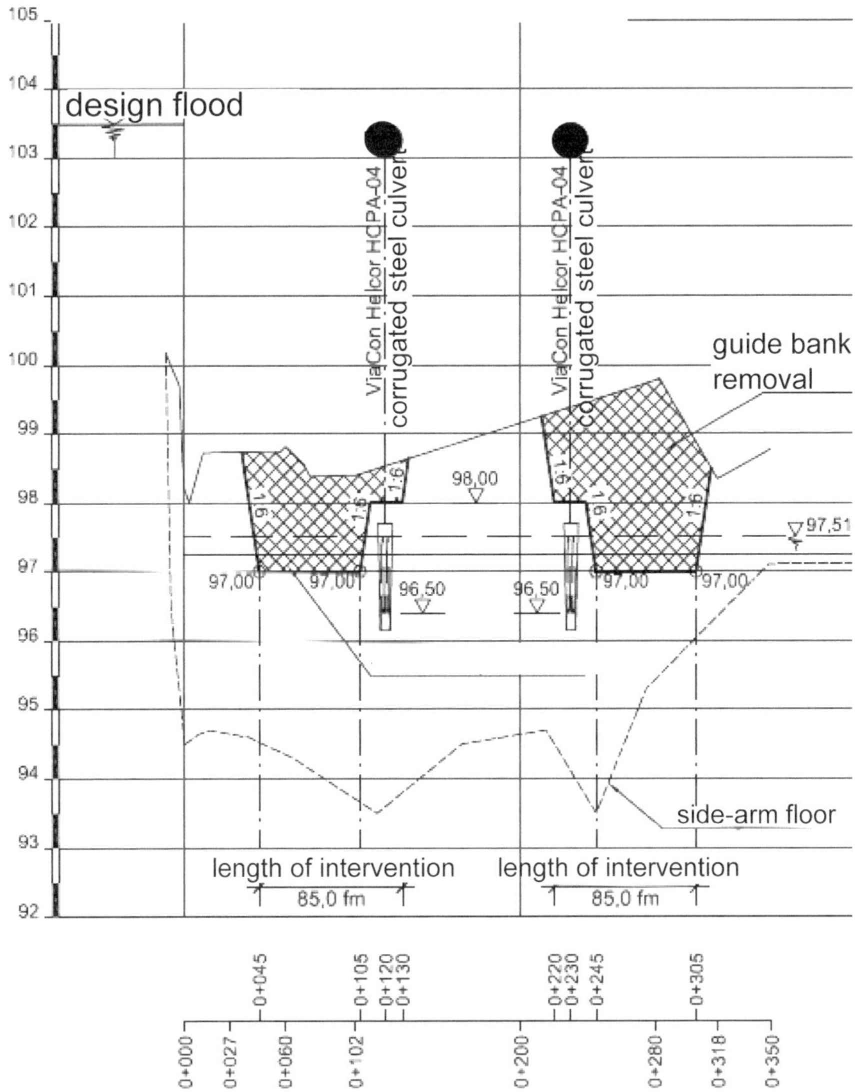

Fig. 20.5 Engineering plan for the partial removal of the closure at the Drávatamási Upper side channel (DDNPI 2010)

Fig. 20.6 The partially demolished stone closure of the Drávatamási Upper side channel (photo by Tibor Parrag)

the chances for revitalization seemed good (Fig. 20.8). The revitalization as implemented in this case was a compromise since the total removal of the closure, most advantageous for nature conservation, could have violated the interests of flood control and navigation route maintenance.

In the five years since the intervention water flow in the side channel intensified and fine deposits were flushed out from the bed. The silt bench in front of the outflow has also shrunk in area considerably in the meantime.

The fish fauna was surveyed after four years and found rich with 25 species present out of the 66 species identified in the Hungarian Drava section over the past 25 years (Ábrahám 2005 and see Chap. 17). Thus, out of the four side-arms mentioned the Drávatamási Upper is the richest in fishes. Protected and Natura 2000 marker species include *Eudontomyzon vladykovi*), *Rutilus virgo*, *Romanogobio vladykovi*, *Rhodeus amarus*, *Cobitis elongatoides*, *Gymnocephalus baloni*, *Aspius aspius* and *Barbus barbus*. Habitat diversity is indicated by the presence of *Rhodeus amarus*, a rather stagnophile species, along with rheophile fishes. In the autumn sample of the Drávapalkonya channel of lentic character *Rhodeus amarus* amounted to 63% of the catch, while in the Drávatamási side channel it remained below 5%.

Fig. 20.7 The lowered closure at high and low water stages (photo by Tibor Parrag)

20.5.1 The Side-Arm at Drávapalkonya

Parallel to the Drávatamási Upper revitalization project, the restoration of the Drávapalkonya arm is underway (Pécsi HIDROTERV Bt. 2010; Fig. 20.9). In contrast to the Drávatamási Upper channel, at Drávapalkonya the mouth was not closed but a spur upstream of the inflow site modified flow. Current velocity was reduced and infilling began along the entire longitudinal profile (Fig. 20.10).

Deposition was further enhanced by a stone weir at 600 m distance from the inflow site built to give access to the island for forestry purposes. However, the DDNPI did not envision commercial forestry in the softwood gallery forest of the island and the weir could be removed as part of the revitalization. The largest-scale intervention was the dredging of the entire, 800-m-long Hungarian section. For the 200-day flow, the bottom level of dredging was set at 88 m elevation. It meant the removal of 50–200 cm sediment (8,000 m^3), mostly sand. Observing an important principle of rehabilitation, the dredged material was disposed in the main channel. The banks of the restored side channel were designed to allow unhindered connectivity between the aquatic and terrestrial ecosystems. Instead of the 1:2 slope, the gradient of the banks was adjusted to the topography: steeper in the vicinity of high banks and gentler on the lower part of the island (Fig. 20.11). Two years after implementation it was impossible to detect the traces of human intervention on the banks any more.

After five years, it is visible that moderate deposition has resumed in the side channel, first of all, on the upstream section. This is most probably due to the

Fig. 20.8 Gravel bar in the Drávatamási Upper side channel

survival of the spur, which still has an unfavourable influence on flow conditions. Another remarkable observation can be made concerning bar formation. The trunk and crown of a tree uprooted and fallen into the channel obstructed flow and a bar appeared downstream the tree. The message is that to avoid large-scale deposition, driftwood has to be removed immediately from the channel.

20.5.2 The Old Drava of Barcs

The Old Drava of Barcs is an overdeveloped and cut-off meander system with advanced succession stages (willow bushes, bulrush swamps, pondweed mats and softwood gallery forests on higher ground) resulting in slow infilling (BioRes Bt 2015—Fig. 20.12). The biota is adjusted to the lacustrine environment and, consequently, the area is part of the Danube-Drava National Park and the Natura 2000 network, both in Hungary and Croatia (see Chap. 13). Among habitats of community importance, there occur 'Natural eutrophic lakes with Magnopotamion or

Fig. 20.9 Location map of the Drávapalkonya side channel. *Source* Pécsi HIDROTERV Bt. (2010)

Hydrocharition-type vegetation' (code 3150), 'Alluvial forests with Alnus glutinosa and Fraxinus excelsior (Alno-padion, Alnion incanae, Salicion albae)' (code 91E0) and 'Riparian mixed forests of Quercus robur, Ulmus laevis and Ulmus minor, Fraxinus excelsior or Fraxinus angustifolia, along the great rivers (Ulmenion minoris)' (code: 91F0). Typical protected lake plants are *Trapa natans* and *Salvinia natans*). Characteristic limnophile fish species are *Rhodeus amarus, Cobitis elongatiodes, Romanogobio vladykovi* recovered from the Fekete-árok canal (see later), where water flow is more rapid. Although the oxbow lake is quite narrow, the

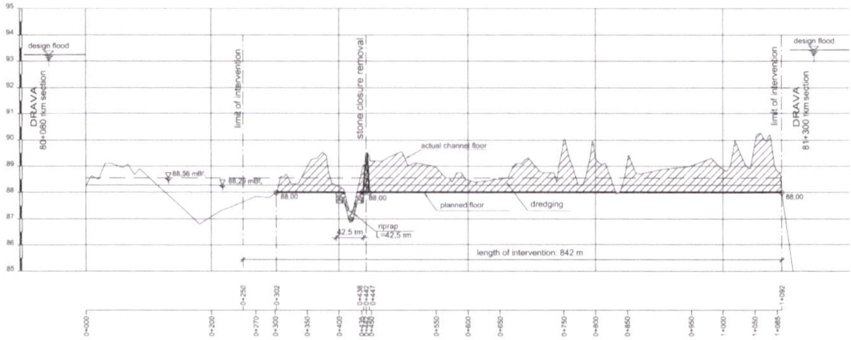

Fig. 20.10 Longitudinal profile of the Drávapalkonya side channel with the section to be dredged indicated. *Source* Pécsi HIDROTERV Bt. (2010)

Fig. 20.11 Banks shaped close to natural conditions in the Drávapalkonya side channel (photo by Tibor Parrag)

densely overgrown reaches accommodate at least 90 bird species, including the protected *Nycticorax nycticorax, Ardea purpurea* and *Aythya nyroca* (Purger 2015).

In contrast to other oxbow lakes (e.g. that of Cún-Szaporca—see Chap. 12), fortunately the Old Drava is connected to the main channel through the Fekete-árok canal. Its water budget is also more favorable due to the constant recharge of water

Fig. 20.12 Slow-flowing downstream reach of the Old Drava of Barcs (photo by Tibor Parrag)

from the north through the Rinya Stream. The wetland area, however, has been shrinking partly because of slow sedimentation and partly because the entrenching Drava main channel drains the lake. The inflow from the Rinya leaves the lake freely through the Fekete-árok outlet. From the Drava, recharge only occurs during flood stages. Thus, regular replenishment from the main channel is not possible. The anglers use the oxbow lake and the draining Fekete-árok. They impounded the latter by a makeshift sand bag dam, which, however, is destroyed by erosion.

Another Croatian-Hungarian LIFE project is designed to solve the problems of the Old Drava of Barcs. The planned revitalization includes the construction of a structure to withhold water on the Fekete-árok, stabilize water level in the oxbow lake, reduce the duration of low-water stages and increase water retention in the channel (DDNPI 2015). Crucial issues are the following: which is the safe water level that does not endanger riparian areas with inundation and whether the discharge of the Rinya Stream is sufficient to fill up the oxbow after the building of the weir. The engineering plan (DDNPI 2015) defined a threshold level of 102.75 m above Baltic Sea level with no inundation hazard for the riparian zone and which is equally favorable for aquatic and riparian vegetation (Fig. 20.13).

As attested by the water gauge in the lake, water levels in August sank to 101.5 m elevation. Therefore, about 1 m water level rise would be necessary to reach the target. Water retention in the lake after raising its level was also calculated (Fig. 20.14).

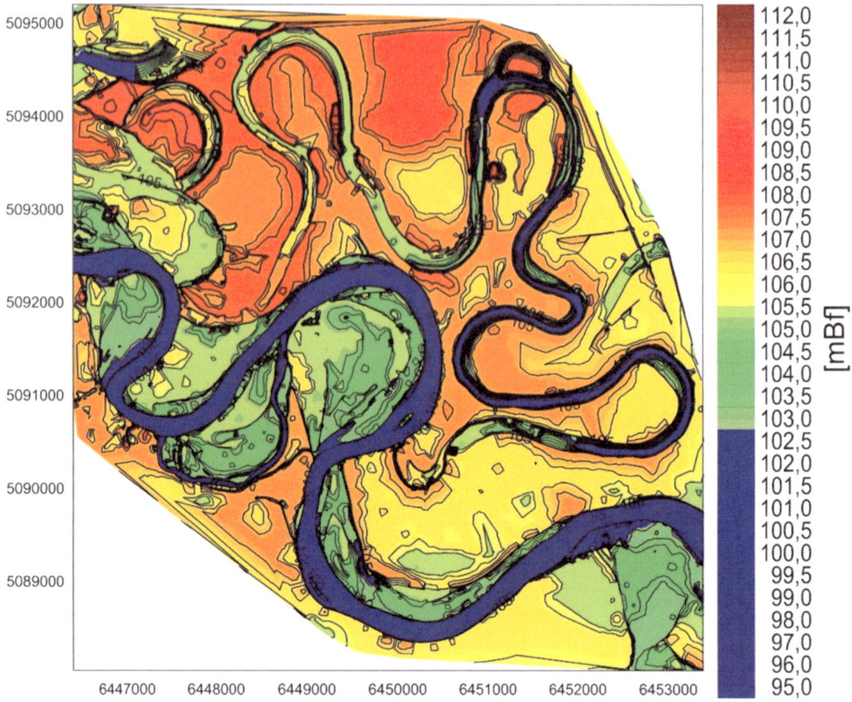

Fig. 20.13 Potentially inundated areas at 102.75 m water level

Fig. 20.14 Relationship between water level and the volume of water storage. *Source* DDNPI (2015)

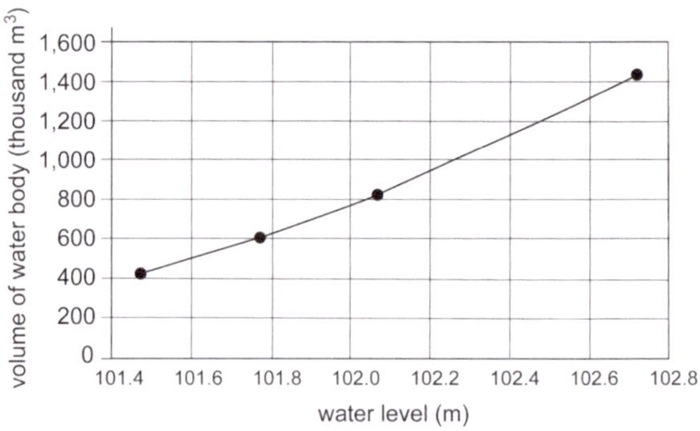

Figure 20.14 clearly shows that even 1 m raising of water level results in several hundred thousand cubic meter water storage. The weir, however, will only be efficient if the water flow of the Rinya can maintain the target lake water level. Calculating water balance, the planning engineers established the monthly discharge of the Rinya amounts to more than 5.000.000 m^3 even in the driest period, in August. This is much more than the 250.000 m^3 evaporation loss. Therefore, the Rinya will be able to maintain higher water level in the oxbow lake.

Naturally, the option of conducting water to the oxbow through a canal directly from the Drava has also been considered. Unfortunately, there are only some days in a year when the water level of the Drava main channel is higher than that of the Old Drava oxbow lake and gravitational water replenishment could be achieved (cf. Chap. 21). Thus, this option was found to be too costly and rejected.

Nature conservation has to cope with the traditional utilization of the lake for angling on both the Hungarian and Croatian banks. The problem is not presented by angling itself, but by the gangplanks built on the shore, which are now unused, dilapidated and nothing more than an eyesore. In cooperation with the local anglers' association, the DDNPI demolished 50 abandoned gangplanks and replaced three of them with new and simple structures, which fit into the landscape.

A further complication of revitalization planning and implementation is presented by the location of the Croatian-Hungarian national border, which crosses the oxbow (Figs. 20.15 and 20.16). Consequently, any intervention on either side directly affects the territory of the neighboring country, too.

Fig. 20.15 The national border mostly follows the centerline of the Old Drava oxbow lake

Fig. 20.16 Old border stone on the shore of the oxbow lake (photo by Tibor Parrag)

The nature conservation organizations of both countries decided to design a LIFE project for the rehabilitation of the Old Drava of Barcs and the related research and dissemination tasks. Agreed on objectives, they can regularly consult on the actions to take and inform each other about the achievements. A good example for international cooperation is that the water retention structure was conceived by Hungarian planners, the technical plans prepared by Croatian engineers, built by the Hungarian partner and operated by the Croatian partner. At present, the engineering plans are completed and environmental and water right permitting procedure is underway.

20.6 Conclusions

The achievements of revitalization can possibly be evaluated after several decades. The evaluation largely depends on the approach to nature conservation objectives: whether we accept that abandoned channels are doomed to disappear due to the natural process of succession or we insist on slowing down or reversing natural processes with all means. Notwithstanding, both hydromorphological and biological monitoring are badly needed for the evaluation of the success of revitalization. However, the financial resources reserved for such activities are usually limited.

In future, thoughts have to be devoted to the opportunities of a larger-scale rehabilitation project, embracing a longer river section and also the main channel, where a (partial) restoration of the natural dynamics of floods would be possible.

References

Ábrahám L (ed) (2005) Biomonitoring along the River Drava in Hungary, 2000–2004. Natura Somogyiensis 7:1–207. http://www.smmi.hu/termtud/ns/ns7.htm

BioRes Bt (2015) Az Ó-Dráva ökológiai alapállapot felmérése (Ecological baseline survey of the Old Drava). BioRes Bt, Pécs. 133 pp (in Hungarian). http://wwf.hu/media/file/1475483468_1-Jelentes-O-Drava-BioRes-2015-12-18_JZs.pdf

DDKÖVÍZIG (2004) A Dráva folyó magyar területeket is érintő mellékágainak fejlesztése (Development of side channels of the Drava River partly located on Hungarian territory). Manuscript report. South Transdanubian Environmental Protection and Water Management Directorate, Pécs (in Hungarian)

DDNPI (2008) A Duna-Dráva Nemzeti Park természetvédelmi kezelési terve (Nature conservation management plan for the Danube-Drava National Park). Danube-Drava National Park Directorate, Pécs (in Hungarian)

DDNPI (2010) Dráva mellékágak rehabilitáció. Drávatamási felső mellékág vízjogi engedélyezési terv (Rehabilitation of Drava side-arms: Water rights permission plan for the Drávatamási Upper side-arm). Manuscript. Danube-Drava National Park Directorate, Pécs (in Hungarian)

DDNPI (2014) Conceptual overview of the revitalization options for the side-arm system of Vízvár-Bélavár. Danube-Drava National Park Directorate, Pécs. 184 pp. http://www.danubeparks.org/files/1985_Drava_vizvar_revitalization_ENG.pdf

DDNPI (2015) A Barcsi-Ó-Dráva ökológiai állapotjavítását célzó lehetséges műszaki beavatkozások egyeztetett koncepció terve (Negotiated conception for engineering actions aiming at improving the ecological status of the Barcs Old Drava). Danube-Drava National Park Directorate, Pécs (in Hungarian)

European Commission (2000) Directive 2000/60/EEC. Establishing a framework for community action in the field of water policy. Official Journal of the European Communities, Luxemburg. L327. 1–71. http://eur-lex.europa.eu/legal-content/EN/TXT/?uri=CELEX:02000L0060-20141120

FISRWG (2001) Stream Corridor Restoration: Principles, Processes, and Practices. Revised edition. Federal Interagency Stream Restoration Working Group, Washington DC. 637 pp. https://www.nrcs.usda.gov/Internet/FSE_DOCUMENTS/stelprdb1044574.pdf

Haraszthy L (ed) (2014) Natura 2000 fajok és élőhelyek Magyarországon (Natura 2000 species and habitats in Hungary). Pro Vértes Közalapítvány, Csákvár. 200 pp (in Hungarian). http://www.novenyzetiterkep.hu/sites/novenyzetiterkep.hu/files/natura-2000-fajok-es-elohelyek-magyarorszagon-ELOHELYEK-L.pdf

Harka Á, Sallai Z (2004) Magyarország halfaunája (Fish fauna of Hungary). Nimfea Természetvédelmi Egyesület, Szarvas (in Hungarian)

Holubova K, Lukač M, Sumbal J, Kališ J, Čapeková Z, Šporka F, Tóthová L, Černy J (1999) Restoration of the Slovak Morava River meanders. Phare project final report. Water Research Institute, Bratislava, Slovakia

Lazowski W, Schwarz U (2011) Auenland. Das Aueninventar als Grundlage einer österreichweiten Auenstrategie. Naturschutzbund, Salzburg, pp 86–89

Life Slovenija (2017) Riparian Ecosystem Restoration of the Lower Drava River in Slovenia. DOPPS, Ljubljana. https://lifeslovenija.si/en/obnova-recnega-ekosistema-nizinskega-dela-drave-v-sloveniji/

Nestler JM, Galat DL, Hrabik RA (2016) Side channels of the impounded and middle Mississippi river: opportunities and challenges to maximize restoration potential. Environmental Laboratory of U.S. Army Engineer Research and Development Center, Vicksburg, MS. 65 pp (Report ERDC/EL CR-16-4). http://www.dtic.mil/dtic/tr/fulltext/u2/1013224.pdf

Pécsi HIDROTERV Bt. (2010) Dráva mellékágak rehabilitáció. Drávapalkonyai mellékág vízjogi engedélyezési terv (Rehabilitation of Drava side-arms: Water right permission plan for the Drávapalkonya side-arm). Pécsi HIDROTERV Bt., Pécs (in Hungarian)

Purger JJ (ed) (2013) Biota and rehabilitation of four Drava river side-branches in Hungary. BioRes Bt., Pécs. 164 pp

Purger J (ed) (2015) Az Ó-Dráva ökológiai alapállapot felmérése (Baseline survey of the Old Drava). Manuscript. Institute of Biology, University of Pécs, Pécs (in Hungarian)

Rákóczi L, Szekeres J (2008) Environmental effects of industrial dredging on alluvial riverbeds. In: International symposium Drava River Vision. 23–25 September 2008, Maribor, Slovenia

Riquier J, Piégay H, Lamouroux N, Vaudor L (2017) Are restored side channels sustainable aquatic habitat features? Predicting the potential persistence of side channels as aquatic habitats based on their fine sedimentation dynamics. Geomorphology 295:507–528. https://doi.org/10.1016/j.geomorph.2017.08.001

Roni P, Hanson K, Beechie T (2008) Global review of the physical and biological effectiveness of stream habitat rehabilitation techniques. North Am J Fish Manag 28(3):856–890. https://doi.org/10.1577/M06-169.1

Roni P, Beechie ATJ, Bilby RE, Leonetti FE, Pollock MM, Pess GR (2002) A review of stream restoration techniques and a hierarchical strategy for prioritizing restoration in Pacific Northwest watersheds. North American J Fisheries Manag 22:1–20. https://doi.org/10.1577/1548-8675(2002)022<0001:arosrt>2.0.CO;2

Sallai Z (ed) (2004) A drávai táj természeti értékei (Natural values of the Drava landscape). Nimfea Természetvédelmi Egyesület, Szarvas. 42–68 (in Hungarian)

Sodhi NS, Ehrlich PR (eds) (2010) Conservation biology for all. Oxford University Press, Oxford, 369 pp

Steiner F, Holubová K, Nemetz S (2014) Revitalisierung der March. via donau, Wien - Výskumný ústav vodného hospodárstva, Bratislava. 108 pp. http://www.etc-more.eu/fileadmin/site/etc_more/MoRe.pdf

Thomas CC, Huber C, Skrabis K, Sidon J (2016) Estimating the economic impacts of ecosystem restoration—methods and case studies. USGS Open-File Report 2016–1016. Reston, VA. 98 pp. https://doi.org/10.3133/ofr20161016 https://pubs.usgs.gov/of/2016/1016/ofr20161016.pdf

UNESCO (2012) Mura-Drava-Danube. UNESCO, Paris. http://www.unesco.org/new/en/natural-sciences/environment/ecological-sciences/biosphere-reserves/europe-north-america/croatiahungary/mura-drava-danube/

VKKI (2010) Vízgyűjtő-gazdálkodási terv. Dráva részvízgyűjtő (Drainage basin management plan: Drava partial catchment). Vízügyi és Környezetvédelmi Központi Igazgatóság, Budapest. 168 p (in Hungarian). http://www2.vizeink.hu/files/Reszvizgyujto_VGT_Drava.pdf

WWF (2001) A Duna-Dráva Nemzeti Park természetvédelmi kezelési terve Dunai szakasz (Nature conservation management plan for the Danube section of the Danube-Drava National Park). World Wide Fund for Nature Hungary, Budapest. 118 pp. (in Hungarian)

WWF (2009) Lifeline Drava-Mura 2009–2020. A Plan for Conserving and Restoring the Drava and Mura Rivers for Nature and People. World Wide Fund for Nature, Vienna. 34 pp. https://www.euronatur.org/fileadmin/migration/uploads/media/DRAVA_Vision_final_06042009.pdf

WWF (2013) Assessment of the river and floodplain restoration potential in the transboundary UNESCO biosphere reserve Mura-Drava-Danube. World Wide Fund for Nature, Vienna. 158 pp

WWF (2016) DRAVA LIFE—Integrated management of rivers. World Wide Fund for Nature, Vienna. http://www.drava-life.hr/en/project/

Chapter 21
Landscape Rehabilitation: The Old Drava Programme

Dénes Lóczy and József Dezső

Abstract Floodplains are highly sensitive to human pressure. The lower sections of the Hungarian catchment of the Drava River, particularly the Drava Plain, have suffered large-scale landscape degradation in recent decades. The negative influences affected both the physical and socio-economic environment. To counter negative impacts from upstream flow impoundment, bed material excavation and other kinds of human impact, a comprehensive government project of landscape rehabilitation, the Old Drava Programme, was launched in Hungary. In the core of the Programme, the water replenishment scheme focuses on the improvement of water availability of the floodplain through replenishment indirectly from the main river channel. The scheme is meant to take advantage of a network of abandoned drainage elements (oxbows, abandoned channels, levee crevasses, backswamps) in the floodplain. On this basis, an ambitious landscape management project is designed with the announced long-term objective of significantly improving economic (employment), social (integration of ethnicities), and cultural (preservation of cultural heritage and its utilization for increasing tourism potential conditions). Rehabilitation potential is used as a measure to express the extent to which the scope of ecosystem services/landscape functions can be broadened. Water availability and the ensuing landscape transformations are monitored with the purpose of assessing the efficiency of the core project of the Old Drava Programme (a water transfer scheme) in the test area of the Cún-Szaporca oxbow. Based on the findings of monitoring the short-term success of the first lake replenishment campaign is evaluated. Through the assessment of expected provision of ecosystem services, the long-term benefits and deficiencies of the scheme are highlighted.

Keywords Floodplain · Rehabilitation · Oxbows · Water replenishment Groundwater · Socioeconomic setting · Ecosystem services

---

D. Lóczy (✉) · J. Dezső
Department of Physical and Environmental Geography, Institute of Geography and Earth Sciences, Faculty of Sciences, University of Pécs, Ifjúság útja 6, Pécs 7624, Hungary
e-mail: loczyd@gamma.ttk.pte.hu

J. Dezső
e-mail: dejozsi@gamma.ttk.pte.hu

D. Lóczy (ed.), *The Drava River*, Springer Geography,
https://doi.org/10.1007/978-3-319-92816-6_21

21.1 Introduction

For the remediation of rivers and their floodplains, i.e. the elimination of degraded conditions, several concepts are employed. River *recovery* is defined as a sequence of stages of geomorphic adjustment governed by the nature of the landscape and its sensitivity to floods following disturbance (Sparks et al. 1990; Fryirs and Brierley 2000; Brierley and Fryirs 2005). The often very limited space available for regulated rivers as geomorphic agents (Schiemer et al. 1999; Piégay et al. 2005), however, does not normally allow recovery.

River *restoration* is conceived as "the complete structural and functional return to a predisturbance state" (Cairns 1991, p. 187). In eastern Central Europe, this concept is formulated in the following way: River restoration improves river quality and allows the recovery of its previous functions (Macura and Izakovičová 2000; Eiseltová et al. 2007). The ecological concept of *resilience* (i.e. a system's potential to recover biophysical properties and processes following disturbance—Niemi et al. 1990) is central to holistic restoration schemes (e.g. along the Missouri—National Research Council 2002). Areas with high potential for ecological recovery and low socioeconomic constraints have the greatest potential for future restoration (Hulse and Gregory 2004). However, complete restoration is a goal not commonly achievable or even desirable (Downs and Thorne 2000).

The introduction of another term seems to be necessary. *Rehabilitation* means "the partial structural and functional return to a pre-disturbance state" (Cairns 1991, p. 187) or, in a holistic sense, "the return of an ecosystem to a close approximation of its condition prior to disturbance" and this can never be perfect (National Research Council 1992, p. 18).

For similar activities, *revitalization* is often the preferred term in Hungary (see Chap. 20 in this volume) and Croatia (e.g., Dragun et al. 2014). The emphasis here is on planning (landscape architecture) with the purpose of re-creating habitats for plants and animals. Wetland *mitigation* or compensatory mitigation, on the other hand, is a legal term, which refers to human interventions to compensate for wetland losses as prescribed by law, for instance, through the creation of constructed wetlands (Kentula 2000).

The concept of rehabilitation, as used in this chapter covers all measures towards improved ecological (environmental) functioning of the system (Lóczy 2013). Rehabilitation potential is central to any rehabilitation scheme as it is a tool to measure the realistic opportunities for re-establishing ecosystem services/landscape functions (Gren et al. 1995; Gilvear et al. 2013). In spite of rather similar formulations of concepts, in this respect, the target of rehabilitation (e.g., with view of future water availability or species composition) is markedly different from that of restoration defined in a strict sense (Jennings and Harman 1999; Lóczy et al. 2017). In addition, *stabilization* is also cited as another type of river remediation, which aims to exclude both aggrading and degrading conditions over time (Jennings and Harman 1999). Whatever approach of river and floodplain remediation is decided on, it is advisable to observe the first 'rule' stated by Leopold (1949): that at least the interventions should do no harm to the system. In addition, the scheme should be satisfactory in the light of social expectations and perception of the landscape (Dufour and Piégay 2009) or 'aspirations of the public' as it is stated in the European Landscape Convention (Council of Europe 2000).

21.2 The Fluvial Environment and Society

Changes in river mechanism, cutoffs and channel shifts, accumulation and degradation, channel broadening or narrowing had been parts of the pre-regulation fluvial systems. All over Europe dam construction and river channelization were major interventions into the life or rivers (Petts 1984), which significantly reduced the space available for river action (which is now restricted to the active floodplain). The geomorphic evolution of the 'protected floodplain' took a new path, instead of fluvial processes, it became governed mostly by the influences of human land use (cultivation—see Chap. 3 in this volume) and natural vegetation.

The 'natural' channel pattern of the Drava River was well-developed meandering and locally anastomosing accompanied by a broad convex floodplain with natural levees, abandoned channels and backswamps. Beginning with 1750, river channelization divided the area into active and 'protected' floodplain (see Chap. 8). Dyke construction ensured increasingly safe conditions for agricultural land use and settlement development even on lower-lying surfaces (Gyenizse and Lóczy 2010).

The environmental problems of the Drava channel and its floodplain are described from numerous aspects in the previous chapters of this book. The main statements are summarized here and some selected additional facts with socioeconomic as well as policy implications are cited.

Commercial *extraction* of sand and gravel are strictly restricted in all countries along the Drava. At Petrijevci, Croatia, some 30 km from the confluence with the Danube, however, Croatian water management authorities and private firms illegally excavated more than 3,000,000 m^3 of sand for motorway construction (Popovič and Mikuska 2010). As an example of conservative water management policy, a 56 km-long stretch of the Drava River is overregulated with some 112 different structures (Popovič and Mikuska 2010). At Osijek, a barrage and a 25-km-long reservoir were established for the purposes of electricity generation, flood protection, irrigation, navigation, tourism and recreation—but despite strong opposition from NGOs and without previous environmental impact assessment.

Flow regulation measures have been repeatedly implemented for the purposes of navigation. In principle, for barges of 400–600 Gross Registered Tonnage (GRT) the river is navigable downstream Barcs, but the actual volume of traffic is negligible. A significant development of navigation would require large-scale interventions which are environmentally not acceptable.

In the 20th century, 22 hydroelectric plants were built on the upper Drava sections, partly with peak-time operation, and caused a huge sediment deficit in the river (see Chap. 9). From this development, severe problems for wildlife, agriculture, forestry, groundwater levels and river stability resulted. The most recent and most downstream of the hydroelectric plants (at Dubrava, Croatia, 17 km upstream of the Hungarian border, completed in 1989) has a reservoir of 150,000,000 m^3 capacity (accommodating three days' mean discharge). A beneficial influence of damming is the mitigation of flood hazard. Floods became rarer and the average annual number of flood days dropped from 18 (before 1976) to 2 days (1989–2013).

The principal problems along the Hungarian Drava section are extreme daily *fluctuation of water level* (up to 1.5 on the uppermost Hungarian section), channel incision, and narrowing as well as intensive bank erosion (Kiss and Andrási 2011 and Chap. 11). Dropping water levels are recorded in groundwater observation wells even at 2–3 km distance from the channel. Natural channel development is not possible for most of the floodplain watercourses, 96% of them are in need of channel rehabilitation (AQUAPROFIT 2010).

Both the climatic and energetic *agroecological potentials* of the arable land of the region (30–32 t ha^{-1} and 30.5–31.5 t ha^{-1} biomass production, resp.) are assessed as slightly above average on Fluvisols and Histosols (AQUAPROFIT 2007a). On the 1 to 100 scores range of the D-e-meter Land Evaluation system, the average score of the Drava Plain is 61.9 (Tóth et al. 2014), i.e. falls somewhat, but not significantly, below the average for the Great Hungarian Plain (63.4). Environmental sensitivity, however, is also above the national average. The overwhelming arable land use is interrupted in lower-lying backswamps occupied by grasslands and pastures with scattered fruit trees (open orchard meadows or in German: Streuobstwiesen), which are a particularly valuable type of seminatural habitat. In spite of the favorable natural conditions, stockbreeding (pigs, sheep, cows, poultry) is of subordinate importance, economically not viable and shows decline.

In Hungary the introduction of sustainable and environmentally acceptable *land use* is promoted by the National Agri-Environmental Programme (NAEP, 2000–2004; renewed for 2007–2013—Nemes and High 2011). The Programme is aimed at the conservation of natural resources, establishment of biotope network, rehabilitation of wetlands, management of derelict land. It was followed by the National Rural Development Plan (NRDP, 2004–2006, with emphasis on support for backward areas, afforestation of agricultural land), the New Hungary Rural Development Programme (NHRDP, 2007–2013, which covers watershed and flood risk management, habitat preservation in the practice of forestry and agriculture) (Gálosi-Kovács 2010) and the Darányi Ignác Plan (National Rural Development Strategy, 2012–2020, focusing on rural employment, preservation of water resources, support to local 'social economy'). The implementation of these programmes, however, progresses very slowly in this region with inadequate human resources.

Ethnically, the floodplain along the lower section of the Drava River, the Ormánság region is an area of mixed population (Hungarians, Romas, Croats) characterized by depopulation with outmigration being the prevalent demographic process (Reményi and Tóth 2009). A general lack of capital and human resources (unskilled labor) are typical. On the other hand, since the region benefited from missing industrial development, the environment is preserved in a healthy state, free of industrial pollution. Historically, the disadvantaged position of the regions springs from the dominance of small-sized, often dead-end, villages with single access roads of poor surfacing and peripheral location next to the Croatian border (Tésits 2012; Tésits and Alpek 2012, 2014). At present, however, economic activities are limited to community service programmes and investments by the government.

For cultural and *ecotourism* the Ormánság region has a wide range of attractions, including genuine curiosities such as the Reformed churches with painted wooden ceilings (their restoration is incorporated into the Old Drava Programme), vernacular traditions and crafts on the one hand and rich wildlife and undisturbed landscapes for hikers, cyclists and fans of water sports on the other. However, all previous development projects in tourism have failed and the opportunities could not be exploited until now (Csapó et al. 2011). To this day, the Ormánság region has remained one of the least developed, peripheral areas of Hungary (Gálosi-Kovács et al. 2011).

21.3 History of the Old Drava Programme

The interventions serving rehabilitation in the degraded landscape of the Drava Plain can be planned observing different (internationally or nationally formulated) requirements. Some important guidelines are presented below as background to action plans.

The Water Framework Directive of the European Union (European Commission 2000) describes tasks to improve aquatic and riparian environments. In addition to the reduction of pollutions of various kind and origin, the related Watershed Management Plan for the partial watershed of the Drava River (VKKI 2010) includes the following general targets:

- To make excess water storage facilities capable of retaining water and to reduce the nutrient load of recipient water bodies; the stored excess water made available for irrigation or induced infiltration; to find solutions for the compensation of land proprietors whose land was used for excess water storage;
- To ensure good fishing and angling practice (management of fish-ponds, dammed reservoirs and natural water bodies);
- To ensure the provision of ecosystem services and public functions of fish-ponds;
- To identify constraints for used thermal water inflow into surface waters;
- To improve the hydromorphological conditions of watercourses and lakes (reducing the impacts of previous regulation measures through channel restoration, slowing down riverbed incision, increasing sinuosity and bank diversity, maintaining seminatural conditions in oxbows enhancing connectivity, building

bottom weir, removing organic and inorganic mud fill from lakes, management of aquatic herbal vegetation etc.);

- To restore riparian vegetation (gallery forests) where sufficient space is available or to create an artificial riparian buffer zone (8–10 m wide if wooded with native trees or wider if bush and grass) along the Drava channel where this space is more limited, to protect riparian zones against the spreading of invasive plants;
- To rehabilitate active floodplains, rationalize their land use and make them suitable for accommodating rising flood discharges (identifying flood reservoirs ['polders'], compensating land owners for losses, removing or shifting back flood-control dykes etc.);
- To preserve wetlands through the regulation (restriction) of surface and groundwater utilization and, if necessary, through water transfer;
- To supervise riverbed structures and operate them in a manner to allow longitudinal river connectivity;
- To design navigation routes in an ecologically acceptable manner (with minimum disturbance);
- To govern water resources economically (retention of floodwater, storage of excess water and use for irrigation in drought periods, additional purification of treated sewage in biological filter zones, prevention of pollutions resulting from accidents);
- To ensure environmental/ecological flow throughout the year (to devise appropriate methodology for identifying environmental flow, restriction of water intake, encouraging ethical water use, etc.).

Naturally, the Watershed Management Plan also identifies a series of legal and technical measures which are necessary to reach the above objectives.

In the international Drava Declaration (Department of Carinthia 2008) ten main tasks in the development of environmental conditions of the Drava region are enumerated. The items most closely related to floodplain functions are the following:

- To enhance flood control through water retention in the floodplain;
- To continue restoration activities in the channel of the Drava River and on its floodplain;
- To re-establish ecological connectivity of the Drava River for migratory fish;
- To create a transboundary recreation area;
- To achieve integrated river basin management;
- To promote further regional development in partnership with the resident human populations.

The changes of the fluvial system coupled with climate change reduced water supply and involved the frequent recurrence of droughts and economic decline in the Ormánság region, i.e. the lower Hungarian Drava floodplain. The Hungarian government recognized that to maintain natural conditions and agricultural activities in the future, water replenishment is indispensable. Therefore, water governance is placed in the centre of a comprehensive development project, called the

Old Drava Programme, which was first proposed in 2004 (AQUAPROFIT 2005) and approved by the Government of Hungary on 17 July 2012. The name itself hints at the reconstruction of old conditions (abandoned drainage network) along the Drava River.

The priorities of the Old Drava Programme (Márk et al. 2006; AQUAPROFIT 2010; Salamon 2014) are

- economic development (agriculture, irrigated horticulture and food industry);
- landscape management (afforestation, creation of water surfaces and grazing lands);
- tourism development (angling, rafting, hunting, horse riding, bicycling, built heritage, nature trails, gastronomy, etc.).

In the *first version* of the Old Drava Programme, a new water governance system is envisaged for the region, a combination of water replenishment ensured indirectly from the Drava River (12 $m^3 s^{-1}$), through gravitational water intake and pumping, distribution of water by a main gravitational canal (at 5–10 km distance from the Drava) using elements of the natural drainage network and floodwater retention in the floodplain (Fig. 21.1). According to the water management plan (AQUAPROFIT 2007a), water replenishment would allow the irrigation of 5,000 ha of agricultural area and the establishment of almost 700 ha total water surface in restored lakes and newly-created reservoirs. In modeling the system, maximum (peak-time) water demands were taken into consideration. The original scheme explicitly claimed that excellent water availability (comparable to the conditions typical before river channelization) can be re-created in the Drava Plain (AQUAPROFIT 2010).

Indirect benefits of the water governance system are envisioned to include improvements in economic structure, safety of harvest, providing touristic attractions, stability of ecosystems and subsistence level of population, leading to a higher carrying capacity of the landscape.

In the proposed land use of the floodplain (Table 21.1) traditional occupations (orchards, reed and willow craftsmanship), mostly strongly dependent on good water availability, are planned to be revived, the extension of forested areas is planned and, at the same time, the water demands of nature conservation and large-scale arable farming are intended to be satisfied simultaneously.

The success of river and floodplain rehabilitation efforts largely depends on reaching the water management targets formulated in the scheme (AQUAPROFIT 2007a), i.e. ensuring water availability for the floodplain and a stable water budget for oxbows. Floodwater storage largely depends on the geomorphology of the floodplain. A recent survey (Schwarz 2014) finds favorable conditions in the Drava floodplain and envisages up to 3,000 ha area with significant water retention potential proposed for floodplain restoration along the lowermost 25-km-long Hungarian section of the Drava River.

In the *new version* (Pécsi HIDROTERV Bt. 2015) the emphasis shifted towards environmentally more acceptable solutions, which promote water retention in the

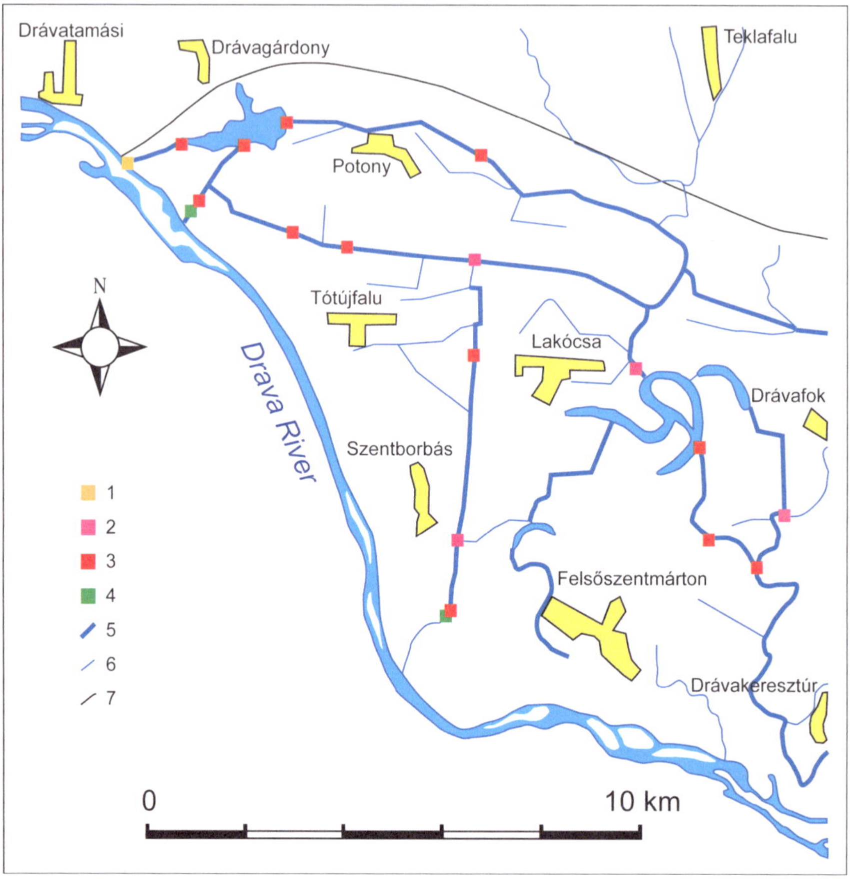

Fig. 21.1 Detail of the water recharge system of the original version of the Old Drava Programme (upper section). Water conducted from the Drava River into a reservoir and distributed through a canal network which partly utilizes abandoned channels. 1, water abstraction from the Drava; 2, distributory structures; 3, pumping station; 4, sluice, dam; 5, main canals; 6, secondary canals; 7, boundary of planning area. *Source* AQUAPROFIT (2007a)

floodplain, taking advantage of natural landforms. The provision of ecological water demands for the ecosystems of floodplain wetlands acquired primary importance. At the same time, the safe drainage of excess water and collection in recipient water bodies has to be ensured. Instead of calculating with maximum water needs, irrigation water demands were resurveyed and found to amount to about 2,767,000 m^3 y^{-1} (2,422 ha area to be irrigated) (Pécsi HIDROTERV Bt. 2015).

The new Old Drava Programme is more prepared to face extreme weather situations (i.e. years with too high or too low precipitation) to be expected under

Table 21.1 Land use proposals for landforms of different elevation (based on AQUAPROFIT 2007b)

Landforms	Elevation range (m)	Frequency of inundation	Proposed land use
Sand dunes	100–110	Flood-free, occasional excess water from precipitation	Built-up, arable, forest, grassland, orchard, hunting, gathering (mushrooms, forest fruits, etc.), apiculture, tourism
Natural levees	98–105	Rare and short-term inundation	Orchard, horticulture, forest, hunting, gathering (mushrooms, forest fruits etc.), apiculture, tourism
Low floodplain level	94–98	Regular (yearly or seasonal) inundation	Pasture, meadow, forest, fishing, growing swamp plants, hunting, gathering (medicinal plants, dried flowers, raw materials for crafts, etc.), apiculture, tourism
Backswamps and infilled abandoned channels	90–98	Waterlogged for most of the year	Fishing, reed-cutting, aquatic plants, waterfowl, hunting, gathering (medicinal plants, dried flowers, etc.), apiculture, water tourism

climate change. For agriculture (and locally forestry), the supply of irrigation water will be crucial in the future.

The pilot action of the Programme was the completion of a feeder canal (length: 1,360 m; capacity: 0.4 m^3 s^{-1}; elevation: 93.5 m above sea level; slope: 0.005) from the Fekete-víz Stream (mean discharge: 4.5 m^3 s^{-1}) to Lake Kisinc of the Cún-Szaporca oxbow in 2016. According to the water management plan (AQUAPROFIT 2007a), a single water replenishment intervention was conceived for March (when the average water flow of the Fekete-víz Stream is 6.379 m^3 s^{-1} and 90%-probability flow is 2.0 m^3 s^{-1}). Later, however, need for summer (June–July) feeding also arose (DDKÖVIZIG 2012). From the impounded Fekete-víz Stream replenishment will require 10–19 days in spring, at 0.4 m^3 s^{-1} (43,200 m^3 d^{-1}) rate, totaling 515,000 m^3 inflow, and 24–28 days in June at 0.33 m^3 s^{-1} (24,512 m^3 d^{-1}) rate, totaling 535,000 m^3 (DDKÖVÍZIG 2012).

The first replenishment campaign took place in spring 2016. Its goal was to fill up the oxbow to at least 91.3 m elevation and merge the individual lakes of into an open water surface of 20.7 ha area. (The actual areas with open water surface are the following: Lake Kisinc: 5.3 ha; Lake Szilihát: 2.7 ha; Lake Inner Hobogy 1.5 ha; Lake Lanka: 1.3 ha.)

21.4 Assessment of the Programme from Nature Conservation Aspect

In the Hungarian Drava Plain ca 25,000 invertebrate species are found. Oak forests with meadows and trees of variable ages show the highest biodiversity. Wetlands and dry grasslands show a somewhat lower diversity but also host rare species (leaches, snails, crayfish, beetles, butterflies, stinging insects etc.). In most habitats, water availability is crucial: amphibian larvae develop in water and desiccation is a major threat for them. Some mammals (such as otters, ermines) are also bound to water.

In general, the experts of the Danube-Drava National Park attribute positive impacts to the Old Drava Programme (Pécsi HIDROTERV Bt. 2015), primarily for the water retention objectives. The raising of groundwater levels (particularly in forests) is also considered a possible favorable outcome of the Programme. Direct flooding is feasible for abandoned channels and backswamps, but wet meadows should only be waterlogged in spring. For bogs, river water inflow may have a negative effect on vegetation. In reed and sedge beds water level raising by more than 20–30 cm could inhibit the nesting of herons. Much higher (50–60 cm) increase in water level is needed to save the desiccating alder groves—but it should be implemented over several years. The oxbow lakes of Bresztik, Old Drava and Lake Fekete (see Chap. 12) would benefit of even 1 m higher water level, while the lakes Verság and Piskó are more sensitive to this kind of change and cannot bear more than 25–30 cm rise.

Anglers (in the Baranya County Association of Angling Clubs) welcome the plans for the establishment of new open water surfaces, replacing some swamps or extending existing lakes (e.g. at Sellye). The sedimentation of oxbow lakes (at Majáthpuszta, Zaláta, Hótedra, Bresztik) endangers their water storage capacity and fish habitats. Deposition is also rapid in the old bed of the Fekete-víz Stream, which used to be an excellent spawning site. Rehabilitation could reverse unfavorable tendencies also here.

21.5 Alternative Approaches to the Environmental Assessment of the Programme

For any river restoration scheme, short and long-term impacts should be evaluated separately. Experience gathered from monitoring and the first replenishment campaign allows us to summarize short-term effects (Dezső et al. 2017).

One option to assess the long-term consequences of water management interventions within the Old Drava Programme was a comprehensive evaluation of floodplain functioning by ecological indicators (Palmer et al. 2005). It can be based on the collection of both archive and actual data and the findings of environmental monitoring (Woolsey et al. 2007; Morandi et al. 2014). Several alternative ways for such an evaluation have been suggested: environmental flow specification (Arthington et al. 2006, 2009), the Floodplain Evaluation Matrix (FEM—Habersack et al. 2015) and analyses following different checklists of ecosystem services (Heal 2000; Ramsar Convention Secretariat 2010).

Kentula (2000) distinguishes between compliance, functional and landscape success of rehabilitation projects. Only the latter ensures the integrity of the region from an environmental aspect. The real challenge is to find an all-embracing set of reliable indicators for judging the success of landscape-scale rehabilitation.

21.5.1 The First Replenishment Campaign

An assessment of empirical data from monitoring may also give an idea of the achievements of water replenishment. Based on modeling seepage from the oxbow lake (see Chap. 14), hydrological scenarios were proposed for both replenishment rate and duration of water retention. In accordance with the water management plan of the Old Drava Programme (DDKÖVÍZIG 2012), the first scenario set the replenishment rate at a 30,000 $m^3\ d^{-1}$, with a water level rise from 90.5 to 91.5 m above sea level.

Our model (for details see Dezső et al. 2017) shows that on day 25 of the replenishment water level reaches an elevation of 91.3 m and this water level remains relatively stable for a long time (Fig. 21.2). However, occasionally, during exceptionally rainy periods, replenishment is capable to raise water level to 91.5 m (DD-KVTF 2013). The rising water level in the oxbow considerably elevates the hydraulic pressure head above the adjacent areas. With the increasing volume in the oxbow, the potential contact surface (from where seepage is possible) also increases.

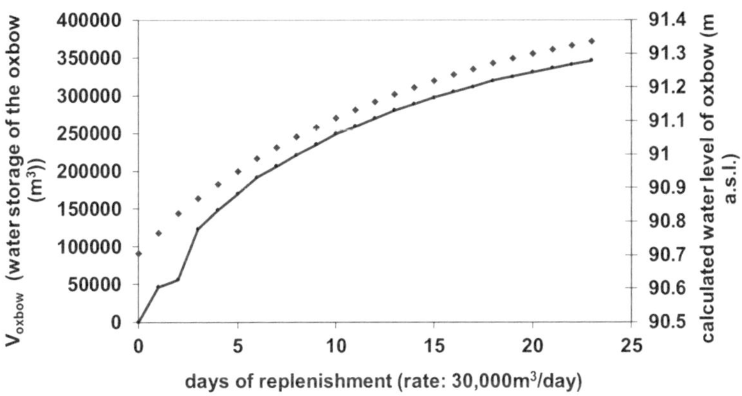

Fig. 21.2 Modelled relationship between water storage (V_{oxbow}) and relative water level (h_{oxbow}) during water replenishment to the oxbow (by József Dezső)

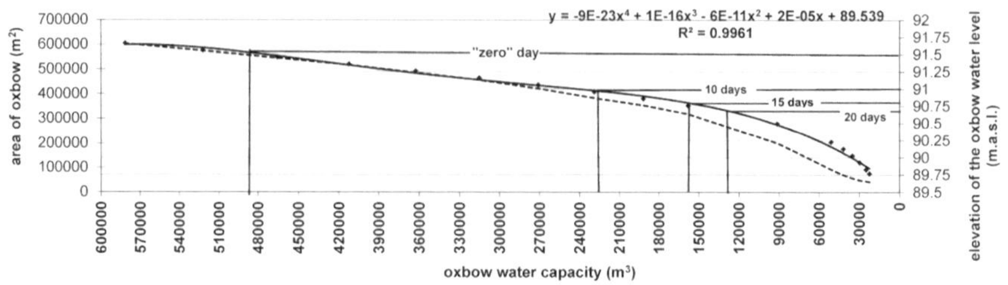

Fig. 21.3 Change of water surface area as a function of volume of water stored in the oxbow at 91.5-m replenished water level (by J. Dezső)

The second scenario was intended to estimate seepage rate from the oxbow and, thus, water retention there. On day 20 of the replenishment water level drops by 70 cm compared to the initial water elevation, while both the volume of the water stored and the area of the water surface shrink to about one third of the initial value (Fig. 21.3). Due to the high seepage rate from the oxbow, the replenishment would last too long, more than 25 days, and thus require improportionately greater amounts of water.

21.5.2 Ecological Indicators

The assessment of the long-term impacts of water recharge requires more comprehensive (but less quantitative) approaches. Palmer et al. (2005) propose five criteria for the ecological assessment of river restoration projects:

1. a guiding image (reference sites, called traditionally in German: Leitbild—Kern 1994), i.e. a more dynamic, healthy river of commeasurable dimensions should be specified;
2. measurable improvements in environmental (hydrogeomorphological and ecological) conditions have to be targeted and achieved;
3. the system has to be turned self-sustaining and resilient to external perturbations with minimal need for follow-up maintenance;
4. during the construction phase, no lasting harm should be inflicted on the ecosystem;
5. both pre- and post-project appraisal has to be completed (Downs and Kondolf 2002) and made available for the public.

In accordance with the above criteria a list of suitable indicators were suggested (Table 21.2), partly based on easily measurable parameters and partly checked on yes-or-no basis. The complete list is assumed capable to demonstrate the environmental success of the project.

Connectivity issues and landscape pattern should enjoy priority in such projects. The restoration of meandering reaches of abandoned river reaches would slow

Table 21.2 Indicators (modified after Palmer et al. 2005) for the ecological evaluation of the river and floodplain rehabilitation scheme

Wetland functions	Weight (w)	Present ESs provision	Exploitation level of potential: rating (s_p)	Planned ESs provision	Rating (s_f)
Hydrological functions					
Short-term storage of surface water	**1.00**	**Flood waves conducted downstream as rapidly as possible, floodwater storage only in side arms of the active floodplain**	**Insufficient: 3**	**Floodwater retention in oxbows and backswamps (ca 10 million m^3 = ca 1 day of absolute maximum discharge**	**5**
Long-term storage of surface water	0.75	Storage in oxbows with appropriate conductivity with the main channel	Insufficient: 2	Storage in better connected oxbows, but significant water losses	3
Storage of subsurface water, moderation of groundwater flow	0.75	Limited infiltration and seepage from oxbows; drawdown by the Drava; dropping groundwater table, low soil moisture content in growing season	Insufficient: 2	Moderate improvement in soil moisture conditions	3
Dissipation of energy at the land/water interface	0.50	Bank erosion precluded by riparian vegetation	Good: 4	No major change expected	4
Biogeochemical functions					
Removal of nutrients and contaminants	0.75	Average primary production, low levels of contaminant loading	Low: 2	Some increase in primary production	3
Retention of particulates	0.75	Sediment load of the Drava only reaches the side-arms	Low: 2	Sedimentation in re-connected oxbows	3
Export of organic carbon	0.25	Carbon storage for floodplain forests along the Danube: 450–500 t ha^{-1}	Medium: 3	Slight increase due to afforestation of some poor-quality arable land	4
Habitat functions					
Maintenance of plant and animal communities	0.25	Many aquatic and riparian habitats, majority of them Community Importance (Natura 2000); riparian forests mostly of native	Medium: 3	Better water provision helps to sustain, even improve the state of aquatic and riparian communities and habitats; some extension of native tree	4

(continued)

Table 21.2 (continued)

Wetland functions	Weight (w)	Present ESs provision	Exploitation level of potential: rating (s_p)	Planned ESs provision	Rating (s_f)
		species, but spreading invasive species		species; raising groundwater table creates wetter habitat types, e.g. sedge beds	
Landscape pattern	0.10	Overwhelming agricultural land use, few ecological corridors, oxbows are valuable refugia	Insufficient: 2	Through conversion of arable land more corridors are created	3

down current velocity and improve groundwater replenishment. Connectivity could also be promoted by grading, breaching dykes, or widening the active floodplain (Palmer et al. 2005). In the case of the Drava such profound interventions are not envisaged. The dredging of oxbow lakes is equally costly and regarded ecologically ineffective on the grounds of the disturbance caused and the need for associated constant maintenance. (However, it is unavoidable in lakes where angling is envisioned as a recreational activity.)

21.5.3 Environmental (Ecological) Flow

Applying the checklist by Palmer et al. (2005), it is a challenge to translate 'natural flow regime' into quantitative environmental flow prescriptions for individual river reaches (Arthington et al. 2006). Hungarian ecologists claim that the ecological water demands of habitats have not yet been specified (Völgyesi 2009). For the water uptake of plants groundwater table depth is of primary significance, but its dependence on instream flow is difficult to establish. River rehabilitation is ineffective ecologically if it focuses exclusively on maintaining minimum instream flow, but fails to re-establish an approximately natural annual surface and subsurface flow regime for the entire riparian zone (Palmer et al. 2005). Our monitoring also proved the view by Sanford (2002) that the groundwater-recharge efficiency of infiltration is highly variable.

A recently elaborated approach for the estimation of 'low-water reserves' of river catchments in Hungary (Szalay 2009) can be helpful since it differentiates water bodies identified in the WFD (European Commission 2000). Low-water reserves are, however, to be distinguished from both environmental and ecological flow. The proposed flow values do not satisfy the criteria of ecological flow as defined in the Hungarian act on nature conservation. Also it does not say anything about the required frequency of medium and high flows—although these could only be vital for the survival of aquatic biota. Therefore, the environmental flow approach has not been considered as a real alternative for floodplain rehabilitation appraisal.

21.5.4 Floodplain Evaluation Matrix (FEM)

Recently three new approaches have been elaborated for integrated flood management in Austria (Habersack et al. 2010):

- The *Floodplain Evaluation Matrix (FEM)* serves the assessment of floodplains along individual river reaches from hydrological/hydraulic, ecological, and sociological viewpoints (Chovanec et al. 2005; Habersack et al. 2010, 2015; Schwarz 2014).
- The indicator *Minimum River Morphological Space Demand* (abbreviated from German as *FMRB*) is defined, based on flood analysis, as three to sevenfold the existing riverbed width, where no construction or cultivation should be allowed.
- The *Spatially Variable Vegetation Management (VeMaFLOOD)* method identifies dynamic, transition and sensitive zones of vegetation.

The FEM method (Chovanec et al. 2005) classifies the oxbows of the Hungarian Drava floodplain mostly to class H2 (water body with limited connectivity to the main river channel). The method includes hydrological (flood peak reduction, flood wave propagation, floodwater retention: floodplain width, slope and roughness, flood risk/inundation depth), hydraulic (water stages, current velocity, specific runoff), ecological (landscape pattern, water regime, connectivity, biodiversity and its conservation) as well as societal factors (land use classes, flow of communication). We perfomed a qualitative FEM analysis separately for the three flood embayments (Ormánság, Kémes-Drávaszabolcs and Old—see Fig. 21.4) of the Drava Plain (Table 21.3).

It is clear from the table that the upper and middle floodplain segments will be significantly affected by the Old Drava Programme and the impacts will be positive, while in the lower segment, which is outside the range of the Programme, no significant change is expected.

21.5.5 Ecosystem Services Approaches

Although any classification system of ecosystem processes and services is laden with redundancy (Wallace 2008), it is useful to compare the present degree of fulfillment of services with that to be attained through the implementation of the Old Drava Programme (Table 21.4). The range of attainable benefits characterizes

Table 21.3 Assessment of the conditions before and after the implementation of the Old Drava Programme using the Floodplain Evaluation Matrix (modified after Habersack et al. 2012) for the flood embayments of the lower Hungarian Drava floodplain (scale: 5, highest priority, 1, lowest priority)

Factors	Assessment of baseline situation in the flood embayments			Assessment of the rehabilitated flood embayments		
	Ormánság	Kémes-Drávaszabolcs	Old	Ormánság	Kémes-Drávaszabolcs	Old
Hydrological						
Flood peak reduction	4	3	3	5	4	3
Flood peak propagation	3	3	3	4	4	3
Floodwater retention	4	4	4	5	5	4
Flood risk/inundation depth	4	4	4	5	5	3
Hydraulic						
Water stages	4	4	4	5	5	4
Current velocity	3	3	3	4	4	3
Specific runoff	2	2	3	3	3	3
Ecological						
Landscape pattern	3	2	2	4	3	2
Water regime	2	2	2	3	3	2
Connectivity	1	1	1	2	2	1
Biodiversity	3	3	2	4	4	2
Conservation	3	3	3	3	3	3
Societal						
Land use classes	3	3	3	4	4	3
Flow of communication	2	2	2	4	4	3
Total	2.9	2.7	2.7	3.9	3.7	2.7

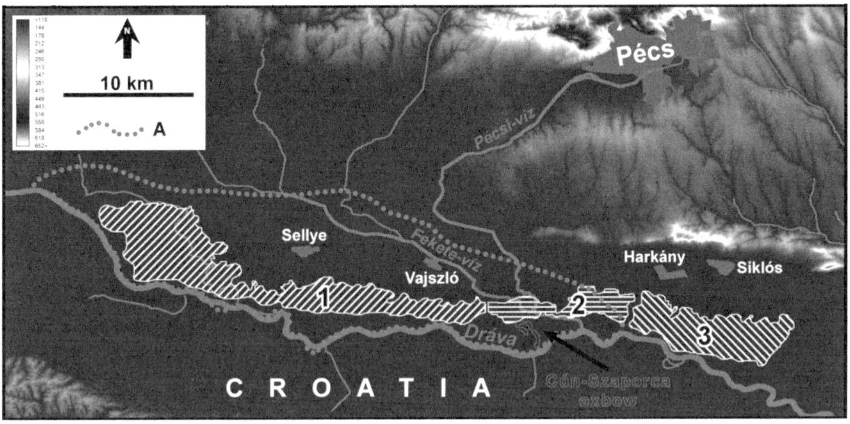

Fig. 21.4 Flood embayments along the lower Drava section. A, boundary of the Old Drava Programme planning area; 1, Ormánság embayment; 2, Kémes-Drávaszabolcs embayment; 3, Old embayment

the dimensions of rehabilitation potential. In landscape ecology, 'active' and 'passive' landscape functions (Konkoly-Gyuró 2011)—interpreted in a broader sense than ecosystem services—are also suggested as a basis for evaluation. Active functions are services provided by human activities, while passive functions are regulating and subsistence functions of the natural systems (environmental regulation, habitat protection, biomass generation, and production, etc.). This approach allows the better consideration of benefits originated from human activities along with natural landscape functions (WWF 2002).

Table 21.4 shows that six items in the Ramsar Convention list show significant growth in services. The predicted improvements are most striking in water-related regulatory (flood control, flood storage, local climate regulation) and cultural services (water sports).

An alternative list was compiled by the United States Army Corps of Engineers (after Brinson et al. 1995). It focuses on water-related ecosystem functions, where new water governance promises remarkable improvements (Table 21.5).

Table 21.4 Evaluation of the Old Drava Programme in the light of production and environment-related ecosystem services as identified by Ramsar Convention Secretariat (2010)

Ecosystem services	Weight (w)	Present-day provision	Exploitation level of potential: rating (s_p)	Expected provision	Rating (s_f)
3.5 Vegetational productivity (S)	0.25	Average primary production: 4.62 t ha^{-1}	Medium: 3	Wetland habitats extended—ca 10–15% increase in primary production	4
4.4 Groundwater replenishment (R)	0.75	Limited infiltration and seepage from oxbows; >1 m drawdown by the Drava River in ca 100 m (max. 300 m) wide zone	Inadequate: 2	Increased groundwater recharge from tributary streams	3
4.6 Food for human consumption (P)	0.10	Food production in ca 39,000 ha of arable land (total area: 54,026 ha), mostly large-scale farming (wheat, sunflower, maize)	Good: 5	To be reduced by ca 26,000 ha through land conversion to pasture, meadow and forest, increased landscape-level diversity, small-scale farming	4
4.7 Food for livestock (P)	0.10	Fodder (legumes, turnips, maize, grass) produced in 5,000 ha of arable land, meadow and pasture	Low: 3	Expansion of meadows and wooded pastures over 10,100 ha area	3
4.8 Wood, reed, fiber and peat (P)	0.10	1,420 ha forest area in previously defined project area	Resources in poor condition: 3	Moderate growth in forest areas	4
4.9 Medicinal products (P)	**0.10**	**Small-scale gathering of medicinal plants (chamomille, lime leaves, nettle, hawthorn, rosehip, etc.), some cultivation (poppyseed)**	**Low: 2**	**120–130 species to be collected or grown on organic farms at commercial scale**	**5**
4.11 Other products and resources, including genetic material (P)	**0.10**	**Organic farming by the Danube-Drava National Park; water melon, pumpkin (oil) production;**	**Insufficient: 3**	**Traditional fruit varieties reintroduced in 295 ha of (apple, pear, plum etc.) orchards, marketing**	**5**

(continued)

Table 21.4 (continued)

Ecosystem services	Weight (w)	Present-day provision	Exploitation level of potential: rating (s_p)	Expected provision	Rating (s_f)
		basket weaving; floodplain orchard gene bank of 400 fruit trees; cattle, sheep, pig, poultry and goat keeping (partly native Hungarian breeds: grey Hungarian cattle, 'racka' and 'cikta' sheep)		**frozen, canned and dried vegetables and fruits, jams, syrups, brandies, candies, honey; development of rabbit, goat and sheep husbandry at household scale keeping of native breeds in the Szaporca Visitor Centre**	
4.12 Flood control, flood storage (R)	**1.00**	**Flood waves conducted downstream as rapidly as possible, floodwater storage only in side arms**	**Insufficient: 3**	**Floodwater retention in oxbows (820 ha area, ca 10 million m^3)**	**5**
4.13 Soil, sediment and nutrient retention (R)	0.75	500,000 tonnes of sand and gravel dredged from the Drava River until now	Low: 2	Sedimentation in re-connected oxbows	3
4.15 Other hydrological services (R)	0.50	Neglected network of drainage ditches (total length: ca 200 km; longest: 28 km)	Resources in poor condition: 2	New facilities to collect and store excess water	3
4.16 Local climate regulation/ buffering of the change (R)	**0.50**	**Actual permanent water surfaces of 605 ha area + 1,500 ha area with forest microclimate, desiccating arable land and meadows**	**Low: 2**	**Additional water surfaces of 449 ha total area; annual evaporation of 458,000 m^3 to increase air humidity**	**4**
4.17 Carbon storage/ sequestration (R)	0.25	Carbon storage for floodplain forests along the Danube: 450–500 t ha^{-1}	Medium: 3	Slight increase due to afforestation of some arable land	4
4.18 Recreational hunting and fishing (C)	0.10	Boar, deer and hare hunting; waterfowl; 11 water bodies available for angling	Medium: 3	Further angling facilities to be established on the new water surfaces	4

Table 21.5 Assessment of the predictable achievements of the Old Drava Programme in the system proposed by the United States Army Corps of Engineers (after Brinson et al. 1995)

Wetland functions	Weight (w)	Present ESs provision	Exploitation level of potential: rating (s_p)	Planned ESs provision	Rating (s_f)
Hydrological functions					
Short-term storage of surface water	**1.00**	**Flood waves conducted downstream as rapidly as possible, floodwater storage only in side arms**	**Insufficient: 3**	**Floodwater retention in oxbows and backswamps (820 ha area, ca 10 million m^3)**	**5**
Long-term storage of surface water	0.75	Storage in oxbows with appropriate conductivity with the main channel	Insufficient: 2	Storage in better connected oxbows, but significant water losses	3
Storage of subsurface water, moderation of groundwater flow	0.75	Limited infiltration and seepage from oxbows; drawdown by the Drava; dropping groundwater table, low soil moisture content in growing season	Insufficient: 2	Moderate improvement in soil moisture conditions	3
Dissipation of energy at the land/water interface	0.50	Bank erosion precluded by riparian vegetation	Good: 4	No major change expected	4
Biogeochemical functions					
Removal of nutrients and contaminants	0.75	Average primary production, low levels of contaminant loading	Low: 2	Some increase in primary production	3
Retention of particulates	0.75	Sediment load of the Drava only reaches the side-arms	Low: 2	Sedimentation in re-connected oxbows	3
Export of organic carbon	0.25	Carbon storage for floodplain forests along the Danube: 450–500 t ha^{-1}	Medium: 3	Slight increase due to afforestation of some arable land	4
Habitat functions					
Maintenance of plant and animal communities	0.25	Many aquatic and riparian habitats, majority of them Community	Medium: 3	Better water provision helps to sustain, even improve the state	4

(continued)

Table 21.5 (continued)

Wetland functions	Weight (w)	Present ESs provision	Exploitation level of potential: rating (s_p)	Planned ESs provision	Rating (s_f)
		Importance (NATURA 2000); riparian forests mostly of native species, but spreading invasive species		of aquatic and riparian communities and habitats; some extension of native tree species	
Landscape pattern	0.10	Overwhelming agricultural land use, few ecological corridors, oxbows are important refugia	Insufficient: 2	Moderate afforestation, conversion of arable land; raising groundwater table creates wetter habitat types, e.g. sedge beds	3

21.6 Conclusions

At present, the Old Drava Programme is the largest-scale landscape rehabilitation project in Hungary. Floodplain rehabilitation should be viewed as a landscape ecological issue. Similarly to most European floodplains, along the Drava River flow regulation disrupted communication between the new straightened river channels and the cutoff oxbows. The latter are doomed to disappear within centuries, while no new meanders develop on the floodplain. Although their natural lifespan is naturally short, from the viewpoint of nature conservation the preservation of existing oxbow lakes as valuable geomorphosites and wetlands through active rehabilitation measures is certainly justified (cf. Tockner et al. 1998).

The environmental benefits of the Programme can be judged in the short-term (for instance, from experience with water replenishment to Lake Kisinc in the Cún-Szaporca oxbow) or in the longer perspective, how the new water governance will promote higher natural potentials.

The evaluation of rehabilitation potential based on ecosystem services/landscape functions, however, does not show a clear picture. Key functions, such as water storage, sustained biodiversity, and land use are explicitly included in the Old Drava Programme, while others, like landscape connectivity are only implicitly or not at all treated. For the efficient protection of wetlands, not only their water supply has to be ensured, but also strictly defined buffer zones have to be established or restored around them. The intensity of land use has to be reduced in areas with low productivity and arable land has to be replaced by a landscape mosaic of woodlands, pastures and open orchard meadows, a centuries-old traditional form of agriculture in the region.

We attempted to approach the problem using several assessment techniques. Although all of them are qualitative, they are partly based on measurable parameters. The outcome of the assessment, however, is jeopardized by the reliability of the available data, which is, unfortunately, still very low.

References

AQUAPROFIT (2005) Az Ormánság komplex rehabilitációja és térségfejlesztése. Ős-Dráva Program (Complex rehabilitation and regional development in Ormánság. Old Drava Programme). AQUAPROFIT, Budapest. 516 pp (in Hungarian)

AQUAPROFIT (2007a) Ős-Dráva Program. Vízügyi műszaki terv (Old Drava Programme: Technical plan of water management). AQUAPROFIT, Budapest – DDKÖVÍZIG, Pécs. 172 pp (in Hungarian)

AQUAPROFIT (2007b) Ős-Dráva Program Tájgazdálkodási Programterv (Old Drava Programme: Landscape management draft plan). AQUAPROFIT, Budapest (in Hungarian) http://www.osdrava.hu/download/tajgazdalkodasi.pdf. Accessed on 11.10.2013

AQUAPROFIT (2010) Ős-Dráva Program – Összefogással az Ormánság fellendítéséért. Vezetői összefoglaló (Old Drava Programme – Cooperation for the Progress of Ormánság: Executive summary). AQUAPROFIT, Budapest. 29 pp (in Hungarian)

Arthington AH, Bunn SE, Poff NL, Naiman RJ (2006) The challenge of providing environmental flow rules to sustain river ecosystems. Ecol Appl 16(4):1311–1318. https://doi.org/10.1890/1051-0761(2006)016[1311:tcopef]2.0.CO;2

Arthington AH, Naiman RJ, McClain ME, Nilsson C (2009) Preserving the biodiversity and ecological services of rivers: new challenges and research opportunities. Freshwater Biol 55 (1):1–16. https://doi.org/10.1111/j.1365-2427.2009.02340.x

Brierley GJ, Fryirs KA (2005) Geomorphology and river management. Applications of the river styles framework. Blackwell Publishing, Carlton, Victoria, 398 pp

Brinson MM, Hauer ER, Lee LC, Nutter WL, Rheinhardt RD, Smith RD, Whigham DE (1995) Guidebook for Application of Hydrogeomorphic Assessements to Riverine Wetlands. Army Corps of Engineers, Waterways Experimental Station, Vicksburg, MS. Technical Report TR-WRP-DE-11

Cairns J (1991) The status of the theoretical and applied science of restoration ecology. Environ Professional 13:186–194

Chovanec A, Waringer J, Straif M, Graf W, Reckendorfer W, Waringer-Löschenkohl A, Waidbacher H, Schultz H (2005) The Floodplain Index—a new approach for assessing the ecological status of river/floodplain-systems according to the EU Water Framework Directive. Archiv für Hydrobiologie Supplement 155:425–442

Council of Europe (2000) European Landscape Convention. Council of Europe, Florence. 9 pp (European Treaty Series No 176). http://conventions.coe.int/Treaty/en/Treaties/Html/176.htm. Accessed on 21.10.2014

Csapó J, Marton G, Aubert A (2011) The aspects of tourism development in the Ormánság (Hungary) with a special attention on the possibilities of internal integration. In: Dombay Ş, Magyari-Sáska Zs (eds) The role of tourism in territorial development. 4th international conference, 7–8 October 2011, Gheorgheni, Romania, vol 4, pp 40–49

DDKÖVÍZIG (2012) Revitalization of the Cún-Szaporca oxbow system. Final Master Plan. South-Transdanubian Environment and Water Directorate, Pécs. 100 pp

DD-KVTF (2013) Jegyzőkönyv közmeghallgatásról: a Dél-dunántúli Vízügyi Igazgatóság „A Cún –Szaporcai holtág vízpótlása az Ős-Dráva Program keretén belül" projekt környezeti hatásvizsgálati eljárása (Protocol of the Public Hearing on the EIS procedure of the project "Water replenishment of the Cún-Szaporca oxbow within the Old Drava Programme" by the

South Transdanubian Water Management Directorate). South Transdanubian Inspectorate for Environmental Protection, Natural Protection and Water Management, Pécs. 3 pp (in Hungarian)

Department of Carinthia (2008) Is the drava river basin management sustainable and well on the way? International Symposium "Drava River Vision", 23–25 September 2008, Maribor, Slovenia. Department of Carinthia, Department of Water Management, Klagenfurt, Austria. 59 p

Dezső J, Halász A, Czigány Sz, Tóth G, Lóczy D (2017) Estimating seepage loss during water replenishment to a floodplain oxbow: a case study from the Drava Plain. Geografia Física e Dinamica Quaternaria 40:61–75. https://doi.org/10.4461/gfdq2017.40.6

Downs PW, Kondolf GM (2002) Post-project appraisals in adaptive management of river channel restoration. Environ Manag 29:477–496. https://doi.org/10.1007/s00267-001-0035-x

Downs PW, Thorne CR (2000) Rehabilitation of a lowland river: reconciling flood defence with habitat diversity and geomorphological sustainability. J Environ Manag 58:249–268. https://doi.org/10.1006/jema.2000.0327

Dragun D, Gavran A, Car V (2014) Hydrographic survey of the River Drava branches in the process of revitalization, flood control and morphological monitoring. In: FIG Congress 2014 Engaging the Challenges – Enhancing the Relevance, Kuala Lumpur, Malaysia 16–21 June 2014. 10 pp. https://www.fig.net/resources/proceedings/fig_proceedings/fig2014/papers/ts06j/TS06J_dragun_gavran_6869.pdf

Dufour S, Piégay H (2009) From the myth of a lost paradise to targeted river restoration: forget natural references and focus on human benefits. River Res Appl 24:1–14

Eiseltová M, Pokorný J, Ripl W, Bodlák L, Pechar L, Pecharová E, Kučera Z (2007) Restoration of water and matter retention functions of a floodplain: ecology and economics. In: Trémolieres M, Schnitzler A (eds) Floodplain protection, restoration, management. why and how. Lavoisier SAS, Paris, pp 190–199

European Commission (2000) Directive 2000/60/EEC. Establishing a framework for community action in the field of water policy. Official Journal of the European Communities, Luxemburg. L327. 1–71

Fryirs KA, Brierley GJ (2000) A geomorphic approach for identification of river recovery potential. Phys Geogr 21(3):244–277

Gálosi-Kovács B (2010) Környezetfejlesztés a kistérségekben (Environmental development in the microregions). Publikon Kiadó, Pécs. 247 pp (in Hungarian)

Gálosi-Kovács B, M Császár Zs, Pap N, Reményi P, Elekes T (2011) Social problems of the most disadvantaged Southern Transdanubian micro-regions in Hungary. Roman Rev Reg Stud 7 (2):41–50

Gilvear DJ, Spray CJ, Casas-Mulet R (2013) River rehabilitation for the delivery of multiple ecosystem services at the river network scale. J Environ Manag 126:30–43. https://doi.org/10.1016/j.jenvman.2013.03.026

Gren IM, Groth KH, Sylvén M (1995) Economic values of Danube floodplains. J Environ Manag 45:333–345

Gyenizse P, Lóczy D (2010) The impact of microtopography and drainage on land use and settlement development in the Hungarian Drava Plain. Hrvatski Geografski Glasnik 72(1):5 20

Habersack H, Schober B, Krapesch G, Jäger E, Muhar S, Poppe M, Preis S, Weiss M, Hauer C (2010) Neue Ansätze im integrierten Hochwassermanagement: Floodplain Evaluation Matrix (FEM), flussmorphologischer Raumbedarf (FMRB) und räumlich differenziertes Vegetationsmanagement (VeMaFLOOD). Österreichische Wasser- und Abfallwirtschaft 62 (1–2):15–21. https://doi.org/10.1007/s00506-009-0153-x

Habersack H, Schober B, Hauer C (2015) Floodplain evaluation matrix (FEM): An interdisciplinary method for evaluating river floodplains in the context of integrated flood risk management. Natural Hazards 75(Supplement 1):5–32. https://doi.org/10.1007/s11069-013-0842-4

Heal G (2000) Valuing ecosystem services. Ecosystems 3(1):24–30

Hulse D, Gregory S (2004) Integrating resilience into floodplain restoration. Urban Ecosyst 7 (3):295–314. https://doi.org/10.1023/b:ueco.0000044041.94705.52

Jennings GD, Harman WA (1999) Stream restoration lessons learned in North Carolina, USA. In: Rutherfurd I, Bartley R (eds) Proceedings, Second Australian Stream Management Conference, Adelaide, vol 1, pp 359–364

Kentula M (2000) Perspectives on setting success criteria for wetland restoration. Ecol Eng 15:199–209

Kern K (1994) Grundlagen naturnahen Gewässergestaltung. Geomorphologische Entwicklung von Fließgewässern. Springer Verlag, Berlin – Heidelberg – New York. 256 pp

Kiss T, Andrási G (2011) A horvátországi duzzasztógátak hatása a Dráva vízjárására és a fenékhordalék szemcseösszetételének alakulására (Impact of dams in Croatia on the water regime and bedload grain size of the Drava River). Hidrológiai Közlöny 91(5):17–23 (in Hungarian)

Konkoly-Gyuró É (2011) A tájfunkció elemzés koncepciójának kibontakozása (Evolution of the concept of the landscape function analysis). Workshop on Ecosystem Services, Corvinus University, Budapest, 24 November 2009. http://www.mta-taki.hu/hu/node/385 Accessed on 10.10.2014 (in Hungarian)

Leopold A (1949) A sand county almanac. Oxford University Press, Oxford, 266 pp

Lóczy D (2013) Hydromorphological-geoecological foundations of floodplain management: case study from Hungary. Lambert Academic Publishing, Saarbrücken. 382 p

Lóczy D, Dezső J, Czigány Sz, Prokos H, Tóth G (2017) An environmental assessment of water replenishment to a floodplain lake. J Environ Manag 202(2):337–347. https://doi.org/10.1016/j.jenvman.2017.01.020

Macura V, Izakovičová Z (2000) Krajinnoekologické aspekty revitalizácie tokov (Landscape ecological aspects of the revitalization of rivers). Slovakian Technical University, Bratislava. 262 pp (in Slovakian)

Márk L, Sindler Cs, Tálos A (2006) Ős-Dráva Program – Víz az Ormánság fellendítésért (Old Drava Programme – Water for progress in Ormánság region). 24th National Meeting of the Hungarian Hydrological Society, Pécs, 5–6 July 2006 vol II. http://www.tankonyvtar.hu/hu/tartalom/arcanum2/HidrologiaiTarsasagVandorgyulesei_2006_024_2/adatok.html Accessed 20.12.2014 (in Hungarian)

Morandi B, Piégay H, Lamouroux N, Vaudor L (2014) How is success or failure in river restoration projects evaluated? Feedback from French restoration projects. J Environ Manag 137:178–188. https://doi.org/10.1016/j.jenvman.2014.02.010

National Research Council (1992) Restoration of aquatic ecosystems: science, technology and public policy. The National Academies Press, Washington, DC, 576 pp

National Research Council (2002) The Missouri River Ecosystem: Exploring the Prospects for Recovery. The National Academies Press, Washington, DC, 175 pp

Nemes G, High C (2011) Implementation of agri-environmental programme in Hungary. Regional studies association annual international conference, 17–20 April 2011. Newcastle-upon-Tyne, UK. http://www.regionalstudies.org/uploads/conferences/presentations/international-conference-2011/nemes.pdf

Niemi GJ, DeVore P, Detenbeck N, Taylor D, Lima A, Pastor J, Yount JD, Naiman RJ (1990) Overview of case studies on recovery of aquatic systems from disturbance. Environ Manag 14 (5):571–597

Palmer MA, Bernhardt ES, Allan JD, Lake PS, Alexander G, Brooks S, Carr J, Clayton S, Dahm CN, Follstad Shah J, Galat DL, Loss SG, Goodwin P, Hart DD, Hassett B, Jenkinson R, Kondolf GM, Lave R, Meyer JL, O'Donnell TK, Pagano L, Sudduth E (2005) Standards for ecologically successful river restoration. J Appl Ecol 42:208–217. https://doi.org/10.1111/j.1365-2664.2005.01004.x

Pécsi HIDROTERV Bt. (2015) Ős-Dráva Program. Aktualizált Területi Vízgazdálkodási Tanulmányterv (Old Drava Programme: Actualized Regional Water Management Plan). Pécsi HYDROTERV Bt., Pécs. 180 pp

Petts GE (1984) Impounded rivers: perspectives for ecological management. John Wiley and Sons, Chichester, 326 pp

Piégay H, Darby SE, Mosselmann E, Surian N (2005) A review of the techniques available for delimiting the erodible river corridor: a sustainable approach to managing bank erosion. River Res Appl 21:773–789. https://doi.org/10.1002/rra.881
Popovič I, Mikuska T (2010) The Drava river – a flowing controversy. Danube Watch 1/2010. http://www.icpdr.org/main/publications/drava-river-flowing-controversy
Ramsar Convention Secretariat (2010) Managing wetlands: frameworks for managing Wetlands of International Importance and other wetland sites. Ramsar handbooks for the wise use of wetlands, 4th edn, vol 18. Ramsar Convention Secretariat, Gland, Switzerland. 96 pp
Reményi P, Tóth J (eds) (2009) Az Ormánság helye és lehetőségei (Position and opportunities of the Ormánság region). IDResearch Kft, Pécs. 56–58:302–355 (in Hungarian)
Salamon B (2014) With collaboration for the advancement of Ormánság. Old-Drava Programme Office, Kémes. 20 pp
Sanford W (2002) Recharge and groundwater models: an overview. Hydrogeol J 10:110–120. https://doi.org/10.1007/s10040-001-0173-5
Schiemer F, Baumgartner C, Tockner K (1999) Restoration of floodplain rivers: The Danube restoration project. Regul Rivers Res Manag 15:231–244
Schwarz U (2014) Restoration potential for floodplains in the Danube River Basin. Presentation at the 1st Danube Regional Workshop on Natural Water Retention Measures, Szentendre, Hungary, 28–29 January 2014
Sparks RE, Bayley PB, Kohler SL, Osborne LL (1990) Disturbance and recovery of large floodplain rivers. Environ Manag 14(5):699–709
Szalay M (2009) A felszíni vizek mennyiségi jellemzése. Kisvízi készlet (Quantitative description of surface waters: low-water reserves). Manuscript report. Öko Zrt., Pécs. 18 pp (in Hungarian)
Tésits R (2012) The nonprofit sector and employment expansion in the rural areas of Hungary. Eastern European Countryside 18(1):125–147. https://doi.org/10.2478/v10130-012-0007-6
Tésits R, Alpek BL (2012) Labour market needs assessment of the Hungarian-Croatian border region. South-Transdanubian Regional Resource Center, Pécs. 106 pp
Tésits R, Alpek BL (2014) Appearance of European employment policy in the rural areas of Hungarian-Croatian border region. Eastern European Countryside 20(1):55–71. https://doi.org/10.2478/eec-2014-0003
Tockner K, Schiemer F, Ward JV (1998) Conservation by restoration: the management concept for river floodplain system on the Danube River in Austria. Aquatic Conservation: Marine and Freshwater Ecosystem 8:71–86. https://doi.org/10.1002/(sici)1099-0755(199801/02)8.1<71:aid-aqc265>3.0.CO;2-d
Tóth G, Rajkai K, Bódis K, Máté F (2014) Magyarországi kistájak földminősége a D-e-meter szántó minősítési eljárás szerint. Tájökológiai Lapok 12(1):183–195 (in Hungarian)
VKKI (2010) Vízgyűjtő-gazdálkodási terv. Dráva részvízgyűjtő (Watershed Management Plan: Drava Partial Watershed). Vízügyi és Környezetvédelmi Központi Igazgatóság (Central Directorate for Water Management and Environmental Protection), Budapest. 162 p (in Hungarian)
Völgyesi I (2009) Ökológiai vízigény vagy megfelelő talajvízszintek? (Ecological water demand or proper groundwater levels). Hidrológiai Közlöny 89(5):53–56 (in Hungarian)
Wallace K (2008) Ecosystem services: multiple classifications or confusion? Biol Conserv 141(2):353–354. https://doi.org/10.1016/j.biocon.2007.12.014
Woolsey S, Capelli F, Gonser TOM, Hoehn E, Hostmann M, Junker B, Paetzold A, Roulier C, Schweizer S, Tiegs SD, Tockner K, Weber C, Peter A (2007) A strategy to assess river restoration success. Freshwater Biol 52:752–769. https://doi.org/10.1111/j.1365-2427.2007.01740.x
WWF (2002) Áttekintés a Dráva alsó vízgyűjtőjének tájhasználatáról (An overview of land use in the Lower Drava River catchment). Report for the Worldwide Fund for Nature Hungary, Budapest. 142 pp (in Hungarian)

Index

D. Lóczy (ed.), *The Drava River*, Springer Geography,
https://doi.org/10.1007/978-3-319-92816-6

MIX
Papier aus verantwortungsvollen Quellen
Paper from responsible sources
FSC® C105338

If you have any concerns about our products,
you can contact us on
ProductSafety@springernature.com

In case Publisher is established outside the EU,
the EU authorized representative is:
Springer Nature Customer Service Center GmbH
Europaplatz 3, 69115 Heidelberg, Germany

Printed by Libri Plureos GmbH
in Hamburg, Germany